Confined Space Entry
and
Emergency Response

Confined Space Entry and Emergency Response

D. Alan Veasey

Lisa Craft McCormick

Barbara M. Hilyer (deceased)

Kenneth W. Oldfield

Sam Hansen
Theodore H. Krayer
Workplace Safety Training Program
University of Alabama at Birmingham
Birmingham, Alabama

WILEY-
INTERSCIENCE

A JOHN WILEY & SONS, INC., PUBLICATION

Library of Congress Cataloging-in-Publication Data is available.

ISBN-13 978-0-471-77845-5
ISBN-10 0-471-77845-1

Printed in the United States of America.

10 9 8 7 6 5 4 3 2 1

To the safety of those who work in confined spaces and the bravery of those who are called up to respond in confined space emergencies

Contents

x Contents

Preface

Welcome to *Confined Space Entry and Emergency Response*. This book addresses the information and training needs of two distinct but overlapping populations: those involved in routine entry and work in confined spaces and those involved in confined space rescue. Examples of routine entrants include industrial workers who enter vessels and public employees who enter utility vaults. Rescuers include private-sector rescue teams and public emergency services personnel who may be called on to perform confined space rescue.

Entrants and rescuers face similar hazards and have interrelated training needs, making an approach that integrates confined space entry and confined space emergency response logical. In teaching confined space entry and rescue for a number of years, we (the authors) never found a book that we believed adequately addressed the needs of both entrants and rescuers, so we wrote *Confined Space Entry and Emergency Response*. We believe that it meets the information and training needs of others as well.

Main Features

The book provides complete information, guidance, regulatory reference, and case studies for all personnel who plan for, supervise, work inside, or provide rescue from confined spaces. The reader is taken carefully through each step from the identification of confined spaces and their hazards, control of and protection from the hazards, equipment and procedures for operations, to complete preparation for rescue.

A CD accompanies the book. On the CD the user will find materials for use in training, using this book as a textbook or reference. The CD includes learning objectives and lesson plans for each chapter, slide presentations for overhead transparencies or PowerPoint presentations, instructions for building and using confined space field simulators, worksheets for classroom hazard analysis and planning, and suggestions for hands-on practice with air monitoring equipment, personal protective equipment, and entry and rescue equipment such as ropes, webbing, harnesses, hardware, and portable anchor systems.

Unique Features

This is the first book to treat both confined space entry and confined space rescue thoroughly with complete illustrated information. The technical information is extensive, and illustrated with over 500 photographs and drawings. The illustrations are especially clarifying in the chapters where ropes, hardware, and other equipment are explained. The accompanying CD includes video clips in addition to the drawings.

The authors' backgrounds are uniquely suited for writing this book. We are confined space rescuers, firefighters, and industrial safety experts with specialized added knowledge of regulatory compliance, toxicology, chemistry, and air monitoring. Equally important, we draw from a combined 75 years of teaching experience to make the information and explanations understandable. We have experienced firsthand the difficulties in performing rescue, writing clear and user-friendly materials, and teaching emergency personnel and industrial workers. We write not from theory, although some of that is included in the training tips on the CD, but primarily from real-world experience. We have taught this material to hundreds of confined space entrants and rescuers, and we know where the difficult concepts and skills are and how to make them understandable without losing the important technical details.

Organization

This book consists of 18 chapters and is organized into four parts. The organization of the book follows the steps suggested for designing and executing a confined space program. For confined space entry, the information follows the logic of an entry permit. For confined space rescue team operations, the information follows the logic of a pre-emergency plan for confined space rescue.

Part I presents all the basic information you will need to judge whether spaces are, by definition, confined spaces, and whether these confined spaces require a permit for entry. You will learn a great deal about regulations covering entry, work, and rescue in confined spaces, and methods to identify and assess the hazards therein. Guidance for writing a permit program and designing an effective permit form is included.

Part II guides you through protective measures that eliminate hazards from the space, or control them if they cannot be eliminated. This part covers preentry procedures such as ventilation, isolation, and personal protective gear and how to use it. Guidelines for the selection of respirators and protective clothing—chemical and otherwise—are described and illustrated.

Part III offers a wealth of information regarding equipment commonly used in confined space entry and rescue. This part describes in great technical detail ropes, webbing, harnesses, hardware, knots, and anchor systems used in basic rigging. This information can be used in rigging retrieval and fall-arrest systems to be used for routine confined space entries. It also serves as the basis for rigging the more advanced systems that may be required by confined space rescue teams and are described later in the book. The last chapter in Part III ties the information in all previous chapters together and sums up all the considerations for making a safe entry.

Part IV covers confined space rescue. Rescue team members must consider all the planning, equipment, and precautions for entry into a confined space, and then must organize and manage a safe and successful rescue. In this part, you will learn basic considerations for performing rescue, how to organize the rescue team, a variety of ways to build and use rescue systems, and how to conduct rescue operations.

Using the Book Effectively

In writing this book, the authors tried to focus the information in such a way that most of it would be valuable both to those involved in routine confined space entry and to confined space rescuers. Those involved in routine work entries should utilize the information in Chaps. 1 though 15. This information focuses on identifying and controlling hazards and conducting safe operations in confined spaces. The role of rescue is restricted to self-rescue and nonentry rescue. Trainees who will enter confined spaces for the purpose of performing rescues need the information in Chaps. 1 though 18. The additional information in Chaps. 16 through 18 is intended to allow them to function as rescue team members to remove a victim from a space. The additional chapters also provide information on how to remove victims from elevated spaces such as tanks or towers after they are rescued from confined spaces. The amount of rope rescue provided is limited to this and is by no means intended to be an exhaustive coverage of rope rescue techniques.

Confined space rescue can be especially dangerous and difficult and deserves special consideration. Rescue training can be conducted safely only by a qualified instructor. This book is definitely not intended as a "self-taught" or "how-to-do-it" approach for the untrained, inexperienced rescuer. It is intended to be used by competent professionals in conducting training. Different schools of thought exist among different agencies and authorities regarding rescue practices. A wide array of techniques can be used to perform confined space rescue. Which techniques will work best depends on the specifics of the situation. In writing this book, we tried to provide information on an array of techniques with examples of how they might be applied and leave it to the trainer to select the practices most appropriate for their situation.

Confined Space Entry and Emergency Response is intended for everyone who writes confined space permit programs; designs permit forms; acts as confined space entrant, attendant, or supervisor; plans or performs rescue; or oversees the safety of entrants or rescue teams. It explains in extensive detail how to plan, execute, and manage rescue operations. It also is designed for those who train entrants or rescuers, including both industrial workers and emergency personnel. It will fill the needs of safety professionals who classify confined spaces, assess hazards, design atmospheric monitoring systems, and implement hazard controls. All personnel who participate in confined space operations will find it helpful. Rescue teams, whether industrial or public-sector, will find this book helpful in planning, coordinating, and executing safe confined space rescues. Used properly, this book will help to keep entrants and rescuers safe, and employers out of trouble with regulatory agencies.

D. Alan Veasey
Lisa Craft McCormick
Barbara M. Hilyer
Kenneth W. Oldfield
Sam Hansen
Theodore H. Krayer

Acknowledgments

The Workplace Safety Training Program at the Center for Labor Education and Research at the University of Alabama at Birmingham provides nationwide training to workers and public responders in chemical emergency response, confined space entry and rescue, hazardous waste site cleanup, and a number of related topics. Funding for training and curriculum development is provided by cooperative agreement #U45ES06155 from the National Institute of Environmental Health Sciences (NIEHS); however, the contents of this text are solely the responsibility of the authors and do not necessarily represent the official views of the NIEHS. Major responsibility for initiating a training grants program was given to the NIEHS under the Superfund Amendments and Reauthorization Act of 1986 (SARA) to fund nonprofit organizations with a demonstrated track record of providing occupational safety and health education in developing and delivering high-quality training to workers involved in handling hazardous waste or in responding to emergency releases of hazardous materials. Since the initiation of the Hazardous Waste Worker Training Program in 1987, the NIEHS has developed a strong network of nonprofit organizations that are committed to protecting workers and their communities by delivering high-quality, peer-reviewed safety and health curriculum to target populations of hazardous waste workers and emergency responders.

Confined Space Basics

1

Introduction to Confined Spaces

In November 1984, a worker in Phoenix, Arizona entered a toluene storage tank to remove sludge from the bottom of the tank. The tank was not tested, evaluated, or ventilated prior to entry, and the entrant used no personnel protective equipment. The entrant became unresponsive soon after entering the tank. The entrant's supervisor called for help from local emergency services personnel. Rescue personnel from the local fire department arrived and immediately placed a high priority on rapid removal of the entrant from the tank. The rescuers found that the only access into the tank was through a 16-inch-diameter manhole on top of the tank, an opening too small for a rescuer to pass through while wearing a self-contained breathing apparatus. The rescuers therefore decided to cut a larger opening in the side of the tank using a rotary saw. While the cutting operation was under way, sparks from the saw blade ignited toluene vapors within the tank. The resulting explosion killed one firefighter and injured 14 others. Autopsy results indicated that the entrant was already dead at the time of the explosion.

The Phoenix tank explosion incident became well known, in part because a camera crew was filming at the time of the explosion and footage from the incident was included in a National Fire Academy video titled *Fire Fighter Safety*. The incident was also well documented in a National Institute for Occupational Safety and Health (NIOSH) publication titled *Worker Deaths in Confined Spaces*. It was one of a number of incidents described in the preamble to OSHA's *Notice of Proposed Rulemaking for the Permit-Required Confined Spaces Standard for General Industry* in order to document the need for such a standard.

Why All the Concern about Confined Spaces?

Stated bluntly, confined work areas have a terrible safety record. Such spaces have long been recognized by occupational safety and health advocates as unhealthy and unsafe places in which to work. Examples of common confined spaces include storage tanks, silos, vaults, vats, sewers, rail tank cars, and tank trucks (see Figs. 1.1 through 1.6).

According to data gathered by NIOSH, at least 670 fatalities occurred in 585 separate confined space incidents in the United States during the 10-year period from 1980 through 1989. It is worth noting that the data gathered by NIOSH included only deaths due to asphyxiation, poisoning, and drowning. Confined space deaths due to electrical energy, explosions, machinery, and other physical hazards were not included, so the actual confined space body count for the decade was probably even higher than these numbers indicate.

It is also worth noting that up to 60% of those who died in confined space incidents entered for the purpose of rescuing others. Some of the would-be rescuers were workers attempting to rescue coworkers, while others were fire, police, or emergency medical personnel who responded to calls.

Figure 1.1   Storage tanks are common confined spaces.

Figure 1.2   Silos, designed to contain granular solid materials, are typical confined spaces.

Figure 1.3   Vaults are common confined spaces in industries involving telecommunications, oil and gas transmission, and electrical transmission.

Figure 1.4   Sewers, including sanitary and storm sewer systems, are very common confined spaces in urban and suburban areas.

Figure 1.5   Open-top structures such as vats, pits, and diked areas around storage tanks may be classified as confined spaces.

Figure 1.6   Transportation vehicles such as railcars and tank trucks are considered confined spaces.

OSHA's Response: The *Permit-Required Confined Spaces* Standard

The large number of deaths in confined space incidents triggered a major push by occupational safety and health advocates for OSHA regulations to protect workers involved in confined space operations in general industry. After many years, this push finally resulted in the promulgation in 1993 of OSHA standard 29 CFR 1910.146, titled *Permit-Required Confined Spaces* (PRCS).

What Are Confined Spaces?

According to OSHA's PRCS standard, a work area is considered a confined space if it meets all three of the following criteria (Fig. 1.7).

- It is of a size and configuration making it possible for a worker to enter and perform work.

- It is not designed for continuous worker occupancy.

- It has limited or restricted means of entry and exit.

A work area is not considered a confined space if it fails to meet these three criteria, despite the hazards that may be present. The characteristics are discussed below.

Size and configuration make entry possible

The PRCS standard regulates operations within confined spaces; therefore, if a space cannot physically be entered by a worker, it is not classified as a confined space. Some

---

CHECKLIST FOR EVALUATION OF SPACES

Is it a Confined Space?
The answer is "yes" if all three of the following requirements are met:
I. ___ It is possible to enter and work within the space
II.___ The space is not intended for human occupancy
III.___ Entry and exit are restricted

Is it Permit-Required?
IV.___The answer is "yes" if any of the following conditions apply:
    A.___Presence of a hazardous atmosphere
        Check "A" if the potential exists for <u>any</u> of the following atmospheric conditions:
           1. Oxygen content less than 19.5%
           2. Oxygen content greater than 23.5%
           3. Flammable gas or vapor in excess of 10% of the LEL
           4. Combustible dust in a concentration that obscures vision at a distance of 5 feet (1.5 meter) or less
           5. Hazardous substance concentrations in excess of PEL and capable of producing significant toxic effects
           6. Any other atmospheric condition immediately dangerous to life and health
    B.___ Presence of an engulfment hazard
    C.___ An entrapping configuration
    D.___ Any significant safety or health hazard

NOTE: The space is a permit-required confined space if I, II, III, and IV are checked.

Figure 1.7 Use a checklist to assess a work area as a permit-required confined space.

spaces may simply be too small for anyone to enter. Other spaces may lack an opening large enough to allow entry. As a way to simplify complying with the PRCS standard, some employers elect to modify confined spaces in order to make entry impossible.

Spaces not designed for continuous worker occupancy

The PRCS standard excludes spaces designed to be occupied continuously by workers. For example, control booths for industrial processes may be of a size, configuration, and location making entry and exit difficult. Since such spaces are designed and intended for continuous occupancy, they are not categorized as confined spaces according to the standard.

Restricted means for entry and exit

Confined spaces are inherently difficult to get into and out of (Fig. 1.8). For example, spaces that require the use of the hands and feet for entry are considered to restrict entry and exit. Spaces are classified into this category if they have an exit pathway that impairs an entrant's ability to perform self-rescue or exit the space without outside assistance, should an emergency occur during entry operations.

Some locations, such as chemical storage rooms, may harbor significant hazards to an entrant, yet not be classified under the standard as a confined space because a well-defined, functional doorway is present (Fig. 1.9). Using the same example, the same location may be classified as a confined space if factors such as poor housekeeping result in obstruction of the doorway.

Nonpermit- versus Permit-Required Confined Spaces

The PRCS standard doesn't regulate work in all confined spaces. It covers only confined spaces classified as permit-required. According to OSHA, *permit-required confined spaces* are confined spaces that contain a potential hazard to an entrant. In addition to meeting all three of the criteria described above, a permit-required confined space must meet the fourth requirement of posing a hazard to an entrant. If a confined space contains no potential hazard to an entrant, then the space is considered a nonpermit-required confined space and operations within it are not regulated by the PRCS standard.

What Hazards Make a Confined Space Permit-Required?

According to OSHA, a confined space is considered permit-required if it contains any one of four major hazard types (Fig. 1.7). These hazards include hazardous atmospheres, potential for engulfment, entrapping configurations, or any other significant hazard to the safety or health of an entrant.

Figure 1.8 Confined space configuration restricts the entry and exit of the entrant.

Figure 1.9 Enclosed areas are not considered confined spaces by OSHA if entry and exit are unrestricted, even though hazards to an entrant may be present.

Hazardous atmospheres

A confined space is considered permit-required if it contains, or has the potential to contain, a hazardous atmosphere. The presence of any one of several types of potential atmospheric hazards triggers permit-required status. Atmospheric hazards include oxygen-deficient atmospheres; oxygen-enriched atmospheres; potentially flammable concentrations of flammable gases, vapors, or dusts; toxic atmospheres; or any other atmospheric hazard representing an immediate danger to the life or health of an entrant.

Oxygen-deficient and oxygen-enriched atmospheres. Permit-required status for a confined space is triggered if the oxygen content of the atmosphere within the space deviates significantly from the normal 20.9%. Oxygen content of less than 19.5% is considered oxygen-deficient; greater than 23.5%, oxygen-enriched. Either condition requires the space to be considered permit-required.

Potentially flammable atmospheres. Confined spaces are considered permit-required if the potential exists for significant concentrations of flammable gases, vapors, or dust to be present (Fig. 1.10). Spaces containing flammable gases or vapors in excess of 10% of the lower explosive limit (LEL), as indicated by air monitoring equipment, are considered permit-required.

Certain finely divided solids or dusts can explode violently when suspended in the atmosphere in flammable concentrations. Air monitoring equipment is not commonly available for measuring flammable concentrations of combustible dust; however, a visual estimate of the concentration of dust suspended in the atmosphere of the work area can be used. OSHA considers a confined space to be permit-required if airborne combustible dust is present at or in excess of its LEL. This may be approximated as a concentration that obscures vision at a distance of 5 feet (1.52 meters) or less.

Toxic atmospheres. OSHA has published dose or permissible exposure limits (PELs) for substances it considers potentially hazardous to workers exposed to atmospheric concentrations exceeding the PELs. PELs are published in Subparts G and Z of

Figure 1.10 A potentially flammable atmosphere within a confined space is a permit-required condition.

Figure 1.11 Toxic and oxygen-deficient atmospheres are permit-required conditions that have produced many casualties in confined spaces.

OSHA's *Occupational Safety and Health Standards for General Industry.* OSHA considers a confined space to be permit-required if atmospheric concentrations of any of these air contaminants exceed the published PELs and are capable of producing significant toxic effects (Fig. 1.11). Significant toxic effects are those resulting in death, acute illness, incapacitation, impairment of ability to perform self-rescue, or injury.

Other atmospheric conditions that are immediately dangerous to life and health. Some air contaminants for which OSHA has not established exposure limits may also be acutely hazardous to workers. For such contaminants, OSHA requires that other sources of information such as Material Safety Data Sheets (MSDSs), published documents, or other forms of communication be used to establish acceptable atmospheric conditions. OSHA considers any confined space containing atmospheric conditions immediately dangerous to the life and health of an entrant to be permit-required.

Engulfment hazards

A confined space is considered a permit-required space if it contains a material that could engulf an entrant. The engulfing material can be in a liquid or a finely divided solid physical state. Materials as seemingly harmless as potable water or sawdust have caused numerous confined space fatalities through engulfment.

Entrapping configurations

OSHA considers a confined space to be permit-required if it has an internal configuration that may trap or asphyxiate an entrant (Fig. 1.12). Examples include spaces having inwardly converging walls or floors that slope downward and taper to a smaller cross section. Workers can easily slide down into structures such as these and become trapped. In some cases, the sides of the entrapping structure cause asphyxiation by compressing the entrant's chest, making adequate breathing impossible.

Other significant safety or health hazards

It is possible for confined space entrants to encounter significant hazards in addition to those discussed above. For example, physical hazards, such as electrical energy sources or machinery that could be accidentally activated, might be encountered during an entry (Fig. 1.13). Under OSHA guidelines, the presence of any serious hazard in a confined space causes the space to be classified as permit-required.

Figure 1.12 Confined spaces having configurations that could trap or asphyxiate an entrant are considered permit-required.

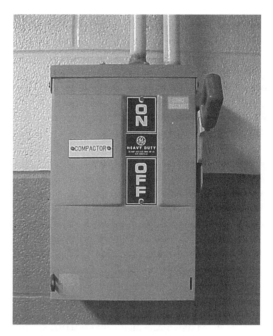

Figure 1.13 Physical hazards such as energy sources and machinery within confined spaces are considered permit-required hazards.

The Importance of Confined Space Recognition

Confined spaces are, and will continue to be, common features of the world in which we live and work. The ability to recognize them and anticipate their hazards is critical to the safety of personnel who may become involved in confined space operations.

Industrial facilities have any number of individual confined spaces such as those cited in the previous examples, plus whole systems of piping or ductwork that qualify as permit spaces (Fig. 1.14). Confined spaces are also common in public facilities and private utility systems such as water treatment plants and distribution systems, natural-gas and electrical transmission facilities, and sewer systems.

Similar hazards may be present in private-sector settings such as local neighborhoods or rural areas (Fig. 1.15). For example, public-sector rescuers may be called on to perform rescues or recovery operations from spaces such as septic tanks, cisterns, wells, root cellars, and manure pits.

Spaces other than those we think of as classic confined spaces may meet the permit space criteria. Such spaces can appear quite harmless, when in fact they are potentially deadly. For this reason, it is important to be objective in applying the evaluation criteria to identify permit spaces (Figure 1.7).

The Intent of OSHA's *Permit-Required Confined Spaces* Standard

In promulgating the PRCS standard, OSHA intended to prevent occurrences such as the Phoenix tank explosion. In researching that incident and other fatal confined space incidents, certain similarities become apparent. Common characteristics of many confined space incidents include

- Ineffective training of personnel involved in confined space operations

- Inadequate safety provisions for personnel involved in the operations

- Lack of preemergency planning and preparation for dealing with emergencies that may arise during entries

- The occurrence of casualties among rescuers as well as entrants

Figure 1.14 Industrial facilities may have any number of individual permit spaces, plus whole systems of piping or ductwork that qualify as permit spaces.

Figure 1.15 Public sector rescuers may face confined space hazards in unexpected settings, such as this abandoned root cellar, in local neighborhoods.

The PRCS standard outlines a proactive approach intended to keep personnel, including potential rescuers, safe during confined space operations. The requirements of the standard are intended to ensure that

- Hazards of spaces are identified before entries are made.
- Adequate safety precautions are taken prior to entries.
- Timely and effective rescue is available if needed during entries.

Summary

Unfortunately, workers continue to become casualties in confined space incidents. Likewise, rescuers continue to become a part of the confined space problem, rather than the solution, during rescue attempts. A quick scan of statistics available on OSHA's Website and the Bureau of Labor Statistics Website reveals that such occurrences have continued with alarming frequency despite the promulgation of the PRCS standard.

The information contained in this book can help to reduce the frequency of confined space casualties by (1) allowing work activities in confined spaces to be carried out safely and (2) providing for safe, effective confined space rescue operations when required.

2

Regulatory and Administrative Aspects of Confined Space Entry and Rescue

Two sewer workers entered an underground pumping station 50 feet below ground, climbing down a fixed ladder in a 3-foot metal shaft. When they removed the bolts of an inspection plate in the station, it blew off and flooded the underground room with sewage. One worker climbed out, radioed for assistance, and climbed back down to help his buddy.

Neither the two men nor two rescuers, a police officer and a sewer system manager, wore respirators or protective clothing. The ventilating fan was not working. A volunteer firefighter removed his self-contained breathing apparatus (SCBA) after falling part way down the ladder and becoming wedged in a position that left him unable to breathe; he lost consciousness but was extricated successfully. The other four men died, two from lack of oxygen and two from drowning in raw sewage.

Everything these men did was wrong, although "right" and "wrong" actions in confined spaces were not well defined when this occurred. This incident and hundreds of other accidents in which entrants and rescuers died in confined spaces led the National Institute of Occupational Safety and Health to recommend and the U.S. Occupational Safety and Health Administration to enact regulations to guide workers as they enter confined spaces.

State and Federal Regulations

The primary regulatory agency for confined space entry and rescue in the United States is the U.S. Occupational Safety and Health Administration (OSHA). States are allowed to set up their own occupational safety administrations with regulations at least as stringent as the federal ones under the guidance of federal OSHA.

In the early days of industry in North America, deaths and injuries were considered part of the industrial process (Fig. 2.1). Common law adopted from the British included the assumption of risk on the part of workers, and allowed employers to blame their workers for accidents. In the nineteenth century, technical and human progress led to industrial safeguarding and financial liability for accidents through insurance and workers' compensation. Financial liability and concern for workers' health led to increased interest in improving working conditions and generated activism aimed at regulatory control. Three-quarters of the way through the twentieth century, the U.S. Occupational Safety and Health Administration formalized controls on workplace

Figure 2.1   With no regulations to protect workers in the early days of industry, death and injuries were considered part of the job. (*Used with permission of the George Meany Memorial Archives.*)

safety, providing a structure and mechanism for the improvement of working conditions for North Americans.

U.S. Occupational Safety and Health Administration

OSHA was established by Congress in the passage of the Occupational Safety and Health Act of 1970 with a mission to

> Assure safe and healthful working conditions for working men and women; by authorizing enforcement of the standards developed under the Act; by assisting and encouraging the States in their efforts to assure safe and healthful working conditions; by providing for research, information, education, and training in the field of occupational safety and health.

The motivation for the act was that "The Congress finds that personal injuries and illnesses arising out of work situations impose a substantial burden upon...interstate commerce." Congress decreed that OSHA would provide research into and develop innovative methods for dealing with occupational safety and health problems, and would offer training programs to increase the number and competence of safety and health professionals. OSHA was charged with developing and promulgating workplace safety standards, enforcing the standards, and setting up appropriate reporting procedures to help describe the nature of problems occurring in workplaces.

Scope and coverage of federal OSHA regulations.   The scope of federal occupational safety and health regulations continues to expand, with new regulations promulgated each year. Each regulation contains a section titled "Scope and Coverage" explaining which employers must comply with the regulation.

OSHA coverage extends to employers and their employees in the 50 states, the District of Columbia, Puerto Rico, and all other territories under federal government jurisdiction. Coverage is provided either directly by federal OSHA or through an OSHA-approved state program.

Some employers are not covered by OSHA: self-employed persons, farms at which only immediate members of the farm employer's family are employed, and workplaces regulated by other federal agencies [such as the Mine Safety and Health Administration (MSHA)] under other federal statutes. Even when another federal agency is authorized to regulate safety and health conditions in a particular industry, if it does not do so in specific areas then OSHA standards apply. For example, cement manufacturing facilities (Fig. 2.2) are regulated by MSHA but OSHA claims regulatory oversight for cement workers who enter confined spaces.

Small employers in certain industries are exempt from some regulations. Exemptions vary from regulation to regulation, and depend on whether the industry is designated as having high risk for the hazards covered by the regulation. No employer having confined spaces is exempt from confined space regulations.

Federal employers must comply with standards consistent with those issued by OSHA for private sector employers; however, OSHA cannot propose monetary penalties against another federal agency for failure to comply with OSHA standards. OSHA conducts federal workplace inspections in response to employees' reports of hazards as part of a special program that identifies federal workplaces with higher than average rates of injuries and illnesses.

OSHA provisions do not apply to state and local governments in their role as employers. The OSH Act provides that any state desiring to gain OSHA approval for its private-sector occupational safety and health program must provide a program that covers its state and local government workers. Several states enacted state OSHA regulations to cover state and local government employees only.

Primary regulations covering confined spaces.   A number of OSHA regulations mandate safe confined space entry and rescue, or include policies for certain types of work in confined spaces. The three regulations with the most application to confined space operations are those covering permit-required confined spaces, controlling hazardous energy, and respiratory protection. See Table 2.1 for a summary of U.S. OSHA regulations covering confined space entry in general industry. Readers should consult the standards to ensure compliance with each regulated aspect of confined space entry and rescue.

29 CFR 1910.146, *Permit-Required Confined Spaces*.   The OSHA PRCS standard covers all aspects of confined space entry, whether for work or for rescue. The standard was published in the Federal Register in May 1994 and revised to include more specific rescue requirements in December 1998. Individual requirements of the confined space standard are covered in detail in relevant chapters of this book.

All employers having confined spaces must comply with the provisions of this standard, except for industries covered by industry-specific standards. Even those employers are covered by the PRCS standard if they have spaces not covered by the industry-specific standard, and must use the PRCS standard to evaluate their confined spaces.

The standard contains requirements for practices and procedures to protect employees in general industry from the hazards of entry into permit-required confined spaces. It does not apply to agriculture, shipyard, or construction employment; however, it does apply when construction workers are engaging in work that meets OSHA's definition of maintenance or modification. The PRCS standard requires a written permit space program if employees will enter permit spaces, and provides guidance for the determination of permit spaces. The written program must be available to employees, and compliant with all sections of the standard.

Figure 2.2   The Mine Safety and Health Administration regulates cement manufacturing plants, but OSHA regulations cover these workers when they enter confined spaces.

TABLE 2.1   U.S. Regulations Covering Confined Space Entry in General Industry

Number	Title	Coverage
29 CFR 1910.146	*Permit-Required Confined Spaces*	All aspects of entry for work or rescue
29 CFR 1910.147	*The Control of Hazardous Energy (Lockout / tagout)*	Servicing and maintenance of machines and equipment with electrical, mechanical, hydraulic, pneumatic, chemical, thermal, or other potentially hazardous energy sources
29 CFR 1910.134	*Respiratory Protection*	Selection, use, and maintenance of respirators
29 CFR 1910.95	*Occupational Noise Exposure*	Protection against noise
29 CFR 1910.133	*Eye and Face Protection*	Eye and face protection
29 CFR 1910.135	*Head Protection*	Head protection
29 CFR 1910.136	*Foot Protection*	Foot protection
29 CFR 1910.137	*Electrical Protective Equipment*	Protection from electrical hazards
29 CFR 1910.138	*Hand Protection*	Hand protection
29 CFR 1910.1000	*Air Contaminants*	Atmospheric concentration limits for toxic and hazardous substances in air
29 CFR 1910.97	*Nonionizing Radiation*	Protection from nonionizing radiation
29 CFR 1910.1096	*Ionizing Radiation*	Protection from ionizing radiation

29 CFR 1910.147, *Control of Hazardous Energy (Lockout/Tagout).* This OSHA standard includes a number of rules that are important to confined space safety. The standard covers the servicing and maintenance of machines and equipment in which the unexpected energizing, startup, or release of stored energy could cause injury to employees. Energy sources covered by the standard are sources of electrical, mechanical, hydraulic, pneumatic, chemical, thermal, or any other potentially hazardous energy. Employers must establish a program and use lockout devices to isolate energy from areas where employees will service or maintain machinery. Employee training and periodic inspections are required.

The standard does not cover construction, agriculture, and maritime employment; they are covered by other standards. Electric utilities are not covered for the same reason, nor are oil and gas well drilling and servicing companies. Servicing and maintenance that take place during normal production are covered only for employees who are required to remove or bypass a guard or other safety device, or place any part of their bodies into a point of operation. Minor servicing activities during normal production operations are not covered if they are routine, repetitive, and integral to the use of the equipment, provided alternative measures are used that provide effective protection.

The identification and likely locations of hazardous energy are discussed in Chap. 3 and control measures, in Chap. 6.

29 CFR 1910.134, *Respiratory Protection.* This important standard, which applies to work in confined spaces, was revised and strengthened in 1998, changing some of the earlier provisions of respiratory protection. Important requirements that impact confined space entrants are that (1) all respirator users must complete a medical questionnaire and achieve an acceptable fit through testing before wearing a respirator in a hazardous environment and (2) employers must have a written respiratory protection program in place, and must explain it to workers and train them annually. The standard and means of compliance are discussed more fully in Chap. 7.

Regulations covering other safety and health hazards. Several OSHA regulations apply to certain kinds of work in confined spaces. These include foot, face and eye, and head and hearing protection, as well as exposure to hazardous substances. Welding operations are covered by a separate standard.

29 CFR 1910.95, *Occupational Noise Exposure.* Elements of protection include the development and implementation of a hearing conservation program, noise monitoring, employee notification of exposure, and audiometric testing of employees. Affected employees must be provided with hearing protection and training.

Personal protective equipment standards. Several OSHA standards cover personal protective equipment other than respirators. They describe requirements for protection of the face and eye (1910.133), head (.135), foot (.136), electrical protective equipment (.137), and hand (.138). Personal protective equipment worn in confined spaces must comply with these regulations.

CFR 1910.1000, *Air Contaminants.* Tables Z-1, Z-2, and Z-3 of this standard list legally enforceable atmospheric concentration limits for air contaminants, including those encountered inside confined spaces and in nonconfined work areas. Parts 1910.1000 through 1910.1051 of the same Subpart Z set limits for chemicals having separate individual standards.

Radiation standards. Two standards covering exposure to radiation have been promulgated: one for nonionizing radiation (CFR 1910.97) and one covering ionizing radiation (CFR 1910.1096). These standards apply to radiation protection in confined space operations.

Welding, cutting, and brazing operations. "Hot work" is regulated under CFR 1910.251-255. Work of this kind in confined spaces must comply with these standards found in Subpart Q of the *OSHA Standards for General Industry*.

Regulations covering confined space work in specific industries. There are quite a few industry-specific OSHA standards that cover work in confined spaces. They are listed here with the recommendation that workers and employers in these industries seek further information. Most of them are found in 29 CFR 1910 Subpart R, "Special Industries." See Table 2.2 for a summary of OSHA standards covering confined space entry in specific industries.

Regarding the coverage of industry-specific standards versus coverage under CFR 1910.146, OSHA makes the following comment: "Confined space hazards in general industry that are not addressed by an industry-specific standard will be covered by 1910.146." OSHA has ruled that employers with spaces covered by a specific industry standard still must do an initial evaluation to determine the status of other confined spaces not covered by the industry specific standard.

29 CFR 1910.261, *Pulp, Paper and Paperboard Mills.* Among its many safe work provisions for the pulp-and-paper industry, the standard includes requirements for personal protective equipment, safe work surfaces, lockout of hazardous energy, and a separate paragraph on vessel entry. 29 CFR 1910.262, *Textiles*, describes requirements for personal protective equipment, ventilation, and other measures to protect workers who enter open surface tanks. Both the textiles and pulp/paper standards incorporate the PRCS standard by reference.

29 CFR 1910.263, *Bakery Equipment.* This standard mentions storage bin entry. 29 CFR 1910.268, *Telecommunications,* mandates protective measures for employees working in manholes or other locations where they must be protected from hazardous atmospheres, vehicles, and pedestrian traffic. The standard requires atmospheric testing of confined spaces and subsequent ventilation.

29 CFR 1910.272, *Grain Handling Facilities.* This standard has the most extensive provisions for special industry employees working in confined spaces. It requires training for employees who enter grain storage structures, a permit system for entry, lockout of energy, atmospheric testing, ventilation, lifelines, observers, and rescue capability.

Shipyards. Requirements for confined and enclosed spaces and other dangerous atmospheres in shipyard employment are found in Subpart B of 29 CFR 1915. Sections of this subpart describe required precautions and order of testing before entering confined and enclosed spaces, cleaning, and hot work (welding, etc.). Compliance assistance guide-

TABLE 2.2   U.S. Regulations Covering Confined Space Entry and Rescue in Specific Industries

Number	Industry	Coverage
29 CFR 1910.261	Pulp, paper, and paperboard mills	Includes confined space work in the pulp-and-paper industry
29 CFR 1910.263	Bakery equipment	Entry into storage bins
29 CFR 1910.268	Telecommunications	Manholes and other confined spaces
29 CFR 1910.272	Grain handling facilities	Entry into grain storage structures
29 CFR 1915, Subpart B	Shipyards and other maritime work	Confined space safety in shipyard employment
29 CFR 1926.21	Construction	Definitions and examples of confined spaces
29 CFR 1926.651	Construction	Entry into excavations
29 CFR 1926.800	Construction	Tunnels, shafts, underground chambers, and passageways
29 CFR 1926.956	Construction	Underground power transmission and distribution lines

lines are included. The standard references U.S. Coast Guard regulations for the determination of Coast Guard Authorized Persons. OSHA has stated that shipyard employees are not covered by the PRCS standard unless there are hazards present that the 29 CFR 1915 does not address.

Construction standards.   Safety in the construction industry (Fig. 2.3) is regulated by 29 CFR 1926, which covers various aspects of confined space construction. This standard includes safety training and education (1926.21); excavations including "deep and confined foot excavations" (1926.651); underground construction such as tunnels, shafts, chambers, and passageways (1926.800); and a brief discussion of construction of underground power transmission and distribution lines (1926.956). The excavation standard includes access control, the establishment of checkin and checkout procedures, safety training, air monitoring with specific limits listed, ventilation, communication, and control of hazards. It also describes required supports and bracing.

Section 1926.21 of the construction safety standard defines confined spaces in construction in the same way as does the PRCS standard, and offers as examples storage tanks, process vessels, bins, boilers, ventilation or exhaust ducts, sewers, underground utility vaults, tunnels, pipelines, and open-top spaces more than 4 feet in depth such as pits, tubs, vaults, and vessels.

The general industry confined spaces standard covers construction industry employees doing maintenance work in confined spaces. In determining when maintenance or modification is defined as construction, and then would be covered under CFR 1926 (construction safety) instead of PRCS (general industry), OSHA defines the difference. "Refurbishing of existing equipment and space is maintenance; reconfiguration of space or installation of substantially new equipment (as for a process change) is usually construction." OSHA provides the following examples of work defined as maintenance rather than construction:

- Relining a furnace or sewer line

- Brick replacement in a manhole

- Lining a tank, including patching or removal and replacement

Even though the employer is a construction company, these jobs are defined as maintenance and covered by the PRCS standard rather than the construction safety standard.

State plans.   The OSH Act encourages states to develop and operate, under OSHA guidance, state job safety and health plans. State plans must provide standards and enforcement programs, as well as voluntary compliance activities, which are at least as effective as the federal program. They must also provide coverage for state and local government employees. Plan states must adopt standards comparable to the federal standards within 6 months of a federal standard's promulgation.

Figure 2.3 29 CFR 1926 protects the safety and health of construction workers and covers confined space safety in construction work.

The states and territories with their own OSHA-approved occupational safety and health plans are Alaska, Arizona, California, Connecticut (for state and local government employees only), Hawaii, Indiana, Iowa, Kentucky, Maryland, Michigan, Minnesota, Nevada, New Mexico, New York (for state and local government employees only), North Carolina, Oregon, Puerto Rico, South Carolina, Tennessee, Utah, Vermont, Virginia, Virgin Islands, Washington, and Wyoming.

Permit program requirements

According to the PRCS standard, an employer who decides that employees will enter permit spaces must develop and implement a compliant written permit space program. The written program must be available for inspection by employees and their authorized representatives (usually meaning organized labor representatives).

The first step is to determine whether there are spaces that meet the definition of "confined" and whether these confined spaces meet the criteria for requiring permits. The employer may decide, following these determinations, that employees will not enter the spaces, in which case they are posted and rendered safe from entry.

In Chap. 1 we explained the determination of nonpermit-required and permit-required confined spaces. An employer having permit-required spaces (usually just called *permit spaces*) must develop and implement a written program that includes certain elements described in the PRCS standard. OSHA requires employee participation in the development and implementation of all aspects of the permit space program. All information required to be developed as part of the program must be made available to affected employees and their authorized representatives.

There are several *alternate procedures*. A confined space where the only hazard is an actual or potential hazardous atmosphere can be declared a nonpermit space if, and only if, the following conditions are met. The employer must demonstrate that atmosphere is the only hazard and that continuous forced air ventilation alone is sufficient to maintain the space safe for entry. The employer must develop monitoring and inspection data documenting atmospheric safety and make these available to each entrant or authorized representative. If entry is made to obtain the data, it must be made in compliance with permit space requirements.

Conditions making it unsafe to remove an entrance cover must be eliminated before the cover is removed; then the opening must be promptly guarded to prevent the accidental fall of people or foreign objects. Before anyone enters, the concentration of oxygen, flammable gases and vapors, and toxic air contaminants must be assessed (in that order) with calibrated direct-reading instrument(s). Periodic testing must take place, and, if a hazardous atmosphere is detected, all entrants must exit and protective measures be implemented. All these precautions are part of a dated written certification.

A permit-required confined space may be *reclassified* as a nonpermit-required space if there are no actual or potential atmospheric hazards and all other hazards are eliminated without entry. Entries to eliminate hazards must be undertaken under permit space conditions. Control of atmospheric hazards through forced-air ventilation does not constitute elimination. Documentation and certification, as well as the signature of the person making the determination of nonpermit-required status, are required and must be made available to entrants. Exit and reevaluation must take place if hazards arise.

Note that forced-air ventilation is not an allowable control for a hazardous atmosphere in a reclassified space, but can be used when alternate procedures, as described above, are implemented.

Guidelines for the written permit program. According to the PRCS standard, the employer must include the following elements in the written confined space program:

Unauthorized entry. Explain the implementation of measures necessary to prevent unauthorized entry. This can be done by posting signs, erecting barriers, or training workers about spaces not to be entered. In a letter of interpretation, OSHA specified that measures could also include permanently closing the space, as well as bolting and locking the space. Before choosing methods, the employer should evaluate the spaces present, since entry prevention means are not necessary for spaces not meeting the definition of confined spaces. Contractors and their employees must be informed about spaces that are not to be entered.

Hazard identification. Carry out identification and evaluation of the hazards of permit spaces before employees enter them. This is done by methods described in Chap. 3 and further explained in Chap. 4. Describe the means and methods for identifying and evaluating hazards in the permit program.

Safe permit space entry provisions. Identify the means, procedures, and practices necessary for safe permit space entry including, but not limited to, the following. Specify acceptable entry conditions and spell out means to allow each authorized entrant to observe air monitoring or testing. State how the employer will isolate the permit space, purge or ventilate the space, protect entrants from pedestrian and vehicle hazards, and verify acceptable conditions throughout the duration of entry. Describe methods to provide equipment for atmospheric testing, ventilation, communications, lighting, rescue, safe entry and exit, personal protective equipment, and any other equipment necessary for safe entry and rescue.

Evaluation of conditions. Describe methods for evaluating permit space conditions before entry and during the course of entry. The standard recommends testing first for oxygen, then for combustible gases and vapors, and then for toxic gases and vapors. Provide entrants or their authorized representatives an opportunity to observe the preentry tests and any subsequent monitoring. If an entrant or authorized representative requests reevaluation because there is reason to believe that the space may not have been evaluated adequately, the employer must test again. Provide results of testing immediately to the authorized entrant or that person's representative.

Attendants. Describe in the program the provision of at least one attendant outside the permit space for the duration of entry operations. If an attendant is to monitor multiple spaces, which is allowed, include the means and procedures that will enable the attendant to respond to an emergency affecting one or more of the spaces without distraction from that attendant's responsibilities.

Personnel roles and duties. Identify the duties of persons with active roles in the permit program. Individuals included in this list are entrants, attendants, entry supervisors, and atmospheric testers. Spell out how these people will be trained.

Rescue. Procedures for summoning rescue and emergency services and for rescuing entrants from permit spaces are important parts of the written program. These requirements were strengthened in the 1998 revision of the standard. This part of the program also must describe procedures for preventing unauthorized personnel from attempting a rescue.

Entry permit. The entry permit is to be outlined in the written program. The program must describe how permits will be prepared, issued, used, and canceled. Entry permits are covered in Chap. 5.

Coordination of multiple employers. Coordination of entry operations is required when employees of more than one employer are working in a permit space. The program must ensure that employees of one employer do not endanger the employees of any other employer. The standard does not require host and contractor employers to use the same permit program, nor does it prohibit a host employer from requiring a contractor to use the host's program. The host employer is obligated to ensure that the contractor abides by a compliant program. Whether contractors develop site-specific programs for different workplaces or use the host employers' programs, the host must advise contractors of permit spaces, inform the contractor of hazards and precautions, compel compliance with a permit program, and conduct a postentry debriefing in which the contractor informs the host of any hazards confronted during entry operations.

Conclusion of entry. Explain the procedures for concluding an entry. The permit program describes the development and implementation of procedures such as closing off a space and canceling the permit after entry operations have been completed.

Program review. Outline and explain the review of the permit space program. Canceled permits must be retained for one year after each entry; these can be used to review and revise the program if necessary to ensure that employees are protected from permit space hazards. If no entry is made during a 12-month period, no review is necessary.

Personnel Roles and Training Requirements

The confined space standard requires the employer to provide training so that all employees whose work is regulated by the standard "acquire the understanding, knowledge and skills necessary for the safe performance of the duties assigned." Training must take place before the first assignment to confined space duties, before there is a change in assigned duties, and whenever there is a change in permit space operations presenting a hazard the employee has not been trained about. At any other time that the employer believes the employee is not knowledgeable about procedures or is deviating from permit program procedures, training must be done to correct the deficiency.

"All employees" includes contractors' employees. Training is the responsibility of the contract employer; however, a wise host employer will verify the contractor employees' training, and may want to evaluate their training programs. Questions regarding the content of training, not just the existence of training, are likely to garner more realistic answers on which to base an evaluation of contract employees' knowledge and skills.

Documentation of training must contain employees' names, signatures or initials of the trainers, and the dates the training took place. Training content should be specific to the assigned duties. OSHA has explained in several letters of interpretation that watching videos and interacting with computer-based programs do not fulfill regulatory training requirements without accompanying discussions with a competent instructor.

Training and duties of authorized entrants

The employer must ensure that all authorized entrants (Fig. 2.4) meet the following performance criteria, as listed in the standard:

Hazards. Entrants must know the hazards that may be faced during entry. This includes the mode (source and route), symptoms, and consequences of the exposure. They must know both physical and chemical exposure hazards.

Equipment use. Entrants must be able to properly use equipment, including equipment for atmospheric testing, ventilation, communications, lighting, rescue, safe entry and exit, personal protective equipment, and any other equipment necessary for safe entry and rescue that may be required of them.

Figure 2.4 Authorized entrants must meet the performance criteria spelled out in the standard.

Communications. Entrants must know procedures for communicating with attendants as necessary for attendants to perform their jobs. Entrants should know how to alert attendants on recognition of a warning sign of exposure to danger or a prohibited condition.

Entrant evacuation. Entrants must know that they are to exit the permit space whenever an attendant or entry supervisor orders evacuation, exposure to danger is recognized, a prohibited condition is detected, or an evacuation alarm is activated.

Training and duties of attendants

The attendant is responsible for all of the duties described below. Attendants' primary duty is to monitor and protect the authorized entrants, and they may perform no other duties that might interfere with these. It is the employer's responsibility to ensure that attendants are trained and competent in all these duties.

Hazards. Attendants must know the hazards of the space, including information on the mode, signs or symptoms, and consequences of exposure to the hazards. They must be aware of possible behavioral effects of hazard exposure. Some chemical exposures result in dizziness, confusion, poor judgment, or declining physical coordination, problems the attendant must be on the lookout for when these chemicals are present in the space.

Count of entrants. The attendant must continuously maintain an accurate count of entrants in the permit space, and ensure the procedures outlined in the permit program for accurately identifying entrants are being used properly. When more than one space is being monitored by the same attendant, it is necessary to identify entrants in each space.

Location of attendant. Attendants must remain outside the permit space until relieved by another attendant. Attendants trained and equipped for rescue operations may enter to perform a rescue, but only after being relieved by another authorized attendant.

Maintaining communication. The attendant must maintain communication with entrants (Fig. 2.5) to monitor entrant status and to alert entrants if there is a need to evacuate. Attendants monitor activities inside and outside the space to ensure continued safe conditions and order entrants to evacuate immediately under any of these conditions:

- Detection of a prohibited condition, as described in the permit program
- Detection of the entrant's behavioral effects of hazard exposure
- A situation outside the space that could endanger entrants
- Any reason why the attendant cannot effectively and safely perform all required duties

Figure 2.5 The attendant must maintain communication with the entrant to monitor status and alert for exit.

Summoning rescue. The attendant summons rescue and other emergency services immediately on determining that entrants may need assistance. Attendants may perform nonentry rescue according to the procedures set forth in the written permit program.

Unauthorized persons. The attendant takes action when unauthorized persons approach or enter a permit space when entry is under way. The attendant warns the persons that they must stay away, advises them that they must exit immediately if they have entered, and informs the authorized entrants and entry supervisor if unauthorized persons enter the space.

Training and duties of entry supervisors

The entry supervisor must know what actual or potential hazards may be encountered in the permit space and verify that all permit program procedures are being followed. The entry supervisor signs the permit.

Hazards. Supervisors are aware of entry hazards, including information on the mode, signs or symptoms, and consequences of the exposures. Hazards include both physical (fire, falls, etc.) and chemical exposure possibilities.

Testing. It is the supervisor's job to verify that testing has been done, equipment specified by the permit is in place, and all required parts of the permit accurately filled out. When this is done, the supervisor endorses the permit and allows entry to begin.

Termination. The supervisor terminates entry and cancels the permit when the entry is completed. The duration of the permitted entry is determined by the permit program and stated on the permit.

Rescue service availability. Verification of rescue services is the responsibility of the entry supervisor. This may require a call to an off-site rescue service whenever an entry is to be made. The supervisor also verifies that the means for summoning rescue are operable.

Unauthorized persons. Unauthorized individuals are removed if they enter or attempt to enter the permit space during entry operations. The entry supervisor is responsible for seeing that they are removed.

Consistency of permit. The entry supervisor determines that entry operations remain consistent with the permit terms whenever responsibility for a permit space entry operation is transferred. This also is done at intervals dictated by the hazards and operations performed in the space.

Requirements for rescue and emergency services personnel

The burden of rescue team competency falls on both the rescue personnel and the employer depending on their services. In the 1998 revision to the PRCS standard, OSHA clarified the employer's responsibilities in evaluating, selecting, and informing rescue personnel. Employers now must prepare for rescue in one of two ways; they must either train and equip an on-site team of company employees or evaluate the training, equipment, competency, and response time of an outside contractor or public agency. "Call 911" is no longer acceptable as a rescue option, unless responders at the 911 dispatch center meet all the criteria required by the standard.

29 CFR 1910.146 *Non-Mandatory Appendix F*—Rescue Team or Rescue Service Evaluation Criteria—is found in App. III to this book. A summary of its requirements follows.

Quick response.   Timely response is critical. "Timely" may vary according to the specific hazards involved in each entry. For example, the respiratory protection standard requires the employer to provide a standby person or persons capable of immediate action to rescue employees wearing respirators while in work areas with "immediately dangerous to life and health" (IDLH) atmospheres. IDLH atmospheric concentrations are available from NIOSH and can be found on MSDS for individual chemicals. For permit spaces where the hazards are mechanical and rescue may be required for such injuries as abrasions and broken bones, 10 to 15 minutes may be an acceptable response time. The employer must verify that the rescue team selected has the capability to reach the victim within a timeframe that is appropriate for the permit space hazards identified. The employer must verify the availability of an off-site rescue team each time a permit space entry is performed.

Proficiency of rescue teams.   Rescue personnel must be equipped for and proficient in performing the anticipated rescue services (Fig. 2.6). Whether they are a company team or an off-site rescue service, the employer must verify their training and proficiency in

Figure 2.6   Rescue personnel must be proficient in performing the rescue services that they anticipate will be necessary.

confined space entry as well as first aid and cardiopulmonary resuscitation. The employer (for in-house rescue teams) or the service (for off-site teams) must provide rescuers with all the necessary equipment, including personal protective equipment, and train rescue personnel on its use.

It is not necessary for an off-site service to have a written confined space permit program. It is required that they have members who are trained, equipped, and practiced for safe entry into the particular kinds of permit spaces from which they will be expected to rescue entrants.

Hazard information. Rescue personnel must be informed of all the potential hazards of the permit space. Information about potential hazards outside the permit space, such as work being done nearby, must also be provided by the employer. Rescue teams should be familiar with the employer's permit spaces, and proficient in rescue from all the configurations of spaces found at the company.

Practice. Rescue practice is essential. Employee rescue teams must practice making permit space rescues at least once every 12 months by means of simulated rescue operations. Rescue personnel should remove dummies, manikins, or actual persons from the actual permit spaces or from representative permit spaces that simulate the types of spaces from which rescue will be performed with respect to opening size, configuration, and accessibility (Fig. 2.7). Employers who select off-site rescue services should ask for the results of critiques of practice rescues that the service has performed in order to identify any deficiencies and their corrections. If the off-site service has not performed a rescue within 12 months, the team must perform a practice rescue.

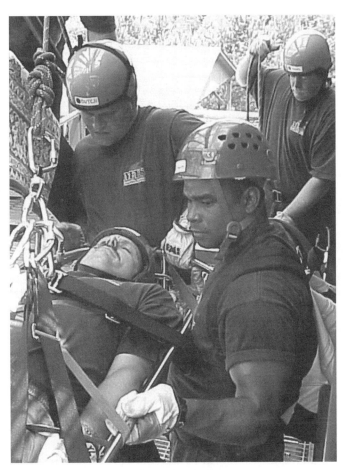

Figure 2.7 Rescue teams should practice removing dummies or actual persons from actual or similar spaces.

Preparation for nonentry rescue. Authorized entrants must use retrieval systems to facilitate nonentry rescue unless the equipment would increase the risk of entry or would not contribute to rescue. Retrieval systems are described in the PRCS standard and will be discussed further in Chap. 13.

Employer's responsibilities. Let us sum up the rescue responsibilities of those who own and are responsible for permit confined spaces, as they have become more burdensome and specific with more recent changes in the PRCS standard. The employer must complete one of these two tasks before an entry is made into a permit-required confined space:

- Train and equip a competent in-house rescue team that meets all the criteria of the PRCS standard, including technical competencies and practice.

- Contract or agree with an outside rescue team or service, public or private, to be on standby for IDLH entries and available during every other permit entry, using nonmandatory Appendix F of the standard (reprinted as App. III to this book) to evaluate the competency and equipment of the rescue service. The evaluation is mandatory, as stated in 1910.146(k)(1)(i), although evaluation methods other than Appendix F can be used.

Appendix F lists questions an employer should ask the outside rescue service regarding response time; availability; communications; equipment; and skills in rope rescue, medical evaluation, patient packaging, and emergency response. The employer should determine whether the necessary training and practice have taken place, if the team can properly test the atmosphere, and whether they perform rescue work safely and efficiently. Using these evaluation guidelines will go a long way to ensure safe and OSHA-compliant rescue plans.

National Fire Protection Association Standards

The National Fire Protection Association (NFPA) is an international nonprofit membership organization founded in 1896. The NFPA's stated mission is to reduce the worldwide burden of fire and other hazards on the quality of life by developing and advocating scientifically based consensus codes and standards, research, training, and education. The organization's activities fall into two broad, interrelated areas: technical and educational. *Technical activity* involves developing, publishing, and disseminating more than 300 timely consensus codes and standards; these are developed by technical committees made up of more than 6000 volunteer representatives. *Educational activities* focus on motivating and enabling the public to be safer.

NFPA standards are recognized by fire departments as national consensus standards; departments voluntarily comply with them in the absence of federal or state standards for many of the topics described by NFPA standards. Because of the detailed nature of NFPA standards, they will be mentioned here only briefly; responders whose organizations comply with NFPA standards should use the standards as a source of further information. Table 2.3 summarizes the five NFPA standards relating to confined space entry and rescue.

NFPA 1670, *Standard on Operations and Training for Technical Rescue Incidents*

This standard identifies and establishes levels of functional capability for safely and effectively conducting operations at technical rescue incidents. Requirements apply to organizations that provide response to technical rescue incidents, including incidents involving confined spaces.

The standard requires the authority having jurisdiction to establish levels of operational capability and to establish written standard operating procedures consistent with one of several operational levels. Subsequent chapters cover rescue in structural col-

TABLE 2.3   National Fire Protection Association Standards Covering Confined Space Entry and Rescue

Number	Title	Coverage
NFPA 1670	*Standard on Operations and Training for Technical Rescue Incidents*	Identifies levels of functional capability for organizational operations at technical rescue incidents, including confined spaces
NFPA 1006	*Standard for Rescue Technician Professional Qualifications*	Describes professional competencies for individuals who perform rescue
NFPA 1983	*Standard on Fire Service Life Safety Rope and System Components*	Inspection, testing, manufacturers' quality assurance programs, certification of equipment, product labeling, and information for rescue equipment
NFPA 1404	*Standard for a Fire Department Self-Contained Breathing Apparatus Program*	Use, maintenance, and inspection of SCBA units; certification, fit testing, and training of users
NFPA 472	*Standard for Hazardous Materials Response*	Knowledge and competencies regarding exposure to potentially hazardous chemicals

lapse, rope rescue, confined space rescue, vehicle and machinery rescue, rescue from water, wilderness search and rescue, and trench and excavation rescue. In each chapter, operational capabilities are described for rescuers at the awareness, operations, and technician levels. The general requirements for all types of rescue include hazard analysis and risk assessment, incident response planning, equipment, and safety.

Chapters on confined space rescue and trench rescue reference NFPA 472, *Professional Competence of Responders to Hazardous Materials Incidents,* and state that organizations that provide confined space and trench rescue shall meet the requirements of NFPA 472 Chapter 2 ("Awareness and Operations Level First Responders").

NFPA 1006, *Standard for Rescue Technician Professional Qualifications*

Where NFPA 1670 provides guidelines for rescue organizations, 1006 describes professional competencies for individuals who perform rescue. It is divided into the same rescue categories as 1670 (omitting wilderness search and rescue) and lists age, medical, physical fitness, emergency medical care, educational, and hazardous-materials (hazmat) requirements before going on to describe job performance requirements for each type of rescue. Job performance requirements are divided into knowledge and skill categories.

NFPA 1006 includes several useful appendices that describe and illustrate administrative organization, first aid and rescue tools, structural drawings, and written forms, as does NFPA 1670. Many of NFPA's professional qualifications and administrative guidelines concerning confined space and rope rescue are incorporated into the techniques covered in later chapters of this book, especially those chapters detailing hazard control, equipment, and rescue techniques.

NFPA 1983, *Standard on Fire Service Life Safety Rope and System Components*

The standard was produced by the committee on fire and emergency services protective clothing and equipment. It provides information on inspection and testing, manufacturers' quality assurance programs, certification of equipment, and product labeling and information. It describes NFPA requirements for the design and construction, performance, and testing of life safety rope, personal escape rope system components, life safety harness system components, belt system components, and auxiliary equipment system components.

Manufacturers whose products meet the requirements of this standard may label their products as follows (shown in Fig. 2.8). "THIS [product type, e.g., "BELT"] MEETS THE BELT REQUIREMENTS OF NFPA 1983, STANDARD ON FIRE SERVICE LIFE SAFETY ROPE AND SYSTEM COMPONENTS, 1995 (or current) EDITION; TYPE (type of belt)." The certification organization's label or symbol is printed, then the name of the manufacturer, the manufacturer's product identification, manufacturer's lot or serial number, month and year of manufacture, and country of manufacture. For products that come in different sizes, the size is indicated. Additional warning labels are required on all equipment, beginning with the attention-grabbing words "YOU COULD BE KILLED OR SERIOUSLY INJURED...."

NFPA 1404, *Standard for a Fire Department Self-Contained Breathing Apparatus Program*

NFPA states unequivocally that "no member should be permitted to enter a confined space for...emergency rescue operations without wearing a SCBA." It outlines procedures for using, maintaining, and inspecting SCBA units and for certifying, fit testing, and training users. This standard is discussed further in Chap. 7.

Additional Standards and Guidelines

There is a plethora of standards describing the construction and testing of equipment use in workplaces, including confined spaces. Their utility is primarily in determining whether tools and equipment are adequate for intended uses.

American National Standards Institute (ANSI)

ANSI serves as the administrator and coordinator of the United States private-sector voluntary standardization system. The institute facilitates development of American National Standards (ANSs) by establishing consensus among qualified groups, and has written over 14,000 standards. OSHA references several ANSs; especially relevant to this book are those describing criteria for items of personal safety equipment. These are discussed in Chap. 9. ANSI also publishes guidelines for safe work; for example, *Safety Requirements for Confined Spaces* is available for purchase from the institute.

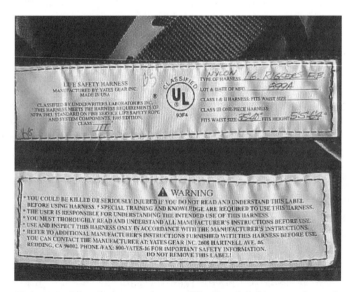

Figure 2.8 Manufacturers place certification labels on equipment meeting NFPA standard requirements.

American Society for Testing and Materials (ASTM)

ASTM writes specifications for conducting the various materials tests referenced in OSHA regulations. The organization develops and provides voluntary consensus standards and related technical information. The stated goal is to promote public health and safety and the overall quality of life; contribute to the reliability of materials, products, systems, and services; and facilitate national, regional, and international commerce. ASTM, for example, has written specifications for testing insulating gloves, matting, blankets, and covers for protection against electrical exposure.

National Institute of Occupational Safety and Health (NIOSH)

NIOSH does not promulgate standards; rather, the institute investigates workplace safety issues and develops criteria documents for OSHA's reference in writing regulations. NIOSH has a large research capability and offers publications of all kinds related to workplace safety. Supported by federal funds, NIOSH is able to provide free copies of scientific documents and articles written for scientists and laypeople.

Summary

Poring over federal, state, and voluntary regulations is not exciting, nor is it easy. Regulations, however, especially those promulgated by OSHA, provide an excellent structural framework for safe work practices in confined spaces. OSHA regulations offer answers to most procedural questions and, with few exceptions, must be followed. Information that cannot be found in the regulations, or is confusing, often can be clarified by reading through letters of interpretation OSHA employees have written in response to questions posed by workers and employers. These are available in writing and online from OSHA. Compliance with OSHA confined space regulations should ensure safe entry and rescue, and prevent employer citations and fines.

Organizations that develop and publish other standards, some referenced by OSHA and therefore mandatory and some nonmandatory, should be consulted whenever additional information is needed to ensure the safety of people who enter confined spaces. NFPA members' familiarity and experience with fire service procedures result in the promulgation of excellent descriptions of rescue operations and competencies for departments and individuals involved in confined space rescue.

3

Identifying Confined Space Hazards

Two workers climbed onto an elevated water storage tank, entered the tank down a ladder from the access portal on top, and began painting the inside of the tank. When they did not exit at the expected time, their supervisor climbed up and entered the tank to check on them. He found them on the tank floor. One man was unconscious; the other was conscious but unable to climb out because of the slope of the floor. The supervisor fell from the ladder, and he and one painter began sliding toward the central pipe in the bottom of the tank, aided by the slope and the slippery wet paint. One was able to arrest his fall, but the other slid into the pipe and fell 110 feet to his death. The two remaining men were rescued from the tank, but one died the next day. One of the three men survived.

The men made a number of mistakes in the light of currently held safe work practices, and they encountered hazards of several kinds: The physical hazards of height, an unprotected ladder, poor visibility, slope and slippery footing, the chemical hazard of toxic paint vapors, potential oxygen deficiency due to displacement by paint vapors, and the possibility of a flammable atmosphere all were factors. Elevated water tanks, such as the one pictured in Fig. 3.1, are only one of the many types of confined spaces where hazards must be identified before entry.

The National Institute of Occupational Safety and Health (NIOSH), in a study of confined space fatalities that led OSHA to write the PRCS standard, reported that the two major factors leading to fatal injuries in confined spaces during the study period were (1) failure to recognize and control the hazards associated with confined spaces and (2) inadequate or incorrect emergency response. Although the standard requires hazards to be identified before entry, deaths resulting from failure to assess potential hazards in confined spaces continue to occur.

The National Traumatic Occupational Fatalities (NTOF) surveillance team reported the number of confined space fatalities between 1980 and 1989 was highest in manufacturing, followed by agriculture, construction, transportation–communication–public utilities combined and the mining-oil-gas industry. The NTOF identified asphyxiation as the leading cause of death, claiming 45 percent of those who died. In 70 on-site fatality investigations, NIOSH found the most common reason for entry into the spaces to be repair and maintenance, primarily into tanks, vats and pits, digesters, and sewer manholes. The second highest number of victims entered the confined space in rescue attempts.

With all the attention on confined space fatalities and a clear focus by OSHA on the prevention of deaths and injuries, why do workers and their would-be rescuers continue to die in confined spaces? As NIOSH did in 1989, we can still lay much of the blame on failure to recognize and control hazards prior to entry.

Figure 3.1 This unusual water tower, in a town where peaches are the principal crop, illustrates all the usual dangers of a water tower structure: height, ladders, sloping floor, and discharge pipe of sufficient size for a worker to fall into.

Physical Hazards

Physical hazards are those things that have the capability to hit, crush, strike, shock, engulf, overheat, or chill a confined space worker or cause that person to fall. The variety of potential physical hazards is great, and the hazards are exacerbated by the small space where work takes place.

Energy

Because of the kind of work done in confined spaces and the reasons for entering the spaces, it may be difficult to separate the worker from hazardous forms of energy such as powered machinery, electrical energy, and hydraulic or pneumatic lines. Control of hazardous energy will be discussed in Chap. 6, where methods of locking out energy will be explained in detail.

Electrical. Depending on the voltage of the circuit and the path of current flow through the body, electricity can have devastating effects on a worker who contacts an electrically energized conductor. Muscles contract involuntarily and cannot be relaxed, in some cases preventing the victim from letting go of the conductor. Even a mild shock causes muscle contractions leading to other injuries if the victim falls. Nerves controlling the heart are affected, and heartbeat is disrupted and may stop entirely if cardiac nerves are paralyzed. Paralysis of other nerves can cause breathing to stop. In cases where death does not occur, electricity moving through the body generates heat that burns tissues on the path of current flow.

Wet surfaces enhance the risk of electrical shock as they serve as additional conductors. Rescuers who touch a victim still in contact with an energized source also are at risk.

Mechanical.   Mechanical hazards in confined spaces are visible. Entrants know that they are there and should be able to prevent injury, primarily by locking out the equipment before entering the space. Still workers die when moving machinery traps, hits, crushes, or cuts them. Luckier victims survive, but often with missing limbs. Mechanical hazards include fixed equipment, such as mixers, blenders, conveyors, and augers, or equipment brought into the space by the entrant to be used in the work being done.

According to OSHA, the most hazardous kind of confined space is the type that combines limited access and mechanical devices. All other kinds of confined space hazards are possible here, with the additional hazard of moving parts. Digesters and boilers usually contain power-driven equipment that, unless properly isolated, may be inadvertently activated after entry. Such equipment may also contain physical hazards such as baffles and narrowing shapes that further complicate the work environment and the entry/exit process.

Pressure.   Confined space work may require the use of pressurized devices, or pressurized lines may already be fixed inside the space. Piping that contains contents under pressure offers both the hazards of the contents, if they are toxic, corrosive, hot, or otherwise dangerous in themselves, plus the hazard of accidentally released pressure. Piping should be isolated using one of the methods described in Chap. 6.

Portable pressurized equipment includes anything the entrant brings inside the space for working. Pressurized lines used in cleaning and spraying are dangerous if they break, or if the pressure increases beyond the worker's ability to control the line. Pressurized cylinders present two kinds of hazards: the hazard of the gas they contain, if it is hazardous, and the danger of the cylinder bursting under pressure and rocketing or becoming flying shrapnel.

Engulfment

Engulfment in solid or liquid material can be fatal within minutes when breathing stops. In the sewer fatality incident described at the beginning of Chap. 2, two of the four deaths resulted from engulfment and drowning in raw sewage, a semiliquid slush. Confined space entrants should be able to predict the presence or potential for inflow of materials capable of engulfing them so protective measures can be implemented.

Liquids.   Even workers who can swim have drowned in water as the water rose too high and filled the breathing space, or no one rescued them before they tired and sank. Liquids that are toxic or give off toxic vapors cause entrants to become disoriented and lose consciousness; unable to self-rescue, they eventually fall into the liquid and die. Engulfment in a liquid causes death either by filling the lungs as the dying person gasps for air, or simply covering the nose and mouth so the victim suffocates.

Sewer system entry is a special situation where surge flow and flooding should be predictable. Sewer crews should develop and maintain liaison with the local weather bureau so that sewer work may be delayed or interrupted and entrants withdrawn whenever sewer lines might be suddenly flooded by rain. It is also possible for flooding to occur following fire suppression activities when the water applied by firefighters moves into the sewer system.

Solids.   Solids like dirt and grain cause death by a slower mechanism. A victim engulfed only to the armpits may die after slowly losing the ability to breathe as the torso is squeezed tighter and tighter by the material. With each exhalation, the grain or other finely divided solid moves into the space created around the torso and prevents expansion of the rib cage for another inhalation. Victims of cave-ins who are completely engulfed may die when they use up all the oxygen in the small space around the mouth and nose.

Flowing materials. Engulfment and suffocation are hazards associated with storage bins, silos, and hoppers where grain, sand, gravel, or other loose materials are stored, handled, or transferred. The behavior of such material is unpredictable, and entrapment and burial can occur in a matter of seconds. Material being drawn from the bottom of storage bins can cause the surface to act like quicksand. When a storage bin is emptied from the bottom, the flow of material forms a funnel-shaped path over the outlet, and the rate of material flow increases toward the center of the funnel. During a typical unloading operation, the flow rate can become so great that once a worker is drawn into the flow path escape is virtually impossible. From the time a bin unloading auger starts, there may be only 2 to 3 seconds to react.

Grain bridging. Engulfment by grain is the most common confined space fatality in grain storage workers. As the grain is drawn into the bottom of the silo onto conveyor belts, a phenomenon known as "bridging" often occurs when grain clings to the sides of the bin. Voids develop in the bin, covered at the top by a solid bridge of grain that must be broken for the flow to continue. Small storage vessel diameters and high moisture content contribute to bridging.

Although there are safe methods for breaking up crust, and wearing fall-arrest lines is recommended, workers continue to die when grain bridges give way under their feet and they are engulfed in grain. Grain accounted for 55% of the fatal confined space engulfments reported in the NTOF study, followed by other agricultural products such as silage or fertilizer, sand, and building materials such as gravel, cement, and clay. Sawdust followed in frequency.

Entrapment

Confined spaces that are funnel-shaped offer the possibility that a worker will fall to the bottom and become trapped in the outlet opening. The opening may be too small for the worker to fall through, but large enough for legs and hips to enter. The worker is unable to pull out of the opening and can become prey to chemical overexposure (if the space has a hazardous atmosphere) to extreme temperatures, or to physical injury from the entrapment. The configuration of some spaces is narrow enough above the bottom outlet to constrict the chest and prevent inhalation (see Fig. 1.12 in Chap. 1).

Gravity

Gravity harms confined space entrants in two ways, by causing them to fall from a height sufficient to cause harm or by allowing objects to fall on them. Many confined space fatalities have been reported where the victims were dizzy because of oxygen deficiency or toxic gas exposure and were killed when they fell from ladders or through central openings. Fall-arrest systems should always be used where the potential for falls exists. These systems are required to be in place when the height of the potential fall exceeds 6 feet, and some kind of protection should be part of ladders used for entry and exit. OSHA lists cages, lifebelts, friction brakes, or sliding attachments as required safety devices for fall protection in general industry (29 CFR 1910.27). For confined space ladders in construction work, see 29 CFR 1916.1053, and 1917.118 for ladders in confined spaces at marine terminals. Fall protection is discussed more fully in Chaps. 13 and 14.

Slippery surfaces contribute to falls, as we saw in this chapter's opening case study that described workers painting the inside of an elevated water tank. Any work that adds liquids to walking surfaces increases the likelihood of falling.

Work in any confined space where there is activity above the space sets the stage for entrants to be struck by falling objects or earth. Multiple entrants endanger workers in lower parts of the space when they carry tools and equipment. Construction workers in trenches and pits are at risk from objects falling from above. Manhole entrants can expect that vehicles and their drivers will present hazards: vehicles by traveling at excessive speeds and drivers by throwing items such as beverage cans and cigarette butts from their windows. Department of Transportation standards diagram protective markings and barriers for utility work zones.

Noise

Sudden loud noise inside a confined space may startle workers and lead to falls. Continuing noise above 85 decibels leads to hearing loss, and is regulated by 29 CFR 1910.95, where permissible noise exposure levels are listed. A noise-monitoring program must be implemented when information indicates that any employee's exposure is 85 decibels or more over an 8-hour period. These workers must be notified of the hazard and the employer must make audiometric testing available to them.

If noise inside the space disrupts communication between the entrant and the attendant outside, the probability of serious accidents increases. This may happen with radios, but is more likely to be a problem when the communication is verbal without mechanical aid.

Temperature extremes

Extreme heat is in the news; as this is being written, an August heat wave is spreading across the midwestern and eastern United States, and five football players, including all-pro Korey Stringer of the Minnesota Vikings, have died from heat stroke. One player's body temperature was 109°F when he died. Heat-related illnesses killed 34 workers in 1999, and 2420 others experienced heat-related occupational injuries and illnesses serious enough to miss work, according to OSHA.

Heat stress.   Heat comes from two sources, environmental heat from sunshine, high ambient temperatures, and nearby equipment and metabolic heat the body generates to maintain a constant temperature of around 98.6°F. When people work, metabolic heat increases. The harder they work, the more heat the body produces. The human body has two built-in mechanisms for cooling when core temperature goes up; under low-stress conditions they work well.

Thermoregulatory systems.   When body heat increases, the "thermostat" inside the brain sets into motion two responses: vasodilation and sweating. The first response causes enlargement of the capillaries in the body surface, reddening the skin and allowing heat brought by the bloodstream from the body core to dissipate. The heat radiates out into the air—unless the air temperature is 95°F or higher, when there is not enough temperature gradient between skin and air to allow radiant heat loss to occur.

Sweating produces moisture on the skin surface, where it cools the body by evaporant cooling as the water changes into vapor. Evaporation cannot take place if the air is already saturated with moisture, as it usually is in the southeastern United States and often is in other places. Sweating will continue even if evaporation cannot take place and may deplete body fluids to a dangerous level.

If neither thermoregulatory mechanism removes heat from the body, as a result of high ambient temperature and/or high humidity, the body temperature continues to climb. The worker is in danger of heat-related illness.

Heat-related illness.   Heat rash is uncomfortable for a time, but not dangerous unless it renders the skin more permeable to workplace chemicals. Heat syncope (fainting) is not serious unless it leads to a bad fall or a head injury. Heat cramps, although very painful, are not a medical emergency. Heat exhaustion and heat stroke, however, are quite serious and possibly fatal. It is important to note that these are not sequential states of heat illness—any one may be encountered without being preceded by the others.

The symptoms of heat exhaustion may be dizziness; nausea; profuse sweating; reddened skin, especially in the face and chest; and a general feeling of exhaustion and malaise. Not all symptoms are necessarily present, although sweating always is, but the victim is quite aware that something is wrong. Treatment is rest in a cool place, plentiful liquids to drink (see text below for preferred liquids), and no return to work that day.

Heat stroke is an immediate medical emergency, and the victim is likely to die even if treatment is administered at once. Heat stroke is fatal in 80% of the people who suffer it, and half of the 20% who survive have permanent organ damage. Treatment is rapid cooling, both outside and inside the body, but do not immerse the victim totally in cold water as the extreme change in temperature may be dangerous when so much blood volume is in dilated skin capillaries.

Symptoms of heat stroke are variable, including hot skin; red face, neck, and chest; rapid heartbeat; rapid and shallow breathing; and confusion or unconsciousness, but the most obvious clue that this is heat stroke instead of heat exhaustion is the lack of sweat. The body has simply run out of fluids, and even though the brain continues to send nerve impulses to sweat glands to keep working, they are unable to do so. Considering the statistics on fatality as a result of heat stroke, prevention is critical.

Heat illness prevention. A variety of prevention methods should be considered; some will be more easily applicable in confined space work than others. Some are individually controlled, and some are employer-driven. All employers should have heat prevention programs. See the NIOSH *Occupational Exposure to Hot Environments, Revised Criteria* for recommendations. Individuals manage heat tolerance through a number of personal decisions and habits. Individual or employer heat stress prevention programs should address the heat protection factors described below.

Drinking large quantities of liquids, whether one is thirsty or not, is probably the most important prevention strategy. Sweating results in extensive fluid loss, leading to internal organs' inability to function. Sports drinks that promise energy contain considerable amounts of sugar, which slows down absorption, and they are not recommended unless diluted with water. It is unlikely that anyone other than an athlete expending extreme and extended effort needs the extra electrolytes sports drinks offer, as all necessary electrolytes are present in a normal healthy diet. Caffeinated drinks and alcoholic beverages increase the rate of fluid loss through frequent urination, and should be avoided when work in hot conditions is anticipated. Water is the best fluid for hydration. Drink 6 to 8 ounces every 15 minutes. Don't wait to feel thirsty; there is a time delay between fluid depletion and a feeling of thirst.

Acclimatizing to heat takes at least a week, and cannot be accomplished in an air-conditioned room. Gradually increasing work hours under hot conditions is the only way to acclimatize, and should be repeated after a restful vacation or an illness involving water loss from vomiting or diarrhea.

Frequent rest breaks in cool places should be written into the heat stress prevention program, increasing in frequency as ambient temperature increases. Provide water at breaks, and monitor workers' temperatures if conditions are stressful. If protective clothing is worn for working in confined spaces, breaks should be even more frequent. Table 3.1 shows recommended rest breaks.

Cooling garments may provide heat relief during brief shifts of work inside confined spaces. Several kinds are on the market, including vests with pockets into which frozen packets of coolant are inserted and garments that move cool water or air through a tube system or into the space inside a totally encapsulating chemical protective suit. Many of these cooling devices are heavy and, except for those that circulate outside air, have short use durations, but should be considered for short workshifts in hot conditions.

TABLE 3.1   Recommended Rest and Monitoring Breaks for Fit Acclimatized Workers during Work in Hot Environments*

Temperature in space, °F	Normal work ensemble[†]	Impermeable ensemble[‡]
≤90	After each 45 minutes work	After each 15 minutes work
87.5–90	After each 60 minutes work	After each 30 minutes work
82.5–87.5	After each 90 minutes work	After each 60 minutes work
77.5–82.5	After each 120 minutes work	After each 90 minutes work
72.5–77.5	After each 150 minutes work	After each 120 minutes work

*For work levels of 250 kcal/h (kilocalories per hour)

[†]Cotton coverall, long sleeves and pants. Firefighter turnout gear is not normal work ensemble.

[‡]Chemical protective fabric, partial or full body coverage.

SOURCE: Adapted from *Occupational Safety and Health Guidance Manual for Hazardous Waste Site. Activities,* NIOSH/OSHA/USCG/EPA, 1985.

The newest generation of cooling garments incorporates new technology in the formulation of a liquid that changes to a semisolid when energized. The garment maintains a temperature of around 55°F to absorb body heat, and can be reactivated repeatedly. The vest encapsulating this material is lightweight, and is more comfortable to use than garments holding ice because of the lower weight and higher temperature of the contents.

Cold stress. Conditions are conducive to cold stress when the ambient temperature is low, there is wind (not likely inside a confined space), and the worker is wet. Physiological adaptations to cold stress are not as effective as those for heat reduction, and acclimatization is not as successful. Workers should avoid getting wet, especially their hands and feet, and should wear layers of insulating clothing. If hard work generates sweat, one or more of the layers can be removed for cooling. Clothing worn next to the skin should be made of fibers that wick away moisture and provide warmth when wet, such as wool or polypropylene fleece. Cotton is not a good choice, as it holds water and cools the body. Breaks in warm locations and liquids to replace body fluids lost to cold dry air are important components of a cold stress prevention program.

Musculoskeletal injuries

Injuries to muscles and joints can occur in any job where hard work is being performed. Inside confined spaces, where there may be little room to move and turn, ergonomic injuries are enhanced by using tools, climbing ladders and scaffolds, and lifting heavy objects without good footing or mechanical aids. Body vibration while using jackhammers and localized vibration to fingers and hands from using handtools can lead to chronic injury. Safe work procedures as identified in Chap. 14 will help prevent musculoskeletal injury.

OSHA's ergonomics standard was proposed after injury reports showed that large numbers of workers were suffering back injuries and repetitive motion strain resulting in high workers' compensation payments and frequent lost work days. At this writing, it has not been promulgated because of objections from employers. One criticism of the standard is that it would be difficult to enforce, since the causes of musculoskeletal injury are so varied they are difficult to determine and address. They are quite varied, as well, in confined space operations. OSHA has said it will enforce the control of ergonomic hazards under the general duty clause of the general industry standard, where it is stated that employers shall provide a workplace free from serious recognized hazards.

Physical Hazards Safety Analysis

A widely used method of analyzing physical workplace hazards is the *job safety analysis* (JSA) developed by the Mine Safety and Health Administration (MSHA). The JSA is a simple five-step procedure that can be applied to a straightforward task such as changing a tire or the complex job of cleaning out a storage tank. Use a worksheet with one column for each step.

Step 1.   Select the job to be analyzed.

Step 2.   Break down the work assignment into individual tasks or procedures that are required to complete the job. Avoid being too general or too specific in listing steps. Aim for 10 to 15 steps. Here are five examples of steps in the order they would be listed; other steps to complete the work would follow:

1. Climb external ladder to top of tank.

2. Enter tank through portal on top.

3. Move tools and equipment through portal.

4. Climb down internal ladder to bottom of tank.

5. Move tools and equipment to bottom of tank.

Step 3. For each step required to complete the job, identify the hazards or potential accidents. Classify accidents into categories. Categories might include the following or others appropriate to the job:

- Falling to same level or to a lower level
- Being struck by or striking against objects in motion, projectiles, or stationary equipment
- Being caught in or between rotating equipment parts, projecting parts, moving equipment and stationary objects, or pinch points on equipment in operation
- Contact with charged electrical equipment or hot equipment
- Exposure to hazardous atmospheres, noise, extreme heat, or cold
- Overexertion by lifting, pulling, pushing, or fatigue

Step 4. For each of the potential accidents identified in step 3, determine safety measures that can be used to prevent the accident. Preventive measures might include eliminating unsafe conditions from the work area, changing work practices to prevent unsafe acts, or a combination of both approaches.

Step 5. Put the safety measures into effect. This requires the cooperation of workers and management. Implementing the new measures calls for information, training, and encouragement for workers to use them. A strong safety culture cannot be achieved without genuine ongoing support from management and buy-in by workers.

Hazardous Atmospheres

A hazardous atmosphere inside a confined space may be oxygen-deficient, contain an asphyxiant or toxic gas, or be flammable. The very nature of confined spaces entraps gases and vapors inside the space. OSHA investigations into deaths in confined spaces in 1984 through 1986, which were a factor in the agency's decision to propose safety requirements, led the agency to state that atmospheric hazards were the leading cause of death. The 10-year NIOSH study mentioned earlier found that 80% of the fatal confined space incidents investigated had hazardous atmospheres (43% were oxygen-deficient, 29% were toxic, and 7% were flammable).

Workers have died in hazardous atmospheres when they entered confined spaces through small portals without their self-contained breathing apparatus (SCBA), intending to put on the SCBA as soon as it was handed through. Some were overcome quickly and never donned the airpacks. Never enter a space unprotected for testing: Assume in every case that an unfavorable atmosphere exists and will be present at the onset of entry.

Tanks were the most common location of confined-space-related deaths from atmospheric conditions, followed by sewers, pits (half of which were manure pits), and silos. Deaths occurred in hazardous atmospheres in vats, wells, bins, pipes, and kilns. The highest number of deaths in hazardous atmospheres was reported in manufacturing facilities.

Sewer system entry is an unusual situation where it may be difficult to isolate the space to be entered, as it is a section of a continuous system. Although some atmospheric hazards in sewers are predictable, such as the gases commonly found there, the atmosphere may suddenly and unpredictably become lethally hazardous from causes beyond the control of the entrant or employer, as when hazardous materials are released into sewers following an industrial or transportation spill. See Chap. 6 for one possible method to isolate areas within sewer systems.

Guidelines are available for assessing dangerous atmospheres; oxygen concentration, gas and vapor concentration, and flammability readings can all be evaluated quickly with the right equipment. Remember that hazardous atmospheric levels are interrelated, and each type hazard should be evaluated together with the others. For example, just because the oxygen-combustible gas meter indicates 20.9% oxygen and no flammable gas is present, don't assume that the atmosphere is safe until other appropriate instruments have been used to test for toxic vapors.

Oxygen concentration

Normal air contains 20.9% oxygen; the rest is made up of other gases. Oxygen is the important gas for our purposes, as it is required by all human cells to function and stay alive. Oxygen is extracted from the air in human lungs, crosses thin membranes into the blood circulating in capillaries around the lungs, combines with hemoglobin molecules in red blood cells, and is carried throughout the body for distribution to cells. On the cellular end of the delivery system, the oxygen is downloaded because the concentration of oxygen is low there as cells continue to use up oxygen. The system operates well under normal conditions where the concentration of oxygen in the air in the lungs is at or near 20.9%.

Oxygen deficiency. When a worker breathes air where the concentration of oxygen is significantly less than 20.9%, the pressure of the oxygen in the lungs is too low to push a complete transfer into red blood cells. Some of the red cells leave the lungs without picking up adequate oxygen for the needs of working cells in the rest of the body. Oxygen-starved cells in organ systems fail to work, especially in the brain, whose need for oxygen is perhaps the most critical—at least the most readily noticeable. The lower the percentage of oxygen in the breathing air, the quicker and more serious are the effects on the worker's body. Disorientation and impaired judgment are among the early signs of oxygen deficiency, and interfere with workers' abilities to respond to threats to their bodies.

Oxygen deficiency by displacement. Oxygen in a top-entry confined space can be displaced by any gas inside the space that is heavier than air. Some of these gases are inert and do no harm to the body except when they deprive it of oxygen. Nitrogen, which makes up 78% of normal air, is one of these. Inert gases are sometimes used to purge flammable or toxic gases from a confined system. Other oxygen-displacing gases are harmful or toxic. Whether inert or harmful, gases that displace oxygen in the confined space atmosphere can be killers.

Oxygen deficiency by consumption. Bacteria use up oxygen as they decompose organic matter, for example, when plant matter rots or ferments. This condition is especially likely to occur in pits and manholes where garbage or leaf litter is plentiful and conditions are wet. Dead plants and moisture provide the best conditions for bacterial action. Decomposition consumes oxygen and usually generates a gas that is hazardous if trapped in a confined space. Manure pits become oxygen-deficient, and may fill with methane gas, due to bacterial decomposition. Grain storage containers where the grain gets wet or is allowed to ferment may be deficient in oxygen; several gases are released during grain decomposition including methane, carbon dioxide, and nitrogen dioxide.

Rust, or oxidation, and rapid corrosion under salty conditions consume oxygen. Oxidation is a chemical reaction between metal and the oxygen in moist air. Confined spaces made from, or containing, old rusty metals may contain oxygen-deficient atmospheres. Metal tanks, vats, and ship holds are especially subject to rusting. Combustion of an open flame consumes oxygen. Space heaters used in confined spaces may result in oxygen deficiency, as might torch cutting or welding inside the space.

Effects of oxygen deficiency. In addition to disorientation and poor judgment, other brain-controlled symptoms of oxygen deficiency include difficulty breathing, rapid fatigue, headache, nausea and vomiting, and dizziness or euphoria. Extreme deficiency leads to complete mental failure, unconsciousness, gasping respiration, and finally to cardiac arrest and death. See Table 3.2 for a summary of the effects of oxygen deficiency on the human body.

OSHA's definition of oxygen deficiency. If atmospheric testing reveals that the oxygen level in a confined space is lower than 19.5%, the space has an oxygen-deficient atmosphere and should not be entered without implementing some kind of control (see Chap. 4 for atmospheric testing and Chap. 6. for methods to control oxygen deficiency and other hazards). Do not assume, however, that an atmosphere with 19.5% oxygen is safe to enter without testing for contaminants. If the oxygen concentration is that much lower than ambient air, ask yourself why. What caused the oxygen concentration to drop? Is there a gas inside the space that displaced the missing oxygen? If the gas is inert (harmless to the body) and nothing else harmful is there, then 19.5% oxygen is adequate to support life. If the displacing gas is toxic, or one that will combine with hemoglobin and

TABLE 3.2 Effects and Symptoms of Oxygen Deficiency in Humans

Oxygen concentration, %	Effects and symptoms
20.9	Normal air concentration
19.5	Minimum safe entry level
16–19	Poor coordination, fatigue
12–16	Rapid pulse, labored respiration
10–12	Very fast and deep respiration, lips begin turning blue, headache
8–10	Fainting, unconsciousness, nausea, vomiting
6–8	Fatal in 8 minutes; 50% fatal in 6 minutes
<6	Coma in 1 minute, convulsions, respiratory and cardiac arrest, death

interfere with oxygen transport and asphyxiate you, or is composed of solvent vapors that may be intoxicating or flammable, then you cannot consider the atmosphere safe just because it contains enough oxygen.

Oxygen enrichment

Conditions leading to oxygen enrichment. Some chemical reactions release oxygen, thereby increasing the oxygen concentration in air. Chemicals classified as "oxidizers" and bearing the oxidizer yellow label release oxygen when they react with certain other chemicals. Oxidizers react, sometimes strongly, with many combustible materials and chemicals. This sets up excellent conditions for fire, as discussed below in the section on flammability hazards.

Several confined space deaths were attributed to the release of oxygen into the space by ongoing or previous work. Three workers burned to death when fire broke out in a barge tank after oxyacetylene hoses, left by a previous welding team, leaked oxygen. Gas welding hoses should not be left inside a confined space when work is not taking place "if it is practicable to remove them" (29 CFR 1910.252); if they are not removed, the torch valves and the gas supply to the torch must be shut off. Incidents have been reported where oxygen was pumped into a space by well-intentioned people who believed that oxygen was the same as air. It is not.

Effects of oxygen enrichment. The biggest hazard posed by an oxygen-enriched atmosphere is the increase in the air's ability to support combustion. Anything combustible—wood, paper, chemicals, or workers' clothing—ignites more easily and burns faster and with greater heat output in an enriched atmosphere. A fire that starts in an enriched atmosphere spreads very rapidly, burns hotly, and is more difficult to put out than it would be under normal conditions.

OSHA's definition of oxygen enrichment. An atmosphere containing more than 23.5% oxygen is considered enriched. As with interpretations of the safety of an oxygen-deficient atmosphere, confined space workers and rescuers should consider the reasons for increased oxygen content even before the atmosphere reaches the OSHA-defined hazard level. What is causing the increase in oxygen? Is the causative condition ongoing? Will the oxygen level go even higher during the entry or rescue? Before entering the space, determine the cause and implement controls, even when the increased oxygen concentration is not above 23.5%.

Toxic and asphyxiating chemicals

Chemicals can cause harm to humans, some immediately following exposure and some later. Although their scientific definitions are slightly different, the words *toxic* and *poisonous* are used to mean the same thing in safety manuals. Chemicals may be toxic, causing sickness or death through a variety of harmful changes in the body as they stop cell processes, prevent or cause enzyme or hormone activities, stop transmission of nerve messages, or interfere with oxygen transport in the bloodstream. Some chemicals asphyxiate workers. Some chemicals are corrosive to skin, eyes, respiratory membranes, and digestive system linings. The most common route of exposure is by inhaling chemicals as we breathe; these may be dusts, but more often are gases and vapors.

Gases and vapors. A *gas* is a physical state in which the chemical expands to fill the space available. *Solid* material keeps its own shape; even if it is formed into powder or granules, when spilled it keeps its shape. *Liquids* are less well formed, and take on the shape of their container. In an open-top container the liquid stays at the bottom, although some liquids evaporate gradually into air.

Gases, on the other hand, disperse immediately and expand to fill all the space available to them. Some gases are lighter than air, and float up and away if the top of the confined space is open. Some are heavier than air, moving down inside a confined space until they can go no lower. In an unconfined area, heavy gases may flow downhill and into basements, trenches, and other low-elevation areas. Table 3.3 provides examples. Look up the density of a gas by name in a Material Safety Data Sheet (MSDS) or other reference to determine if it is heavier than air (density greater than 1.0) or lighter than air (density less than 1.0). We have already seen how heavier-than-air gases displace breathing air in confined spaces and lead to oxygen deficiency.

A *vapor* is a type of gas that is released from a liquid as it boils or simply evaporates. Vapors behave as gases, and their properties can also be used to predict behavior. The term *density* is applied to both gases and vapors; usually when the gas in question is a vapor, the term *vapor density* is used.

Asphyxiant gases. Simple asphyxiants displace breathing air and may kill a worker simply by not leaving adequate air to support life. The asphyxiating atmosphere contains too much of the displacing gas and not enough oxygen. Even though the gas is itself not harmful, the lack of oxygen causes cells, and perhaps the worker, to die. Acetylene, carbon dioxide, and methane are simple asphyxiants. Symptoms of simple asphyxiation are the same as those experienced by workers in oxygen-deficient atmospheres. Note that in addition to their potential for asphyxiating workers, acetylene and methane are flammable; in fact, acetylene would likely ignite before it asphyxiated. Of the three examples only one, acetylene, has an odor. (Pure acetylene does not, but commercial-grade acetylene gas a distinctive garliclike smell.) Carbon dioxide and methane are odorless, and do not warn of their presence.

Carbon monoxide, hydrogen sulfide, hydrogen cyanide, and sulfur dioxide gases are chemical asphyxiants. They use various means to halt oxygen delivery to cells. Carbon monoxide (CO) gas binds to hemoglobin in red blood cells, filling the binding sites that normally would transport oxygen. It commonly is found in gases used for heating and cooking, accumulating in dangerous concentrations when combustion is incomplete or heaters and stoves do not exhaust properly. Carbon monoxide can be formed from microbial decomposition of organic matter in sewers, silos, and fermentation tanks. It is odorless and colorless. Fatalities due to CO poisoning are not confined to any one industry.

Hydrogen cyanide gas prevents cells from using the oxygen delivered to them, causing death due to respiratory arrest. Hydrogen sulfide poisoning produces symptoms similar to those produced by hydrogen cyanide and is irritating to lung surfaces. This gas, which smells like rotten eggs, is found in some natural-gas deposits, in sewers (where it is sometimes called "sewer gas"), and in the vicinity of industrial paper plants using the kraft process. Inhalation of very high concentrations of hydrogen cyanide and hydrogen sulfide is lethal within minutes. Sulfur dioxide, used as a disinfectant in breweries and food factories and as a bleach for fibers, is odorless and nonflammable but is strongly irritating to eyes and the respiratory tract, so its presence in confined space atmospheres is immediately known.

Warning properties. Gases that have a strong odor, such as hydrogen sulfide; have a visible color, such as chlorine; or are very irritating, such as sulfur dioxide, have good

TABLE 3.3 **Examples of Heavier-than-Air Gases**

Chlorine	Hydrogen sulfide	Nitrous oxide
Carbon dioxide	Isobutane	Phosgene
Butane	Ketene	Propane
Freon	Methyl mercaptan	Sulfur dioxide
Hydrogen chloride	Nitric oxide	Vinyl chloride

warning properties. Those that are colorless, odorless, and nonirritating do not warn of their presence. Even though odor is a warning property for some gases, three problems with relying on odor make this reliance unsafe: (1) not every nose detects odors at the same concentration—some people have a better sense of smell than others; (2) the odor threshold (concentration at which the odor is detectable) may be too high, as some chemicals are dangerous at levels below the odor threshold; and (3) a phenomenon called "olfactory fatigue" occurs when the nerves transmitting the message of odor to the brain exhaust their supply of neurotransmitter and stop sending the signal. When this happens, the worker stops smelling the odor even though the chemical is still present.

Chemical vapors. Vapors are gases that are released when liquids boil or evaporate. In confined spaces, vapors may be present from evaporation of the liquid formerly stored there. Workers cleaning out "empty" storage tanks often encounter hazardous vapors along with the sludge and residue. In the toluene tank explosion described in Chap. 1, only residue toluene remained but toluene vapors lingered inside the tank to poison the entrant and explode on the intrusion of air and a spark. Many of the chemicals commonly stored, used, or left behind in confined spaces release toxic, flammable, corrosive, or otherwise harmful vapors into the atmosphere inside the space. Hazardous vapors are harmful if inhaled, or if they dissolve in the moisture on a worker's skin or eyes.

Vapor density. As with other forms of gases, vapors may be heavier or lighter than air and therefore move downward or upward in and around confined spaces. On the MSDS or other reference, look up the vapor density of the chemical. If the listed number is greater than 1.0, the vapor is heavier than air. If the number is less than 1.0, the vapor is lighter than air. When vapor density is not listed, look up the molecular weight of the liquid. If the molecular weight is greater than 29, the vapor is heavier than air. If the molecular weight is less than 29, the vapor is lighter than air.

Hazard assessment. The examples given above are not the entire list of potentially harmful gases and vapors. Others are hazardous because of their properties and human health effects. Characterize all atmospheres and analyze them for hazardous properties before entry. Carry out a combination of air monitoring and reference research for a complete hazard analysis. When seeking information about hazardous atmospheres, complete these three steps: (1) determine the potential for hazardous atmospheres, identify the agent(s) by name if possible; (2) look up the chemical(s) by name to learn their properties relative to oxygen displacement, flammability, corrosivity, and toxicity; and (3) measure the atmospheric concentrations of the chemicals present in the confined space. Assess the chemical hazards before entry, as part of the permit process. Figure 3.2 shows one example of a chemical hazard assessment sheet. Only with this knowledge can safe work practices be implemented according to the written permit plan.

Routes of entry. For chemicals to harm you, they must get onto or into your body. Experts can predict the route each chemical will take to do this, and these are listed as routes of entry for the chemical in various references. Knowing the likely route of entry for a chemical in a space about to be entered allows entrants to cover or otherwise protect that route and prevent exposure.

Inhalation. Everybody breathes. Working harder leads to breathing harder, both faster and deeper. If the air inside a confined space contains chemical gases, or solids in the form of dusts or particles, they will be inhaled as the worker breathes. Chemicals that are corrosive or irritating damage the delicate linings of every part of the respiratory tract. Depending on the solubility of the gas or the size of the particle, chemicals are deposited in the nose and pharynx, the bronchi and bronchioles, or the millions of tiny alveoli that make up the lungs (see Figs. 3.3 and 3.4).

Some chemicals directly damage the delicate surfaces of the respiratory tract. Corrosive chemicals such as chlorine gas or the vapors given off by some acids and strong bases dissolve in the wet surfaces, forming compounds that damage the tissues. Some of these tissues, especially in the deeper parts of the system, do not repair themselves. Permanent damage results, leading to chronic bronchitis or pulmonary edema. Dusts and particles may clog the tiny passageways in the bronchioles, leading to chronic obstructive pulmonary disease. Workers have inhaled grains while working in silos,

CHEMICAL HAZARD ASSESSMENT

Chemical Name_____

Physical State   solid___ liquid___ gas___ dust___
Properties:
      Gas/vapor density_____
      Flash point_____
      LFL/UFL _____
      Reacts with_____
      Fire Hazard?   Yes___   No___
      Water Reactive? Yes___   No___
Health Information:
      PEL(TWA)_____ ST_____ C _____
      Route of Entry_____
      Symptoms_____
      _____
      Target Organs_____
      _____
      Asphyxiant?   Yes___   No___
      Toxic?   Yes___   No___
First Aid:_____

Other Comments:_____

_____

Signature_____
Date_____

Figure 3.2   A worksheet aids complete chemical hazard assessment.

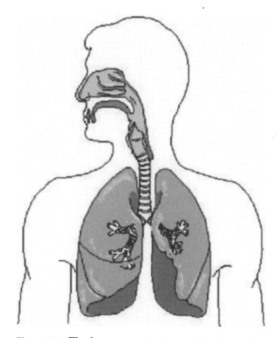

Figure 3.3   The human respiratory system branches repeatedly into ever smaller passages, each of which is vulnerable to chemicals.

Figure 3.4   At the ends of the repeated branchings in the respiratory system are millions of tiny alveoli with very thin walls that are easily crossed by certain chemicals.

and died from clogged breathing passages when the grain sprouted inside the damp lungs and bronchi. A pediatrician reports working for several days to remove sprouting grain from a teenager's lower airway, and her sadness and frustration when he died.

Some gases, including the vapors released by solvents, easily cross the linings of the lungs and enter the bloodstream, where they are carried all over the body. The lungs contain a huge surface area (about the size of a singles tennis court!), well supplied with capillaries, where air contaminants are only a few thin membranes away from the blood supply. Since blood circulates to all the body's cells, chemicals that are absorbed into the lung capillaries can cause health effects far from the point of entry.

Ingestion. Each organ in the digestive system has evolved features to improve its ability to perform its task (Fig. 3.5). The stomach is strongly muscular to churn and mix its contents and break up the larger pieces. Its acids and digestive enzymes influence digestion. Stomach contents move into the small intestine, a long narrow tube where most digestion and nutrient absorption into the bloodstream take place. The large intestine reabsorbs much of the water used in the digestive process and stores feces.

Swallowing chemicals results in the same two kinds of tissue damage as does inhaling them. Some damage the linings of all parts of the digestive system, and some (once again, solvents do this easily) cross the cells lining the system and move into the bloodstream for transport to other organs. Consider how quickly one begins to feel the effects of swallowing the solvent known as *ethyl alcohol* ("drinking" alcohol), indicating that it has been absorbed and reached the brain.

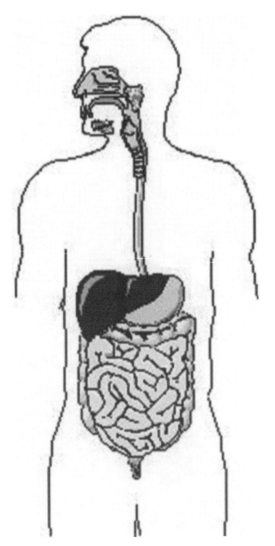

Figure 3.5 The human digestive system allows absorption of swallowed chemicals into the bloodstream through the walls of the small intestine. Water is absorbed in the large intestine.

It is rare that anyone ingests hazardous chemicals purposely. It is not unusual for chemicals to enter the digestive tract by splashing into the worker's face, spreading from contaminated hands into the mouth, or dripping into the back of the mouth after capture by the respiratory system's protective mucous.

Skin contact and absorption.   Chemicals can damage skin on contact, and some permeate the skin and enter the bloodstream. Corrosive chemicals are most likely to cause surface damage, but solvents also cause dryness and painful cracking as they dissolve protective skin oils and waxes. Solvents readily permeate healthy skin, especially skin that is damaged from previous chemical exposure, moving into the bloodstream for bodywide transport. Chemicals that cross skin easily are indicated by the (skin) notation in many references, most notably the *NIOSH Pocket Guide to Chemical Hazards.* Although there are no exposure limits for skin contamination as there are for inhalation, the skin notation is an indicator of potential overexposure by this route of entry.

Exposure limits.   *Exposure limits* are guidelines to safe levels for the inhalation of hazardous substances. Exposure limits should not be considered a hard and fast cutoff between safe and unsafe inhaled concentrations; rather, they should be used as guidelines to make changes in the confined space atmosphere before entry. More than one agency sets exposure limits, and within each agency's list there are several types of limits.

OSHA exposure limits.   The only exposure limits that are legally enforceable are those set by OSHA. OSHA was forced by a court to roll back its newest limits in 1992, retreating to earlier—and less safe—limits for a number of chemicals. Other agencies' limits are more protective, and should be used whenever possible.

NIOSH exposure limits.   These are listed in NIOSH documents and often on MSDS. They are based on research carried out by NIOSH and the American Conference of Governmental Industrial Hygienists (ACGIH). These are recommended limits, and are not enforced by OSHA. Frequently they are more protective than OSHA limits.

ACGIH limits.   The ACGIH publishes exposure limits, which they call *threshold limit values,* in their booklet *Threshold Limit Values (TLVs) for Chemical Substances and Physical Agents.* These limits describe lower concentrations in breathing air, and are more protective than those set by OSHA.

Types of limits.   The basic exposure limit is the permissible exposure limit, often abbreviated PEL. The ACGIH calls them TLV. These describe the concentration in breathing air that the average healthy worker can inhale for a full work day (8 hours) every day for 50 weeks a year (a vacation is assumed) without harmful health effects. Limits for an eight-hour working day are time-weighted averages (TWAs), with the concentration in air measured for a full or partial shift and averaged out over the 8-hour period. For longer or shorter shifts, calculations must be performed by someone who understands the extrapolation. (See Chap. 4 for measuring techniques and equipment.)

Exposure for periods shorter than 8 hours may be covered by a different kind of limit, the short-term exposure limit (STEL or ST). With the chemicals that have listed STEL, a worker might go into the confined space for only 15 minutes in the concentration listed as STEL four times in a working day if each entry is separated from the previous one by an hour in uncontaminated air. The average concentration for all the entries cannot exceed the listed TWA for the day.

A few chemicals have a listed ceiling (C) limit. This means it does not matter what the overall average is as long as the concentration in the breathing air never exceeds the listed ceiling.

Short-term and ceiling limits are listed for certain chemicals only, on the basis of their molecular structure and how they interact with body cells. These are chemicals whose health effects are generally limited to acute effects (those that are apparent almost immediately after exposure—see text below) and unlikely to cause chronic (long-term) impact.

Almost all chemicals have a listed IDLH (immediately dangerous to life and health) limit. It is higher than the other exposure limits, and describes the atmospheric concentration at which someone inside the space without a respirator would suffer within

30 minutes effects severe enough that he would be unable to self-rescue. When the IDLH concentration is detected, a worker must put on the proper respirator or exit the area immediately.

Exposure limits and permit spaces. One condition that makes a confined space a permit-required space is a hazardous atmosphere with concentrations above PEL, but, says OSHA, only for chemicals "capable of causing death, acute illness, incapacitation, impairment of ability to perform self-rescue, or injury." A competent person, one who can research and interpret toxicologic data, must make the determination of whether the chemical(s) in any confined space meets (meet) this definition. Contamination above PEL inside any confined space, permit- or nonpermit required, should trigger control actions to reduce the concentration or protect entrants by some other means.

Interpretation of exposure limits is confusing. Consult an industrial hygienist or other safety professional for good advice. Subsequent chapters will cover the means to protect confined space entrants from spaces where the atmosphere contains chemicals.

Target organs and harmful effects. Chemicals that make their way into the bloodstream are beyond the body's voluntary control at that point. All organs and tissues are available for damage, as all are supplied with blood vessels. Each human organ system has its own specialized functions, and each requires certain nutrients and body chemicals to perform its task. Cells select what they need from the bloodstream, often by means of specially formatted entry ports on the cell surfaces that fit the shapes of body chemicals in the blood. The cells cannot distinguish between helpful and harmful molecules; they simply allow whatever fits the portal to enter. A workplace chemical whose structure mimics a needed body chemical is invited in.

Because each organ has specialized entry gates for its nutrient and chemical requirements, harmful chemicals that appear to match the required shapes are more likely to enter certain organs than others. It follows that harmful chemicals are more likely to cause damage in some organs than others. Research in laboratory animals and studies of workers and other people exposed to chemicals provide information that leads to knowledge about the organs targeted for damage. These target organs are listed on MSDS and in references describing the health effects of overexposure to chemicals. Table 3.4 gives several examples of target organs.

Symptoms of exposure. Following an event that may have resulted in exposure to chemicals, workers should be alert for signs and symptoms indicating overexposure has occurred. Different kinds of chemicals cause different groups of symptoms because they target different organs. Noticing the symptoms is frequently the first sign of overexposure, so it is important to look up the chemicals present in the confined space and be on the lookout for their symptoms. Table 3.5 lists examples.

Some symptoms of chemical exposure leave the worker unable to self-rescue from the confined space, or to make other decisions regarding safety. The primary target organ for these chemicals is the central nervous system (CNS)—the brain and the spinal cord. Table 3.6 lists some of the chemicals causing these kinds of symptoms.

TABLE 3.4 Selected Chemicals and Their Target Organs

Chemical name	Target organ
Arsenic	Skin, respiratory system, kidneys, central nervous system, liver, gastrointestinal system, reproductive system
Butyl alcohol	Eyes, skin, respiratory system, central nervous system
Gasoline	Eyes, skin, respiratory system, central nervous system, liver, kidneys
Lindane	Eyes, skin, respiratory system, central nervous system, blood, liver, kidneys
Nitric acid	Eyes, skin, respiratory system, teeth
Phenol	Eyes, skin, respiratory system, liver, kidneys
Sodium hydroxide	Eyes, skin, respiratory system
1,1,1-Trichloroethane	Eyes, skin, cardiovascular system, central nervous system, liver
Turpentine	Eyes, skin, respiratory system, central nervous system, kidneys

TABLE 3.5 **Symptoms of Overexposure to Selected Chemicals**

Chemical name	Symptoms of overexposure
Acrolein	Irritated eyes, skin and mucous membranes, decreased pulmonary function, pulmonary edema (fluid in lungs)
Ammonia gas	Irritated eyes, nose, and throat; bronchospasms; chest pain; pink frothy sputum; skin burns
Methyl ethyl ketone	Irritated eyes, skin, and nose; headache; dizziness; vomiting; dermatitis
Methyl isocyanate*	Irritates eyes, skin, nose, and throat; cough; chest pain; shortness of breath; asthma; eye and skin damage
Portland cement	Irritated eyes, skin, and nose; cough; spitting; wheezing; bronchitis; shortness of breath on exertion
Sulfuric acid	Irritated eyes, skin, nose, and throat; pulmonary edema, emphysema; dental erosion; skin burns; tracheobronchitis
Toluene	CNS symptoms plus dermatitis, liver, and kidney damage

*Killed an estimated 10,000 people in Bhopal, India, in 1984.

TABLE 3.6 **Chemicals Affecting the Central Nervous System (CNS)**

Chemical name	CNS symptoms
Acetone	Headache, dizziness, slowed CNS activities
Butane	Drowsiness, narcosis, asphyxiation
Chloroform	Mental dullness, nausea, confusion, headache, fatigue, anesthesia
2,4-D herbicide	Weakness, stupor, muscle twitch, convulsions, overactive reflexes
Ethylene glycol	Nausea, vomiting, weakness, dizziness, stupor, convulsions, slowed CNS activities
Naphtha (coal tar)	Lightheadedness, drowsiness
Nicotine	Nausea, salivation, headache, dizziness, hearing and visual disturbances, atrial fibrillation, convulsions
Propane	Dizziness, confusion, excitation
Tetrachloroethane	Weakness, restlessness, irregular respiration, lack of muscular coordination
Trichloroethane	CNS slowdown
Toluene	Weakness, confusion, euphoria, dizziness, dilated pupils, running eyes, insomnia
Xylene	Dizziness, excitement, staggering walk, drowsiness, nausea, vomiting

Acute health effects. Symptoms that appear soon after chemical overexposure are referred to as *acute symptoms* or *acute health effects.* They may be serious or relatively minor, and in some cases disappear when exposure ceases. Examples of acute effects are burning eyes, coughing, dizziness, nausea, headache, or any other symptom that manifests quickly. Acute effects also include very serious outcomes such as rapid coma and death.

Chronic health effects. Some chemicals cause harm over a long period; some of these lead to poor health for a lifetime. Chemicals that cause human cancer (carcinogens), liver disease, respiratory dysfunction, or anemia are examples. Table 3.7 shows some examples of chemicals and their chronic health effects. Some chronic effects appear soon after exposure, but some, such as asbestos-related cancer, lie hidden for many years.

Diagnosis and repair.   Unfortunately, the diagnosis of chemical overexposure is difficult and the repair usually impossible. For the majority of chemicals, there is no one test showing that exposure has occurred. A few chemicals can be found by laboratory analysis of blood, urine, or exhaled air, but most cannot. In some cases the resulting organ damage is diagnosed, but the causative agent can only be speculated. Liver damage, for example, is diagnosed through atypical lab results, but the cause of the damage might be alcoholism, prescription drug use, infection, or chemical exposure. Lung disease may result from bacterial infection, cigarette smoke, asbestos particles, or inhaling a corrosive gas. Repair, in most cases, is equally difficult.

It should be noted that many physicians are not formally trained about the health effects of chemical exposure. A visit to a physician for suspected illness as a result of chemical exposure should include a review of potential exposures. Take along information about the chemicals in the workplace or home in the form of MSDS or other scientific reference.

TABLE 3.7   **Examples of Chronic Symptoms of Overexposure to Selected Chemicals**

Chemical name	Chronic symptoms of overexposure
Acrolein	Chronic respiratory disease
Benzene	Leukemia
Chlordane	Liver and kidney damage, possible liver cancer
Chloroform	Enlarged liver; possible liver and/or kidney cancer
Coal dust	Chronic bronchitis, decreased pulmonary function, emphysema
Coke oven emissions	Cough; shortness of breath; wheezing; cancer of skin, lung, kidney, and bladder
Diesel exhaust	Pulmonary function changes; possible lung tumors
Formaldehyde	Nasal cancer
Methyl alcohol	Optic nerve damage, blindness
Silica, crystalline	Cough, wheezing, decreased pulmonary function, progressive respiratory symptoms (silicosis)
Vinyl chloride	Enlarged liver, liver cancer

Fire and Explosion Hazards

Three factors are required for fire: something that will burn (fuel), oxygen (almost always present in air), and enough heat to start combustion (an ignition source). The fire triangle (Fig. 3.6) illustrates the requirements. The fourth aspect of fire, the chemical reaction included in the fire tetrahedron, will be ignored here as beyond our control in assessing confined space fire safety.

Think of the fire triangle as a three-legged stool. If one corner is missing, the stool fails to support weight, or the fire fails to occur. Fire hazard analysis in a confined space seeks to identify the three corners and assess their presence in the space.

Oxygen

Wherever there is air, there is oxygen. Normal air contains 20.9% oxygen. Enriched air may contain even more, increasing the fire danger. Practically speaking, there is no good way to remove this corner of the fire triangle from a confined space. Purging with an inert gas displaces air and removes oxygen, but eliminates breathing air and leads to the required use of air-supplying respirators. Some extinguishing agents that apply a foam blanket eliminate air from the surface of the combustible material, but this generally occurs after a fire has started. Hazardous materials responders may have foams that do the same thing. These normally are not used for confined space entry, but might be used during rescue by trained hazmat responders.

The oxygen corner of the fire triangle is not the easiest one to eliminate from a confined space, and removal probably is not desirable since entrants need breathing air. Consider the other two corners of the fire triangle; one of these may be easier to eliminate from the confined space prior to entry.

Ignition source

Sparks, hot equipment and machinery, welding and cutting, open flames, or the heat from a chemical reaction can all serve as ignition sources. Other sources are static electricity, radios, lights, and other electrical equipment. This is by no means a complete list. Control of most ignition sources will be discussed in Chaps. 6 and 14. High charges of static electricity, which rapidly accumulate during periods of relatively low humidity (below 50%), can cause certain substances to accumulate electrostatic charges of sufficient energy to produce sparks and ignite a flammable atmosphere.

A *chemical reaction* may serve as an ignition source. Heat release is one of the potential outcomes when incompatible chemicals combine and react with each other. Some reactions give off enough heat to serve as an ignition source, particularly when the reaction also evolves a flammable gas. For example, the reaction between calcium carbide and water generates acetylene gas, which is highly flammable, and provides enough heat to ignite the gas. Other undesirable outcomes of chemical reactions include the production of toxic gases or a mixture that boils and splatters.

FUEL

HEAT

OXYGEN

Figure 3.6 All three corners of the fire triangle must be present for ignition to occur: fuel (something that will burn), heat (an ignition source), and oxygen (always available in air).

Some chemicals are incompatible with each other. To learn what chemicals should not be allowed to mix and what will happen if they do, look under the section on incompatibilities on the MSDS or in the reference source. One especially dangerous reaction is the one between oxidizers and flammable or combustible materials. The reaction generates heat and extra oxygen in the presence of the material serving as a fuel. All three corners of the fire triangle are present in this reaction.

Some chemicals are incompatible with water and react with it to produce toxic or flammable vapors. These reactions almost always are accompanied by the release of heat, which can ignite the vapor cloud if it is flammable. Check the reference source for this information. Usually the source mentions water reactivity only if the chemical is water-reactive, and does not mention the lack of reactivity if it is not. This is one case where the absence of information is reassuring, but several sources should be checked if there is any suspicion the chemical is water reactive.

Fuel

Without something to serve as a fuel, there will be no fire. Fuels may be solid or liquid in the original state; however, it is the gases released from solids and liquids at certain temperatures that actually ignite. Although it is difficult to remove the fuel source after fire has started, there are several ways to eliminate it prior to a fire. Checking for these conditions should be part of the hazard assessment process.

Keep the temperature of flammable and combustible liquids well below their flash points. (See discussion below for an explanation of *flash point.*) Prevent evaporation from liquid containers by keeping them sealed, or remove the containers from the space prior to entry. Ventilation, discussed in Chap. 6, removes gases from confined space atmospheres. Use only inert gases to purge a space before entry.

Prevent the buildup of materials and trash inside the space. Clean up after working inside the space and remove any previously used equipment, especially equipment related to welding or cutting. Include safe work practices in the written confined space permit plan, and enforce them. Outlaw smoking in and around the confined space.

Chemical properties

Chemicals and other materials that will support combustion are classified as combustible or flammable according to the temperature at which they will ignite in the presence of an ignition source. They serve as a fuel in the presence of oxygen and an ignition source.

Flash point (liquids). Liquids do not burn, but they may give off vapors that will burn. The temperature at which a liquid gives off enough vapors to support fire is called the *flash point.* Since this term is defined as it is mentioned in the last sentence, solids and gases do not have flash points. When you look up information about a chemical and see no flash point listed, there may be several reasons: (1) it is not a liquid, (2) it will not burn, or (3) the reference is incomplete and you should consult another. Usually when the liquid has been tested and the lab testers could not make it burn, the letters "NA" will be seen in the section on fire hazards. Some references state what test method was used; for example, CC (closed cup). This information is not needed for hazard analysis.

A chemical with a tested flash point below 140°F (Fahrenheit) is classified by the U.S. Environmental Protection Agency (EPA) and the U.S. Department of Transportation

(DOT) as a flammable material. Practically speaking, this indicates that the presence of such a chemical is a good reason to find out more about the chemical and the conditions surrounding it. It means that if the temperature of the liquid chemical reaches the flash point, and if there is an ignition source present, fire will result.

The lower the flash point, the more likely that the temperature of the chemical will reach the flash point and present a potential fire hazard. Gasoline, for example, has a flash point of around minus 45°F. A match held above an open container of gasoline may ignite the vapors at any temperature normally experienced in the United States, since in most regions of the United States the temperature is almost always above −45°F. Kerosene, on the other hand, cannot be ignited with a match until its temperature reaches 100 to 162°F (depending on the recipe—there are several ways to make kerosene), so the presence of kerosene causes a fire hazard only when ambient temperatures approach 100°F or higher.

Conditions affecting a confined space that may send temperatures within higher than the air outside the space include heat gain from sun shining on the space, especially if it is painted a dark color, or the industrial process or work taking place—and thus serving as a heat source—inside the space. A fire in another material nearby can increase the temperature of a flammable liquid to its flash point. The temperature surrounding the chemical should be measured.

Temperature and vapor pressure. Liquids evaporate at a measurable rate. Vapor pressure, the push to evaporate and the pressure a vapor puts on the inside of its container, has been measured and listed for each liquid chemical. Some chemicals evaporate more quickly than do others; these have higher vapor pressures. *Vapor pressure* is an indication of how much vapor a liquid will generate, and can be used to predict the size of the vapor plume or the likelihood of vapors being present in a confined space. Vapor pressure inside a closed container may cause the container to rupture.

Look up the vapor pressure of chemicals of concern. The listed pressure was measured at a specific temperature, and the reference will so state. For example, vapor pressures listed in the *NIOSH Pocket Guide to Chemical Hazards* were measured at 68°F, except where otherwise indicated. It is important to know what the temperature was at the time of the pressure test. If the temperature in the space is higher than that at which the vapor pressure was measured, the pressure will be higher than that listed. Vapor pressure can go up dramatically when temperature increases, and may lead to container failure or a much larger vapor concentration than expected.

The vapor pressure of liquids is measured in millimeters (mm) of mercury (Hg) inside a calibrated tube. The easiest reference point for vapor pressure is water, since water is common in our experience. The vapor pressure of water at standard temperature is around 20 mm Hg, so if the vapor pressure of a chemical is 180 mm Hg (the vapor pressure of acetone), it will evaporate around 9 times as fast as water. A chemical with very low vapor pressure, for example, kerosene (5 mm Hg measured at 100°F), evaporates very slowly. Water, acetone, kerosene, and all other chemicals evaporate faster at higher temperatures.

Gas pressure. Gases that are shipped and stored in pressurized containers such as cylinders also have listed pressures; however, the units of measurement are different. These are chemicals that are in the gas state at normal temperatures, so in order to store them in small containers, they are squeezed to force their molecules closer together to take up less space. Pressurization turns some gases into liquids until they are let out of their cylinders, and some remain in the gas state under pressure. The pressures of these gases are usually listed in atmospheres (atm). When the pressure of a chemical is listed in atmospheres, that is a signal the chemical is a gas and stored under pressure. One atmosphere is equal to 760 millimeters of mercury (mm Hg).

Flammable limits and range (gases). Since the definition of flash point applies only to liquids, flammable gases do not usually have a listed flash point (although occasionally one is listed). The reference will state whether the gas is flammable. Temperature is used as an indicator of fire hazard only when the chemical is a liquid. Remember that a vapor is a type of gas, as mentioned earlier, and flammable limits apply to vapors. The term *gas,* as used in this section, includes vapors as well.

To assess the fire hazard of a gas, the concentration of the gas inside the confined space must be determined through the use of monitoring equipment. Here's how this works.

Each gas has a known range in which it forms a flammable mixture with air. This is called the *flammable range* (or the *explosive range*). At the low end of the range is the lower flammable limit (LFL or LEL), below which there is not enough of the gas in the gas-air mixture to support fire. At the upper end of the flammable range is the upper flammable limit (UFL or UEL), above which there is too much gas present, and not enough air, to support combustion. See Fig. 3.7 for a graphic illustration of flammable limits.

This concept is more easily understood in reference to tuning the carburetor of a vehicle. Just the right mixture of fuel and air must be achieved, or it will not ignite. Below the LFL the mixture is "too lean" to burn; above the UFL the mixture is "too rich" to burn.

OSHA's rule on flammable atmospheres. The PRCS standard states that if the chemical concentration of an atmosphere inside a confined space is above 10% of the LFL of the chemical, this constitutes a hazardous atmosphere. This atmosphere may be entered, but only in compliance with a comprehensive written confined space program meeting all the requirements outlined in the standard. See Chap. 4 for a discussion of monitoring for flammable atmospheres.

Inside a tall tank, or in a sewer system, the atmosphere may become stratified with lighter-than-air gases at the top or heavier-than-air gases at the bottom. Several miles of a municipal storm sewer system exploded in Kentucky after heavier-than-air hexane vapors were ignited by a spark from car traveling over an opening. Monitors had been lowered into manholes, but not deeply enough to detect the heavy flammable vapors.

Solids and dusts. Workers and their would-be rescuers die each year in confined spaces when dusts and other solid particles explode. These include coal dust, grain, wood chips, and hazardous dusts containing metals such as magnesium or aluminum. Materials rarely considered flammable, like flour, create explosive mixtures with air if they are dispersed in the space in the proper concentrations. The more finely ground the solid, the more surface area it offers to air contact and the easier it is to ignite. 29 CFR 1926, *Safety and Health Regulations for Construction,* includes in Subpart K the requirements for electrical equipment in locations where combustible dust is in suspension in the air under normal operating conditions.

There is no direct-reading instrument available that measures flammable concentrations of dusts and particles. Combustible gas monitors do not measure solids. Air sampling and laboratory analysis measure the concentration of dusts and particles in air, but this procedure does not give immediate feedback, and there is a delay in getting the results. Sampling and analysis usually are done to determine whether a dust is present in amounts greater than its listed exposure limit, for health protection rather than explosion prevention.

World record grain elevator explodes. The DeBruce grain elevator in Kansas was reported in the *Guinness Book of Records* as the largest grain elevator in the world. This grain elevator contained 246 silos, each measuring 30 feet in diameter and 120 feet in height, with 164 interstice spaces between the silos, and its 310 storage bins held 20.7 million

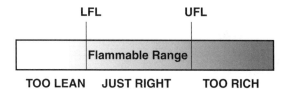

Figure 3.7 Below the lower flammable limit (LFL) the air-fuel mix is "too lean" to burn; above the upper flammable limit (UFL) it is "too rich" to burn. In the range between the limits, the mixture is just right and will ignite if an ignition source reaches it.

bushels. The central 216-foot headhouse stood 197 feet above ground; the overall length of the elevator was well over one-half mile. Four 1300-foot underground tunnels contained belts conveying grain discharged from the silos toward the elevator legs in the headhouse (Fig. 3.8).

In 1998, grain dust in one of the tunnels ignited, setting off a series of explosions propagating throughout the elevator and setting off a blast wave with a force sufficient to pulverize much of the structural concrete and kill seven workers. Ten others were seriously injured. An OSHA investigation determined that the most probable ignition source was created when a roller bearing, which had seized because of lack of lubrication, caused the roller to lock as the conveyor belt continued to roll over it. Witnesses reported that during elevator operation the cloud of suspended grain dust was often so thick that workers could not see their hands in front of their faces. OSHA blamed corporate decisions to (1) allow massive amounts of fuel (grain dust) to be created and distributed throughout the elevator, (2) forgo repair of long-failed grain dust control systems, and (3) abandon preventive maintenance of elevator equipment, declaring that these three factors "made the disaster an inevitability."

After several preexplosion eyewitnesses stated in more than one accident investigation that "I couldn't see my hand in front of my face," a working rule of thumb was developed that a combustible dust atmosphere was likely to be present when dust obscures objects at a distance of 5 feet.

Special Hazards
Radioactivity

Ionizing radiation sources are not likely to be present in confined spaces; however, entry into spaces in certain industries may give cause for concern. Medical and research facilities, mines, and companies engaged in the manufacture of televisions and computer monitors or items painted with luminous paint may have low-dose radioactive sources on hand. Special facilities like those producing or storing nuclear fuel or weapons must conduct training and assessment for these special hazards. Mixed-waste storage tanks, where waste materials are both radioactive and hazardous by EPA criteria, cannot safely be entered without complete hazard analysis and protection. The level gauges inside some tanks contain small radioactive sources that are safe when closed if they are operating properly. Nonionizing radiation, such as ultraviolet radiation, may be present during heliarc welding or operations using lasers.

Biological hazards

Danger from living or dead biological organisms is a possibility in some confined spaces. The hazards of grain explosion and engulfment and oxygen depletion by plant or microbial

Figure 3.8 The DeBruce grain elevator, and world's largest, exploded when dust in the underground tunnels was ignited by a spark from an unlubricated conveyor belt roller.

activity have been discussed. Confined spaces are unlikely to contain human pathogens classified as biohazards and recognized by the familiar biohazard symbol unless the space is associated with a hospital or research lab. (Sewer systems, of course, are the exception to this, as human feces contains a variety of pathogenic organisms.) Rescuers working with injured entrants must always take standard biohazard precautions.

The most likely harm from living things in confined space operations comes from toxic plants, such as poison oak and ivy, or venomous insects or spiders living in or around the space. Occasionally someone encounters a snake inside a confined space, especially during operations in sewers and other spaces below ground. The best protection from these kinds of hazards is recognition and avoidance.

Poison oak and ivy have three-leaflet leaves, while the similar nontoxic Virginia creeper has five. Stinging insects can be removed from a space with insecticide sprays, which then must be purged because of their toxicity. Confined space entrants who are allergic to insect venom should consult a physician and prepare for immediate emergency care in case of a bite. There are just two poisonous spiders in the United States, and neither will kill a healthy adult human, but both the black widow and the brown recluse prefer dark secluded places such as tunnels and sewers. They are leggy spiders; one is black with the characteristic red abdominal hourglass; the other brown with a violin-shaped back marking.

Other than the coral snake, which has brightly banded red/yellow/black skin, all the poisonous snakes indigenous to the United States are pit vipers. These are distinguished by their wedge-shaped heads, with the jaw wider than the neck. Don't let a harmless snake make you hurt yourself. One harmless mimic of the coral snake has similar coloration, but the red bands do not touch the yellow bands. An anonymous poet contributed this helpful reminder: "Red on yellow, kill a fellow. Red on black, friend of Jack."

Label and Placard Hazard Recognition

The first step in assessing hazards presented by chemicals is recognizing their presence. Clues to the presence and identification of chemicals include knowledge of the process taking place or the work being done and what chemicals are involved. Labels on small containers and placards on large containers and vehicles such as tankers are visible indicators of the presence, the potential hazards, and in some cases the names, of chemicals that are stored or in use. Three classification and labeling systems are commonly used in industry and transportation.

DOT hazard classification system

The U.S. Department of Transportation (DOT) classifies chemicals according to their hazardous properties, and requires hazardous chemicals to be labeled or placarded during transportation. OSHA requires that the DOT labels remain on containers in the workplace, or be applied if the container arrives without them or the chemical is recontainerized (with a few exceptions). Nine hazard classes constitute the system, each with a unique label that is easily recognized by color, symbol, class number, and text. If a four-digit identification number is on the label, as is required most of the time, the chemical can be identified by name. Even if the chemical name cannot be determined, recognition of the hazard class informs about the kinds of hazards present.

Chemicals are classified according to their primary hazard. Several of the classes include multiple divisions. Each class has a unique symbol at the top of the label or placard. The nine DOT hazard classes and their label colors are

- Class 1: explosives, orange
- Class 2: gases, various colors depending on primary hazard
- Class 3: flammable liquids, red
- Class 4: flammable solids, red and white stripe, and some require blue labels
- Class 5: oxidizers, yellow
- Class 6: poisons, white
- Class 7: radioactive materials, white or yellow above and white below

- Class 8: corrosive materials, white above and black below

- Class 9: miscellaneous, black and white stripes above, white below

Hazardous materials transported or stored in bulk loads (more than 119 gallons) must be identified by their four-digit DOT identification number. The placard shows the number across its face, or the number is added on an orange rectangle. Labels on non-bulk containers have a space along one side of the label where the four-digit number is printed. Since identification numbers designate a single chemical in many cases, or a group of chemicals with similar properties and hazards, they are very useful for identifying chemicals by name or at least by small group with similar hazards. Identification by name is the first step in researching the properties and health hazards of a chemical. See Fig. 3.9 for examples of DOT labels.

NFPA hazard ranking system

The National Fire Protection Association (NFPA) classifies chemicals according to the hazards they present, but the system does not allow identification of the chemical by name. The labeling system is voluntary, not required by OSHA or any other agency, but is widely used in industry and may be required by local ordinances. The label is a diamond divided internally into four smaller diamonds colored blue, red, yellow, and white (Fig. 3.10).

Figure 3.9 The DOT label gives information in four ways: color, symbol at the top, words stating the hazard class, and class number at the bottom.

Figure 3.10 The NFPA label does not aid identification of a chemical by name, but shows its hazardous properties in color-coded diamonds using a numerical ranking system of 1 to 4.

A number can be marked inside each of the smaller diamonds where appropriate to indicate level of hazard in four categories: blue for health, red for fire, yellow for reactivity, and white for any other special hazard such as radioactivity or oxidizer. Rankings are from 0 to 4; a 4 indicating the highest hazard level.

HMIS hazard labeling system

The Hazardous Materials Information System (HMIS) voluntary label uses the colors and ranking numbers of the NFPA label arranged in blocks or horizontally in stripes. The label may contain additional information showing a respirator icon or other symbol or letters.

Neither the NFPA nor the HMIS label leads to the identification of chemicals by name, an important piece of information when using references such as MSDS in hazard assessment. For this reason, the DOT label and placard system may prove a better tool for evaluating specific hazards in confined space work.

Summary: Prevention through Assessment

Hazard assessment is vital to ensure the health of every entrant, whether for work or rescue. Knowing what the space is used for and the reason for entry is the first step in assessing the potential hazards. Looking for clues to chemicals' presence is next, followed by monitoring the space for actual hazards. Air monitoring and hazard control are discussed in subsequent chapters.

4

Air Monitoring in Confined Spaces

Confined spaces are typically small or crowded and have limited ventilation or air movement. Consequently they offer excellent opportunities for gases and vapors to accumulate, often to explosive or lethal concentrations. Even at these high concentrations, many chemicals give little or no evidence that they are there at all. In fact, for many chemicals, the saying "You'll be dead long before you smell it" holds true because they cannot be detected by smell. Our senses simply are not reliable indicators of the presence or absence of toxic or explosive chemicals in a confined space.

Spaces must be evaluated by air monitoring to determine if it is safe for workers to enter and work. OSHA requires employers to use air monitoring equipment to test the atmosphere inside the space before entrants enter and while they work there. Air monitoring equipment technology is always improving, and today's instruments are simple and easy to use; however, there are significant considerations and limitations regarding what the equipment can measure. This chapter discusses what the PRCS regulation requires, common equipment and practices for confined space monitoring, and challenges to the accurate interpretation of the results.

Regulatory Requirements

Air monitoring is vital to the success of all aspects of a confined space entry. Most work in confined spaces is regulated under OSHA's PRCS standard. Paragraph (d)(5) of the standard requires employers to test the atmosphere in a space to see if acceptable conditions exist before and during a normal entry by personnel. They also must test spaces to document decisions to make entry under the alternate procedures described in Paragraph C(5) or to reclassify a space as a nonpermit confined space according to Paragraph C(7). The PRCS provides minimum requirements for monitoring in these work situations.

Some industries have their own comprehensive regulations that include requirements for working in confined spaces. These regulations take precedence over the general industry PRCS standard, as discussed in chapter two. It is beyond the scope of this text to address the air monitoring requirements of each of these regulations.

Normal permit-required entry

The employer must test the atmosphere inside the space before any worker enters (Fig. 4.1) to determine if a hazardous atmosphere exists. Paragraph (d)(5) does not describe detailed requirements for testing the space, but allows the employer to use any appropriate equipment and procedures such as those described in this chapter. The nonmandatory Appendix B of the standard (described below) provides general guidelines and considerations that may be helpful.

Figure 4.1 The atmosphere of a confined space must be assessed from outside before entry is made.

Since entry occurs any time any part of the body passes through the opening, the tester must be able to extend the air monitor or its probe to all parts of the space from outside. Many spaces are too large for the tester to evaluate completely from outside. Other spaces are part of a continuous system and cannot be isolated and evaluated. In both of these cases, OSHA allows the employer to test to the extent feasible prior to entry and, if entry is allowed, to monitor continuously where authorized entrants are working.

Besides testing before entry, the employer must test or monitor the space as necessary to determine if acceptable conditions are being maintained. The standard leaves the frequency of this testing to the discretion of the employer, except as noted above for large spaces where continuous monitoring is required. Entry supervisors must understand the chemical and physical properties of the hazardous atmosphere, the unique characteristics of the space, and the work to be done in order to predict how the atmosphere will change during the entry. Testing should be more frequent for work involving

- Volatile chemicals
- High temperatures
- Highly toxic chemicals
- Extensive physical action, such as scraping, that disturbs chemical deposits
- Cleaning solvents or other chemicals that may react with chemicals in the space
- Frequently changing tasks

Today's small, durable air monitoring instruments (Fig. 4.2) make continuous monitoring during the entry easy and practical for oxygen levels, combustibility, and a few specific toxic gases. For other hazards that cannot be monitored continuously, entry supervisors who are uncertain about the appropriate frequency should check with industrial hygienists, chemists, or others familiar with chemical behavior to determine how concentrations will change.

OSHA specifies the order of testing for different hazards before and during entry. Testers must measure oxygen first, then combustible gases and vapors, and then toxic gases and vapors. The rationale for this order is discussed below in the description of Appendix B to the standard.

Figure 4.2 Small personal monitors make continuous monitoring during the entry easy.

The results of the testing are to be listed on the permit posted at the entrance to the space or available by other means to affected employees. The permit also shows the name or initials of the person performing the testing and when the testing was performed. Since the permits and notes regarding any problems during the entry are to be retained for one year, entry supervisors can use previous permits for a space to indicate appropriate monitoring for later entries.

Entry using alternate procedures

Some employers choose to make entries under the alternate procedures described in Paragraph C(5) of the standard. Entry under these procedures can occur only in spaces where the only hazard present is that of a hazardous atmosphere and where continuous forced-air ventilation alone is sufficient to eliminate the hazardous atmosphere. For such spaces, the employer does not need to comply with large portions of the standard [Paragraphs (d) through (f) and (h) through (k)] but can use the simpler guidelines set out in Paragraph C(5)(ii).

The key requirement to the use of this procedure is that the employer develops the data to support the demonstration that the conditions are met. Such data can come only from air monitoring performed in the space. The monitoring must be performed over a long enough period to demonstrate the range of expected concentrations of contaminants. In addition, the variety of tasks and conditions expected during the entry must be represented in the data.

Testing to obtain the supporting data must be completed before any entrant enters the space. If an initial entry is necessary to obtain the data, it should be a permitted entry in compliance with the full standard. Again the standard specifies that the atmosphere must be tested for oxygen content, for flammable gases and vapors, and for toxic gases and vapors in that order.

The atmosphere in the space must be tested periodically to ensure that ventilation is preventing the accumulation of a hazardous atmosphere. Just as in a normal entry, the frequency of the periodic testing depends on the hazards and the conditions in the space. Since the potential for a hazardous atmosphere has been established and the employer is relying on ventilation to control it, it is very important that monitoring be performed frequently enough to catch changes before entrants are exposed.

Reclassifying a space as a nonpermit-required confined space

The employer may determine that a space once classified as a permit-required confined space no longer poses a hazard to workers entering the space. Controlling a hazardous atmosphere through ventilation does not constitute elimination. If the space once contained a hazardous atmosphere, air monitoring will be necessary to demonstrate that the hazard truly has been eliminated.

A reclassified space can be entered by workers without the protections of the permit system. To be sure that the hazardous atmosphere will not return, air monitoring should be performed for a prolonged period and under the worst-case circumstances. All monitoring results must be documented. An instrument with datalogging capability can be useful in this demonstration.

Employee observation of monitoring and results

The revision of the PRCS standard in 1998 added requirements that employers allow any employee who enters a space or that employee's representative to observe the preentry and subsequent testing of the space [Paragraphs C(5)(ii)(C and F), (d)(3)(ii), and (d)(5)(iv)]. In addition, the employer must "reevaluate the space in the presence of any authorized entrant or that employee's authorized representative who requests...such reevaluation because the entrant or representative has reason to believe that the evaluation of that space may not have been adequate." The certification that a space has been reclassified also must be made available to employees or their representatives.

The addition of these requirements provides two benefits:

1. Employees observing the monitoring as it's being done increase the likelihood that human errors will be caught. Entrants into the space are often familiar with the hazards of the space and may recognize when inappropriate or ineffective equipment or procedures are used.

2. Employees who observe the monitoring will be both more aware of the hazards of the space and more confident in the testing that was done. In both cases, they will be more likely to be aware of their surroundings and changes that may increase the hazard of the space. Confidence in the testing can allow them to focus on their work and not be overly worried about whether they are working in a hazardous atmosphere.

Nonmandatory Appendix B

OSHA included Appendix B to the PRCS standard as guidance on several important topics. The appendix does not make any requirements of employers and the topics it addresses do not necessarily have to be addressed in an employer's confined space permit program; however, the issues addressed potentially will affect entries into many confined spaces.

Evaluation and verification testing. Two purposes for testing are discussed. *Evaluation testing* is the process of determining what hazards are or may be present in a given space. Preliminary information about a space can be used to select instruments that will measure specific atmospheric parameters. In some cases, specific contaminants may not be identified and equipment that responds to a broad range of hazards can be used to determine if unknown materials may be present. In other cases, residues or air samples may need to be collected and sent to a laboratory to be analyzed by sophisticated equipment to identify the contaminants of a space. Evaluation testing should be done or reviewed by a technically qualified professional so that appropriate testing and entry procedures can be developed.

Verification testing is the process of determining if unacceptable levels of contaminants identified by evaluation testing exist at the time of entry into the space. It is performed with equipment and procedures selected on the basis of evaluation testing data. The results of verification testing should be posted on the entry permit along with the acceptable levels or conditions for entry. While evaluation testing may involve sampling and laboratory analysis, verification testing must always be done in "real time" with direct-reading equipment.

Duration of testing. The instruments used to measure the atmospheric parameters prior to entry are referred to as *direct-reading instruments* (DRIs), implying that they give instant, direct results. There is actually a certain time period required for the instrument sensor to interact with the atmosphere and produce a response. This time period, called the *response time,* varies for different types of instruments or hazards. The instrument response time is specified by the instrument manufacturer and is usually provided in the instruction manual. It is often reported as the time (in seconds) that it takes the instrument to reach 90% of a full-scale reading.

Appendix B points out that the test for each parameter should last at least as long as the minimum response time for the instrument measuring it. Testing the atmosphere of a space for a period less than the response time could result in an underestimate of the hazard.

Testing stratified atmospheres. Confined spaces with little or no ventilation may develop stratified atmospheres. This happens when gases of different densities settle into layers. If the space is too large to be completely evaluated before entry, the appendix suggests that the atmosphere is to be tested at least 4 feet in the direction of travel and to each side during entry.

Appendix B further suggests that, if the instrument uses a sampling probe, the progress should be slowed. This makes a point related to the instrument response time. If the instrument has a probe with sampling line and either a hand-operated or a battery-powered pump, it will take time for the air being tested to travel to the sensor. This "travel time" is in addition to the response time of the instrument. An entrant moving through a space too quickly while testing may pass right through a hazardous atmosphere without being warned. Just as the test must last at least as long as the response time of the instrument, the entrant must not move faster than the air can travel to the sensor.

Order of testing. Elsewhere in the standard, OSHA specifically requires that testing of the atmosphere follow the order

1. Oxygen testing

2. Testing for flammable gases and vapors

3. Testing for toxic gases and vapors

In the appendix OSHA explains the rationale. Oxygen is measured first because instruments measuring flammable gases and vapors are oxygen-dependent and will not function properly in oxygen-deficient atmospheres. Flammable gases and vapors are tested next because the threat of fire or explosion is deemed both more immediate and more life-threatening, in most cases, than exposure to toxic atmospheres. This does not mean that toxic atmospheres are less dangerous than flammable atmospheres. For some chemicals, the threat of lethal toxic exposures may be more likely than fire or explosion. The order is specified because a fire or explosion occurs more rapidly and without warning. Toxic gases and vapors still must be measured if they may be present.

Employers increasingly are using multigas instruments (Fig. 4.3) that have separate sensors to measure oxygen, flammable gases or vapors, and two or more toxic gases all in the same instrument. The instruments monitor all of these hazards simultaneously and will alarm if any or all of the parameters move out of acceptable ranges. These instruments satisfy the order of testing requirements of the standard.

Appendix E: "Sewer System Entry"

Appendix E is a nonmandatory appendix addressing the unique nature of work performed in sewers. OSHA acknowledges that work in sewers differs from work in other types of confined spaces because sewers are typically difficult to isolate completely (Fig. 4.4), so hazardous atmospheres may develop very rapidly from causes beyond the control of the

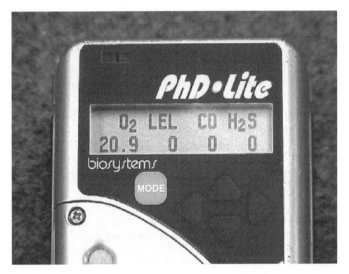

Figure 4.3   Today's multigas monitors simultaneously measure the hazards listed in OSHA's order of testing requirements.

Figure 4.4   Sewers are difficult to fully assess prior to entry.

entrant or the employer. Experienced sewer workers are knowledgeable about entering and working in their permit spaces because they enter more frequently than do entrants in other permit-required spaces. OSHA has provided special guidelines for work in sewers.

The appendix suggests that sewer workers be equipped with instruments that will sound an alarm when one of these conditions is encountered:

- Oxygen less than 19.5%

- Flammable gas or vapor at 10% or more of the lower flammable limit

- Hydrogen sulfide at 10 ppm as a time-weighted average

- Carbon monoxide at 35 ppm as a time-weighted average

The appendix goes on to state that such instruments should be carried and used by the entrant throughout the sewer-line work. Where several entrants are working in the same area, one instrument, used by the lead entrant, is acceptable. It further discusses the use of broad-range sensors and substance-specific sensors in sewers. The characteristics of these types of instruments and how to use them are discussed in detail later in this chapter.

Special industries

Several industries have specific regulations covering health and safety in their unique workplaces. Chapter 2 discusses the relationship of these regulations found in Subpart R of the PRCS standard. Several of the industries' regulations do not have sections specifically covering entry into confined spaces, so the requirements of the PRCS would apply.

Others, most notably telecommunications (1910.268); electric power generation, transmission, and distribution (1910.269); and grain handling facilities (1910.272) address work in confined spaces. The air monitoring requirements of these standards differ slightly from those of the PRCS standard, but still require some basic atmospheric testing before entry. Employers must follow at least the requirements of these standards, but may choose to do additional testing in some situations, such as when unanticipated hazards or unusual conditions are encountered.

Air Sampling in Confined Spaces

The nature of confined space entry requires the ability to get immediate measurements of the atmosphere and to monitor changes that may occur rapidly. Air monitoring with a DRI provides such real-time information.

DRIs have limitations, such as the inability to specifically and selectively measure one chemical of interest or to separately measure the components of a mixture of contaminants. Real-time monitoring equipment cannot identify an unknown contaminant; therefore, it is worthwhile to take a brief look at air sampling as an additional means of evaluating the atmosphere of a space. Air sampling differs from air monitoring in that a sample of the atmosphere is collected and sent to a laboratory for analysis.

Using air sampling data

The benefits of using air sampling techniques come primarily from their ability to produce accurate and selective measurements of chemicals. For instance, air sampling can document entrants' actual exposure to trichloroethylene, even if other chemicals are also present. The air sampling results obtained during entry into a space can be used to plan for protecting entrants during future entries into the same space. An entry supervisor can use air sampling to identify the specific organic chemicals present in the atmosphere of a storage tank with an uncertain history or a reaction vessel where different reactions have occurred.

Air sampling results should be compared to the appropriate exposure limit for the chemical sampled as described in Chap. 3. The length of the sample period determines whether the result is compared to the 8-hour time-weighted average (TWA) exposure or

15-minute short-term exposure limit (STEL). Results that indicate an overexposure to a chemical can be used by a safety and health professional to determine if follow-up action is needed to protect the health of the entrant. The results should be attached to the entry permit and retained for the program review.

Limitations

The obvious limitation to using air sampling data for confined space entry is that results are not available for days or weeks after the sample is collected and the entry has occurred. The decision to allow entry will have to be made with real-time air monitoring readings.

Air sampling results report only the average air concentration during the whole sample period. The highest (peak) exposure and the changes in concentration during the sample period won't be known; therefore, they don't describe the danger from chemicals that cause acute or short-term health effects.

These limitations explain why air sampling is not performed frequently for confined space entry work. Nonetheless, for situations like those described above, air sampling can provide useful information and is worth the investment of resources.

General procedures

Air sampling techniques involve the collection of contaminant onto a sampling medium, such as a filter, that is sent to a laboratory for analysis. The analytic equipment in the laboratory is sensitive and often able to isolate and identify specific contaminants. The analysis accurately measures the total amount of contaminants collected. Information from the sampling period is used to calculate the average concentration of chemical that was in the air when and where the sample was taken.

Air sampling must be performed according to a sampling and analytic method. The method guides two phases of the process—collecting the sample from the work environment and analyzing the sample in the laboratory. OSHA, NIOSH, and EPA develop, test, and publish these methods for specific chemicals and for groups or families of chemicals. The method lists the required equipment and describes the procedures for collecting an accurate sample and to ship the sample to the laboratory. The laboratory follows the analytic method to remove the chemical from the sample medium properly and to use sophisticated equipment to measure the amount collected. Selecting and following the proper sampling and analytical method will reduce error and yield a more accurate result.

A discussion of analytic equipment and procedures is beyond the scope of this book. Training in how to conduct air sampling projects may be needed to prepare someone to collect good data. Basic information about active and passive sampling equipment and procedures is given as an introduction.

Active sampling

Active air sampling techniques involve drawing air through a sampling medium by means of a pump or other air moving device (Fig. 4.5). The contaminant is trapped and held by the sample medium while the air moves through tubing to the pump and is exhausted back to the space. The sampling medium is then packaged and shipped to the laboratory.

Pumps. The critical feature of active sampling is that the air is moved through the sampling medium at a constant rate of flow for a known period. The rate of flow in volume (liters, milliliters, or cubic centimeters) of air per minute is multiplied by the period over which the sample was collected in minutes to calculate the total volume of air sampled. For instance, a pump calibrated to one liter per minute (1 L/min) will draw a total of 480 liters during an 8-hour shift.

This example is true only if the pump maintained the 1-L/min flow rate for the entire shift. Pumps used for air sampling can be adjusted to a desired flow rate and maintain

Figure 4.5 Active sampling provides an accurate measurement of an entrant's exposure to toxic chemicals.

that rate using flow sensors. If something blocks the airflow or if the battery power fails, the pump shuts down or "faults" and warns the user.

The pump flow rate should be checked by calibration before and after each use. Calibration equipment is available to check the flow against a known standard. Even pumps controlled by microprocessor with a digital flow rate display must be checked to be sure the flow rate displayed is accurate.

Sample medium. The air passes through a sampling medium that traps the contaminant. Different types of sampling media are used to trap different contaminants. Glass tubes containing granules of solid sorbents often are used to collect gases and vapors. The contaminant is adsorbed onto the surface of the granules and held there by molecular forces. In the laboratory, the gases or vapors are "desorbed" off the granules and run through the analytic equipment. Solid sorbents are made of different materials with different properties and are specified by the sampling and analytic method.

Some gases or vapors are absorbed in liquid media. The air is bubbled through the liquid, often an acid or base solution, which is contained in a bubbler or impinger. The liquid is packaged and sent to the laboratory for analysis. Because of the potential for breakage or spillage, liquid-media samplers are used less frequently.

Solid particles are collected using filtering media. Filters are either a fiber mesh or a thin membrane with tiny pores and are mounted in cassette holders. The sample is drawn through the cassette, and the particles are impacted onto or blocked by the filter while the air passes on through. The whole cassette is sent to the laboratory for analysis.

Passive sampling

A chemical in the air as a gas or vapor generally will move from areas of high concentration to areas of low concentration until it is evenly distributed throughout the space. This movement, known as *diffusion,* is the basis for passive air sampling.

A badge or holder with fresh sample medium (low concentration) is worn by an entrant in a contaminated atmosphere (high concentration) for a period of time. Contaminant from the air diffuses into the passive sampler and adsorbs to the medium. At the end of the sample period, the sampler is sealed and sent to the laboratory for analysis. The amount of chemical analyzed, the rate of diffusion of that chemical into the badge, and the sample period are used to calculate the average exposure concentration.

Passive samplers do not involve pumps or batteries (Fig. 4.6). This makes them safer to use in areas where a flammable atmosphere may be present. Because they can be used without special training, they provide a simple and inexpensive way to measure many gases and vapors during an entry.

Passive samplers have some limitations. Particles do not diffuse in the same way as gases, so they cannot be sampled by passive samplers. In addition, passive samplers require air movement across the face of the sampler to prevent "starvation" of the sampler. Normal movement by a worker wearing a badge typically produces enough air movement; however, mounting a sampler in a fixed position inside the space would not provide for such air movement. The sampler must be worn by a worker and should not be used for fixed-area sampling.

Real-Time Monitoring with Direct-Reading Instruments

Direct-reading instruments provide information about the atmosphere directly to the user without need of laboratory analysis. Entry cannot occur unless air monitoring confirms that the atmospheric conditions are acceptable immediately before the entry. DRIs can respond to and report changes in the conditions during entry, warning of a dangerous condition as it arises. While air sampling *can* be performed as a part of an entry, air monitoring with DRIs *must* be done for each entry. A discussion of DRIs and their use follows.

Basic operation

Many manufacturers develop and sell DRIs. As in other technologies, instruments are becoming smaller, lighter, smarter, faster, and, in some cases, less expensive. All instruments have at least these components:

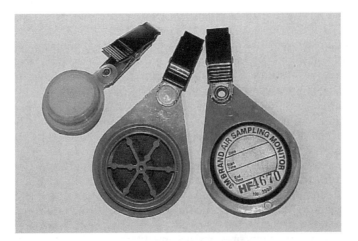

Figure 4.6   Passive sampling is an easy and effective way to measure many toxic gases and vapors.

- Sensor(s)

- Circuitry

- Display

- Battery power

- Audible and visual alarm

These are the basic elements of any piece of monitoring equipment. The sensor interacts chemically or physically with the chemical in the air to produce an electronic signal. Circuitry in the instrument converts the signal to an appropriate displayed reading and typically stores information. If readings are outside allowed ranges, an audible and visible alarm warns the user of the dangerous condition. Some other features that may be added to various instruments are discussed later in this section.

Sensors may respond to a particular gas or to a broad range of chemicals with a common characteristic, such as flammability. Thanks to miniaturization of components, instruments typically are equipped with as many as five different sensors installed in a single meter. These multigas meters provide readings from all the sensors at the same time. If the alarm point of any sensor is crossed, the alarm sounds and the dangerous condition is highlighted in the display.

While multigas meters consolidate several sensors into a single instrument, it is helpful to discuss the different types of sensors separately. The following sections describe the main types of sensors used in confined space entry: oxygen sensors, combustible-gas sensors, chemical-specific sensors, and survey sensors for toxic gases and vapors.

Measuring oxygen

Oxygen sensors are basic electrochemical sensors in which oxygen in the atmosphere passes through a permeable barrier and interacts with an electrolyte solution and electrodes to create an electric current. Specifically, oxygen reacts with a sensing electrode (usually of lead or zinc) in potassium hydroxide (electrolyte) solution and electrons are released. These electrons migrate to a counter electrode (usually gold or platinum) and an electrical current is produced that is directly related to the amount of oxygen in the atmosphere.

Interpreting readings. Oxygen meters display readings as percent oxygen (by volume) in air. Normal air has about 20.9% oxygen and about 79% nitrogen with trace amounts of other gases. OSHA considers an atmosphere hazardous if the oxygen concentration drops below 19.5% or rises above 23.5% in air.

The concentration of oxygen is constant across the earth's entire surface, even at higher elevations. What is different at higher elevations is the atmospheric pressure and, thus, the partial pressure of oxygen. This lower partial pressure results in the perception that there is less oxygen; however, the actual volume-to-volume ratio of oxygen to nitrogen is the same unless something unusual acts to add or displace oxygen.

While the primary purpose of the oxygen sensor is to ensure a proper oxygen concentration in the atmosphere, it may warn of other kinds of hazards. For instance, the oxygen meter reads 19.9%—down 1%, but still above the minimum allowable oxygen level in air. Since oxygen represents one-fifth (20%) of the air, a drop of 1% of oxygen may mean that there is 5% of some gas or vapor in the air. Now consider that a 1% concentration equals 10,000 ppm. A drop of 1% oxygen may not be an oxygen-deficiency problem, but it could warn that there are 50,000 ppm of some chemical in the air. That's likely to be a dangerous concentration of any chemical!

Limitations and considerations. Oxygen meters are among the simplest to operate and read. There are some things to keep in mind. First, the electrolyte in the sensor is typically a base (or alkaline) solution that can be neutralized when exposed to acid gases, such as carbon dioxide in exhaled air. Other chemicals may also poison the sensor, and these would be indicated by the instrument manufacturer. Oxidizers may cause an artificially high reading. If the instrument is used in a space that contains such chemicals, it should be calibrated afterward to ensure that it is still functioning properly.

Excess moisture and humidity also can affect the permeation of oxygen into the sensor. Temperature extremes also affect the electrochemical reaction. Instruments are typically able to operate in a range of 32 to 104°F because they are temperature-compensated. Measuring in temperatures outside this range may require special procedures, such as recalibrating the instrument at that temperature.

Oxygen sensors should be calibrated at the elevation or pressure conditions at which they will be used. The change in the partial pressure of oxygen can affect the permeation of oxygen through the protective barrier of the sensor. This will lead to low readings, even if the actual concentration is normal.

Measuring combustible gases and vapors

Combustible or flammable gases and vapors are measured with instruments that commonly are called *combustible-gas indicators* (CGIs). Several types of sensors may be used in CGIs, and in this section we will look at the most common types.

Catalytic sensor. Most CGIs use a form of catalytic sensor to detect the presence of a potentially flammable atmosphere. This type of sensor has two filaments that carry electric current through the sample chamber. One of these filaments is treated with a catalyst that will burn flammable gas even at low levels, while the other filament is untreated and will not burn the gas. When the treated filament, called the *sensing filament,* burns the gas, it heats up and the electric current it carries decreases, while the other filament is unchanged. The difference in electric current between these two filaments, measured by a Wheatstone bridge circuit, indicates how much flammable vapor is in the air.

Thermal conductivity sensors. Thermal conductivity sensors also have two filaments; however, only one is exposed to the contaminated air. The other is sealed away from the air. The filaments are heated to a high temperature and carry an identical electric current. As flammable gases pass over the exposed filament, they cool it and increase its electric current. The other wire is not affected. Once again, the difference in current between the two filaments is measured by a Wheatstone bridge circuit to indicate how much vapor is present.

The thermal conductivity sensor is not selective for flammable gases and vapors. It doesn't respond to low concentrations of gases and vapors. It is used primarily to measure high concentrations of gas or vapor and reports the result as percent gas in air.

Metallic oxide semiconductor sensors. *Metallic oxide semiconductor* (MOS) sensors are a type of solid-state sensors. The sensor consists of two electrodes and a heating element embedded in a ceramic bead that is coated with a metal oxide semiconductor such as tin oxide. The sensor surface is heated to a high temperature, and the semiconductor material absorbs atmospheric oxygen, establishing a baseline conductivity through the sensor. As a combustible gas contacts the bead and reacts with the absorbed oxygen, this changes the conductivity of the sensor. As the gas leaves, the conductivity returns to normal. The change in conductivity indicates the amount of gas or vapor present.

CGI readings. CGIs display the sensor response as a percent of the lower explosive limit (percent LEL). As discussed in Chap. 3, LELs differ from one flammable material to another. Each CGI manufacturer recommends a calibration gas for the instrument as described later in this chapter. CGIs are not able to distinguish one flammable gas from another, so the displayed percentage of the LEL is relevant only to the calibration gas. If different flammable gases or vapors are present, the readings do not tell you the actual percentage of the LEL of the actual gas present. This is discussed further in the section on relative response later in the chapter.

Atmospheres that provide any response on the CGI must be treated carefully. OSHA considers the atmosphere hazardous if the reading exceeds 10% of LEL. This conservative limit is appropriate because if there is enough flammable material present to give a reading of 10% LEL when and where the testing is performed, the concentration may be higher in another location or at a different time.

Limitations and considerations. There is currently no DRI available that will measure combustible or explosive solids. CGIs do not respond to combustible dusts at all, even in high concentrations. The OSHA standard states that the LEL of a flammable dust has probably been exceeded if it is difficult to see something within 5 feet. Entries into spaces with flammable solids must be made with caution since the hazard can't be measured.

The catalytic and MOS sensors require adequate oxygen to function properly. An advantage of the multigas instrument is that the user is always able to confirm that there is adequate oxygen for the proper operation of the combustible-gas sensor. Combustible-gas meters that do not have an oxygen sensor included should be used only if it is confirmed that they do not require oxygen.

Silicon compounds, leaded gasoline, and some other chemicals can "poison" CGI sensors. Corrosive gases can damage the sensor. Halogenated compounds exposed to high temperatures can degrade to corrosive materials, as can other compounds. It is a good practice to check the function and calibration of a CGI after exposure to a high concentration of any material.

Measuring toxic atmospheres

Toxic atmospheres offer the greatest challenge in confined space testing because of their potential variety. Unlike oxygen and flammable atmospheres, the exact identity of toxic chemicals that may be in the space must be known. In addition, each chemical must be measured individually and compared to the appropriate exposure limit or IDLH level.

The preentry hazard assessment is critical in identifying toxic gases or vapors that may be present. If the confined space is a storage or reaction vessel in an industrial facility, process records usually can identify what chemicals and/or reaction products to expect. In sewers, methane, hydrogen sulfide, and other breakdown products of organic material should be suspected. Entry supervisors must use whatever resources are available to identify all potential toxic gases or vapors before deciding on air monitoring instruments to be used.

DRIs that monitor toxic gases and vapors may be chemical-specific or broad-range survey instruments. Appendix E of the PRCS standard discusses the uses of each of these types in monitoring sewer atmospheres. If a few, well-identified chemicals are expected, instruments specific for those chemicals provide the most accurate readings. If the actual or potential gases or vapors have not been identified, the broad-range survey instruments should be used.

Chemical-specific sensors. A number of instruments are available to measure the concentration of specific chemicals in the air. The most common examples of these are carbon monoxide and hydrogen sulfide, but others include

- Ammonia
- Chlorine
- Formaldehyde
- Carbon dioxide
- Oxides of nitrogen
- Hydrogen cyanide
- Sulfur dioxide
- Ozone

The sensors in these chemical-specific instruments are most often electrochemical or MOS sensors similar to those described previously. These sensors are made specific for a particular chemical by changing the electrolyte solution, the semiconductor material, the electrode metals, the operating temperature, or other factors. This means that the sensors will measure the concentration of a specific chemical in the space and generally will ignore other chemicals.

Chemical-specific instruments are used to measure toxic chemicals in low concentrations. They display the results in parts of chemical per million parts of air (ppm). These are compared to exposure limits for the chemical to determine if the space will pose a health hazard to entrants.

There are three types of exposure limits (discussed in Chap. 3) related to the duration of exposure: ceiling limit (peak, instant exposure): short-term exposure limit (STEL 15 minutes), and time-weighted average (TWA 8 hours). Most of the instruments available today are able to keep a "running average" of the accumulated exposure to compare to the different exposure limits. Users can check these 15-minute and 8-hour averages to be sure that they are complying with each limit.

The sensors are designed to measure a specific chemical; however, other chemicals also may produce a response in the sensor. These other chemicals are referred to as *interferences* and are reported in the instruction manual from the manufacturer. If these are present in the space, they will cause a false reading that can create problems when preparing a permit. Careful review of the use of the space should identify the potential for interferences before entry measurements are taken.

Sensors have a limited service life. This is because they contain electrolytes and other chemical constituents that may be consumed. Manufacturers currently guarantee their sensors for 1 to 3 years, depending on the type. Since a depleted sensor would read zero even if measuring a hazardous atmosphere, these instruments must be calibrated each time they are used as discussed later in the chapter.

Survey instruments. Survey instruments use sensor technologies that respond to many different chemicals. They give one reading that is the total response to all of the detectable chemicals present. The sensors are not able to identify unknowns or measure the individual chemicals in a mixture. Since they respond differently to different chemicals, the meter reading can be very different from the actual concentration present. This is discussed in the next section, on calibration and relative response.

Photoionization detectors (PIDs) are survey instruments that use ultraviolet (UV) light to break the gas or vapor molecules into ions (charged particles). The ions move to electrodes and create an electric current that is proportional to the amount of chemical in the air. The PID will detect many organic and some inorganic compounds at concentrations as low as one part per billion (ppb) and up to 2000 ppm.

The amount of energy necessary to ionize a chemical is called its *ionization potential* (IP). The IP is expressed in units called electron volts (eV) and is different for different chemicals. The UV lamp in the PID must be stronger (have more eV) than the IP for the chemical(s) in order for the instrument to respond. Most PIDs use a 10.6-eV lamp, although other-strength lamps, such as 9.8 and 11.7 eV, are available.

High humidity and high electromagnetic energy may interfere with the PID's response. If the UV lamp is dirty, the UV light will be blocked and unable to ionize the chemical, causing the instrument to give an inaccurate reading. Charged particles in the air other than the ions of the contaminant (dusts, particulates from diesel engines, etc.) will collect on the electrode and cause a false reading.

A *flame ionization detector* (FID) uses a hydrogen flame to ionize gases or vapors, resulting in the release of charged ions. Like the PID, the ions are collected on electrodes and cause an electric current that is proportional to the amount of chemical in the air. It responds to a broad range of organic chemicals in concentrations from one to 1000 ppm. Despite its hydrogen flame, the instrument is intrinsically safe and can be used where flammable atmospheres may arise.

The FID ionizes organic compounds only and does not respond to inorganic chemicals such as carbon monoxide, hydrogen sulfide, and ammonia. Because the flame requires oxygen, the unit may not function properly in oxygen-deficient atmospheres. As with the PID, other charged particles in the space may collect on the electrode, interfering with the instrument readout.

In addition to the PID and FID, MOS sensors are available as broad-range survey instruments. Infrared spectrophotometers use the absorption of discreet wavelengths of infrared light to measure chemicals in the air. Other relatively expensive technologies sometimes are used in special circumstances to evaluate a space. Such instruments offer some attractive features, but simply are not economically practical for occasional confined space entries.

Survey instruments respond to a variety of chemicals. This is both an advantage and a disadvantage. It is helpful when trying to determine if any of a number of possible chemicals is present; however, it is difficult to selectively measure one chemical. A positive response with a survey instrument simply indicates that *at least* one detectable material is present unless other information rules out other possibilities.

Where a single, identified chemical is of concern in a space, a survey instrument can be used to assess the space. This may require calibrating the instrument to read that chemical directly or converting the displayed reading with a relative response factor. These topics are discussed later in the chapter. The results from this type of assessment can then be compared to the appropriate exposure limit for the chemical.

Instrument calibration

DRIs have electronics, sensors, and, sometimes, moving parts that may wear down, become dirty, or stop functioning properly. Over time and with use, the instruments may lose their accuracy or begin to malfunction. *Calibration* is the process of checking or adjusting the instrument readout to correspond to a known concentration of a gas (Fig. 4.7).

Calibration serves two important purposes necessary for the proper use of DRIs:

1. Calibration confirms that an instrument is functioning. Except for oxygen, nearly all instruments would be expected to read zero when started in normal fresh air. A malfunctioning instrument may also read zero when turned on. Unless the instrument is exposed to a chemical that would cause a response from the sensor, there would be no way to be sure the instrument is even functioning.

2. Beyond simply functioning, it is important for the instrument to give an accurate reading in the hazardous atmosphere. Calibrating the instrument involves adjusting the electronics that interpret and display the signal from the sensor. Adjusting the instrument display to agree with a known concentration of a chemical gives confidence that the instrument will accurately measure a hazardous atmosphere.

Procedures. There are two steps in the calibration procedure. First, except for oxygen sensors, all instrument sensors should be *zeroed*. Each sensor is exposed to clean or "zero" air that contains no detectable material and the display is adjusted to read zero.

Figure 4.7 Air monitoring instruments must be calibrated against a known concentration of gas, such as from a calibration cylinder.

This establishes the baseline and removes the effect of electronic "noise" and other factors not related to the actual measurement.

Second, the accuracy of the instrument's response can be tested in two ways:

1. A *calibration check* or "bump test" involves simply exposing the instrument to an atmosphere that will cause a response from the sensor. In a strict sense it is not critical that the exact concentration be known as the purpose is simply to cause any reading greater than the baseline reading. The test is most useful when the instrument is exposed to a known concentration and the reading is compared to that concentration. A bump test differs from a full calibration discussed below (in paragraph 2) in that no adjustment to the reading is made. The user simply confirms that the instrument will react in a hazardous atmosphere.

2. A *full* or *span calibration* entails connecting the instrument probe to a source of calibration gas at a known concentration and adjusting the reading to match. The user usually must enter a special calibration mode of the instrument to be able to adjust the reading. The instrument display is allowed to stabilize once it is connected to the source, and then the reading is raised or lowered to match the expected reading.

Some instruments use a *one-button calibration* procedure. In this case, a standard calibration concentration is programmed into the instrument by the manufacturer. When the calibration is performed, the user attaches the instrument to a calibration gas cylinder with the same formula of gas and presses one button to initiate calibration. The instrument then automatically sets the programmed readout to correspond to the amount of signal it is receiving from the sensor. The user must use the exact formulation of calibration gas from the manufacturer since there is no opportunity to manually adjust the reading to a different concentration. Used as directed by the manufacturer, this feature can simplify the calibration process for the user.

For instruments that have an internal or attached pump, the flow rate of the regulator on the calibration gas cylinder must be compatible with that of the pump. A flow rate higher or lower than that of the pump could cause the sensor to be flooded or starved and therefore give a false reading. If the instrument has no pump, the flow rate may still be important. In any case, purchasing the calibration kit from the manufacturer will insure the proper regulator is used.

Most calibration checks can be made in a laboratory or office environment prior to deployment to the field. Instruments have a range of conditions of temperature, pressure, and humidity in which they operate properly. In extreme conditions or at significantly different altitudes, the calibration should be under conditions similar to those of the space in which the instrument will be used.

Calibration should not be done in the hazardous atmosphere. An entrant who suspects that the instrument is malfunctioning while in a space should exit immediately and have the instrument checked.

Calibration gases.   The instrument manufacturer specifies a calibration gas for each instrument. A chemical-specific instrument uses a known concentration of that chemical for calibration. The concentration used is generally lower than the exposure limit for the chemical so that the person calibrating the instrument is not overexposed during the calibration.

For CGIs or survey instruments that respond to different chemicals, the calibration gas chosen is one that will cause the instrument to respond to most other gases. The properties of other flammable gases may cause more or less response than the same amount of the calibration gas. For example, many CGIs use pentane as their calibration gas. If a CGI calibrated to pentane is used to measure xylene, another flammable material, it will give a lower reading than the concentration that is actually there. The reading can be converted by means of a relative response factor described later in the chapter.

A survey instrument can be calibrated to directly and accurately read any gas that will cause its sensor to respond. If a space is known to contain only one chemical and if that chemical has been identified, an employer can use a known concentration of that chemical to calibrate an instrument. Once calibrated to that chemical, the instrument will display the actual concentration of that chemical when measuring the air in the space.

Frequency. Instruments should be calibrated before each day's use. There has been some confusion about whether this was actually necessary. It seems that some instrument manufacturers have instructed that their instruments do not need field calibration; rather, the instrument should be sent back to the manufacturer once or twice each year to be factory-calibrated. An informal review of major instrument manufacturers by the author did not reveal any that advocated this. Users should treat any such instruction with a healthy dose of skepticism.

Many factors affect the way any sensor responds to atmospheric hazards. These range from atmospheric conditions such as temperature, pressure, and humidity to the presence of interfering chemicals. Some chemicals poison or degrade sensors, such as the effect of acid gases on oxygen sensors. Then there is the wear and tear of normal usage that can cause components to fail without notice. It should be obvious why it is important to verify the function and accuracy of the instruments at least before each day's use and more frequently if needed.

Calibration checks also can be done after each use. Such a check demonstrates that the instrument continued to function properly while it was in use. Although not as critical as the calibration before use, calibration after use enhances the confidence that the entrants were not overexposed during the entrance and identifies equipment that may need repair before the next use.

Common features of DRIs

There are many different instruments on the market. For the most part, these instruments have many common features and do the same thing: monitor the level of gases in the air. Each instrument does this in a slightly different way and may have unique options available. Users must familiarize themselves with whatever instruments are purchased by their employer. This section looks at some features and characteristics common to all or most DRIs.

Portability and ease of operation. Instruments should be easy to operate and the results easy to read. Instruments that are difficult to operate likely will require too much of workers' attention, perhaps causing them to miss important visible information during an entry. Any buttons, switches, and knobs should be easy to find and operate, even while wearing gloves.

A bulky instrument may limit the activities of the worker during an entry. A small, compact device is preferable, especially if it will be worn on the worker. The equipment must also be durable. Confined spaces are often tight, hot, wet, or humid places. The instrument must be able to withstand a certain amount of abuse and still function properly.

Selectivity. *Selectivity* is the ability of an instrument to monitor one chemical or group of chemicals and ignore others. It is useful when only a few chemicals may be present or when one specific chemical poses a higher hazard than do others. The user can focus on a particular hazard with a greater degree of confidence; however, the inability to detect other chemicals is a potential limitation if multiple hazards may be present. Users must also be aware of interference and cross-sensitivity in the presence of other chemicals.

Sensitivity and operating range. An important feature of an instrument is its sensitivity. This is the lowest concentration of chemical in air that will cause a response by the instrument. The full range of concentrations that can be measured by an instrument is its operating range. The sensitivity and operating range requirements for an instrument will depend on what hazard you want to monitor.

For example, a combustible-gas meter, which measures in the range of percent (parts per hundred), would not be appropriate for monitoring toxic gas or vapors. This is because most gases and vapors are toxic at concentrations in the range of parts per million (a 1% concentration equals 10,000 ppm). In other words, a concentration of the gas that is high enough to be detected by the combustible-gas meter would be much higher than the toxic exposure limit, perhaps higher than the IDLH level. On the other hand, an instrument designed to monitor toxic exposures may not have a sufficient upper range to monitor the flammability of a gas.

Some instruments are capable of reporting measurements in more than one range. For instance, some CGIs can report as percent of LEL or percent of methane. Some older instruments that use an analog (needle) display use a switch to change the scale value. The user must be sure what units the display represents.

Intrinsic safety.   DRIs must be safe to use, even in hazardous environments. Because an instrument contains electronics and possibly other ignition sources, it must be constructed in such a way that it will not cause ignition of an explosive atmosphere. The National Electrical Code by the National Fire Protection Association describes minimum criteria for an instrument to be considered "intrinsically safe." Instruments typically are tested by Underwriters' Laboratory (UL) or Factory Mutual (FM) and must be marked as to the hazardous atmosphere for which they are certified. The code classifies hazardous atmospheres by class, group, and division.

Class and group are used to describe the type of flammable material present. Class I, including flammable vapors and gases, is further divided into groups A, B, C, and D, based on similar flammability characteristics. Examples include gasoline and hydrogen. Class II, combustible dusts, is divided into groups E, F, and G. Examples include coal, grain, or metals such as magnesium. Class III includes ignitable fibers such as cotton.

Divisions are used to describe the likelihood that the flammable contaminant will be present in a concentration sufficient to pose an explosion or combustion hazard. Division I atmospheres are considered most likely to contain the hazardous substance in flammable concentrations. Division II atmospheres have flammable or combustible substances present, but they typically are handled or contained in closed systems that are not likely to generate hazardous concentrations under normal conditions.

A typical marking on an instrument (Fig. 4.8) is that it is "intrinsically safe for Class I, Division I, Groups A, B, C, and D as approved by FM." This means that the instrument can be used in an atmosphere that potentially contains flammable concentrations of combustible or flammable gases or vapors. Approval of an instrument for use in one hazard class does not mean it can be used in all hazard classes. This approval assumes that the instrument will be used according to the manufacturer's directions and that the user does not modify it.

Datalogging.   *Datalogging* is the ability of an instrument to record the results of its measurements for later printout or downloading to a computer (Fig. 4.9). This ability

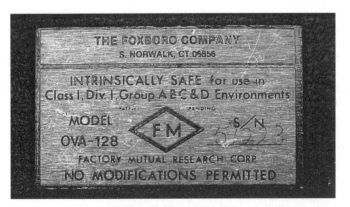

Figure 4.8   Instruments that have been tested and marked as intrinsically safe will not ignite flammable atmospheres.

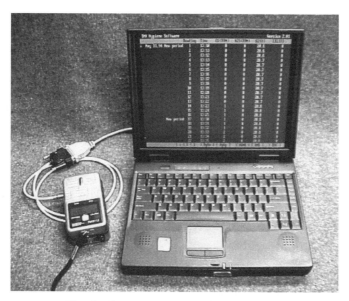

Figure 4.9 The datalogging feature on air monitoring instruments allows readings to be stored and later downloaded to a computer.

usually is built into the instrument, although it is typically an optional feature costing extra. The basic operation is that the datalogger records the electronic output from the sensor and stores it in memory in the same way as does a computer.

Datalogging in instruments typically takes one or all of three forms: (1) the datalogger may be programmed to record the instrument readings for each sensor at specific intervals of time, (2) only readings that exceed preset levels or alarm points may be recorded, or (3) the datalogger may record readings only when switched on and off manually by the user. For many instruments, the user can choose the desired mode in the field.

Recording results with a datalogger has many benefits. Much of the writing burden for users is removed. The user is free to record visual observations and work activity descriptions. Users still may need to assign sample or location numbers in their notes to tie the readings to their observations. The datalogger can capture and store peak readings that the user may not see.

Instruments that have one or more toxic gas sensors usually have an averaging function that keeps up with the percentage of the allowed TWA exposure experienced by the worker. In other words, a worker may have spent a sufficiently long portion of the day at such a high concentration that even if the rest of the day's exposure was zero, the TWA result would be over the exposure limit. At that point, the instrument would display 100 percent of the TWA. This is not the same as datalogging since it doesn't record specific readings at specific times.

Alarms. In confined space work, a critical function of air monitoring is to warn entrants when dangerous conditions arise suddenly. The warning usually comes from an audible and/or visible alarm built into the instruments. The alarms come on once readings pass a preset level. Typical alarm points for instruments match the limits in the definition of a hazardous atmosphere in the standard. The alarm points usually can be changed by following the manufacturer's guidelines.

Alarms are either latching or nonlatching. A *latching* alarm locks on if an alarm condition is reached and the user must push a button to unlock or turn off the alarm. *Nonlatching* alarms sound while the instrument readings are in alarm conditions, but turn off automatically when readings return to normal. Latching alarms ensure that workers are made aware of temporary excursions into alarm conditions. These excursions, however brief, should serve as warnings of the potential for hazardous atmospheres and they may be missed if nonlatching alarms are used.

Detector Tubes

Detector tubes are widely popular because of their ease of use and quick results. They consist of a glass tube containing a granular solid that has been coated with a specific chemical reagent. As contaminated air passes through the glass tube, the contaminant reacts with the reagent on the carrier solid to produce a color stain. The air is drawn through the tube by either a hand pump or a battery-powered pump; thus, the volume of air sampled can be controlled.

Detector tubes provide a means of quantifying air contamination with a reasonable degree of selectivity. The tubes may be specific for a certain gas or vapor or may detect groups of chemicals, such as alcohols or aromatic hydrocarbons. Tubes designed to detect one chemical may react with certain other chemicals (known as *interferences*) to produce a similar color change. The manufacturer provides a list of known interferences.

Modes of operation of detector tubes

The tubes may operate in one of three ways (Fig. 4.10); it is critical that the operator of the tube be familiar with the manufacturer's directions and which mode of operation is used:

1. The pump may be operated until the length or degree of the stain is complete. The number of pump strokes (i.e., the volume of air) required to reach this full stain is compared to a chart to determine the concentration. A high concentration of contaminant in air would require fewer strokes to reach the full stain.

2. Length of stain is the most common mode of operation. A set number of pump strokes (volumes) of air are drawn through the tube. The length of stain is compared to a calibration scale, often printed on the tube, to determine the concentration. For a set number of pump strokes, a high concentration causes a longer stain.

3. In the third possible mode, a fixed volume of air is drawn through the tube and the degree or tint of the color change is compared to a chart to determine the concentration. For these tubes, a high concentration causes a deeper or darker color change after a set number of pump strokes.

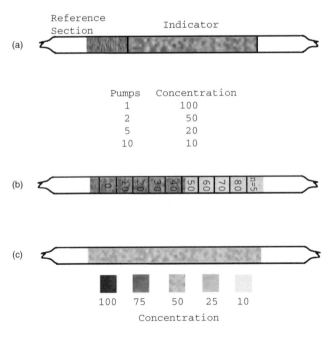

Figure 4.10 Detector tubes provide information in three ways: (*a*) color matching, (*b*) length of stain, and (*c*) degree of color change.

Figure 4.11 Air is drawn through short-term detector tubes with either bellows or piston pumps.

Pumps for use with detector tubes

Two kinds of pumps are used to take detector tube samples (Fig. 4.11). A detector tube, with both ends broken open, is inserted into the inlet for the pump. The *bellows pump* has a rubber collapsible bellows that is compressed by the user. As the bellows expands, air is drawn through the tube. Each complete stroke draws 100 ml of air and the number of strokes determines the volume of air sampled.

The other type is a *piston pump*. An opened tube is inserted into the pump inlet. A spring-loaded handle is pulled on the pump. This action draws a piston to expand the cylinder inside the pump body. The air to fill this expanding cylinder is drawn through the tube. It is the reverse action of a bicycle tire pump.

The tube manufacturer's pump must be used to draw air through the tube. Using a detector tube with another manufacturer's pump, even if the pump volume is the similar, can lead to inaccurate results.

Sample pumps must be checked for leaks to ensure that the appropriate volume of air is drawn through the tube. The pump must be allowed to fully complete every pump stroke. Incomplete strokes or leaks in the system will cause less than the appropriate volume of sample air to pass through the tube, potentially resulting in a lower reading than is actually present.

Limitations of detector tubes

Some limitations must be considered when detector tubes are used to test atmospheres in confined spaces. A significant limitation is that detector tubes do not measure changes in concentration over time as do DRIs. A sample taken with a detector tube is like a snapshot of the concentration at the time of the test. This can be a problem if the chemical must be monitored continuously.

Detector tube systems have varying accuracy, with errors ranging from 15 to 25% for many tubes. If the tube is being used only to verify the presence of a suspected contaminant, this error is not a major issue; otherwise, safety factors must be used in interpreting the data. Manufacturers provide accuracy information with the tubes.

Because a chemical reaction is involved, detector tube accuracy may be affected by such factors as temperature, humidity, and atmospheric pressure. Where temperature significantly affects the performance of the tube, the manufacturer will include compensation factors in the instructions. Colder temperatures usually slow down the reaction. If detector tubes are to be used in cold weather, they should be stored in a warm place and carried next to the body. High temperatures may affect the rate of the chemical reaction and may reduce the detector tube's shelf life (described below).

Another limitation to the use of detector tubes is the visual interpretation of the length or degree of color change. The leading edge of the stain may be uneven or may be lighter than the rest of the stain. This calls for judgment on the part of the operator to determine where the stain ends. The same difficulty applies to judging the degree or tint of color change, even when comparison charts are provided. When in doubt, it is advisable to use the most conservative (i.e., highest) reading so that more protection is provided to the worker.

Detector tubes have a specific shelf life. Chemical reagents will deteriorate over time, even if the tube is not opened and exposed to air, and high temperatures may cause the degradation of the reagent. The manufacturer stamps an expiration date on each pack of tubes. Storing the tubes in a refrigerator may maintain or extend the shelf life, but expired tubes should not be used.

Challenges to Interpreting Air Monitoring Results

The appropriate interpretation criteria for each type of monitoring device have been discussed above. A user can compare the readings in most situations directly to the action level or exposure limit to determine if a hazardous atmosphere exists. There are circumstances in which interpretation of the results is more challenging. Some are discussed below.

Relative response

Sensors that respond to many different chemicals are calibrated to a single calibration gas. When the instrument is used to measure a chemical other than that gas, the reading that is displayed is not necessarily the actual concentration of the chemical in the air. While inconvenient, this is not an insurmountable problem if the chemical present in the space has been identified. Manufacturers of such instruments usually have determined relative response factors for many common chemicals. These factors are applied to the reading to calculate the actual concentration of the chemical from the reading.

For example, a manufacturer of a PID (Fig. 4.12) includes a list of response factors in the operating instructions for the instrument. For n-hexane, the response factor is 6.2 and readings of the instrument measuring only hexane should be multiplied by this factor. A reading of 10 ppm means that the actual concentration of hexane is 62 ppm (6.2 × 10 ppm). In the same way, an actual 100-ppm concentration of hexane would be expected to cause a reading of about 16 ppm on the instrument.

A relative response factor can be used with confidence only in a space where there is just one chemical in the air and it has been identified. Obviously it is impossible to choose the proper response factor if the chemical has not been identified. If a mixture of two or more chemicals is known to be in the space, the reading could not be multiplied by more than one response factor. Readings of atmospheres containing unidentified chemicals and mixtures cannot be converted to an actual concentration(s) of the chemical(s) present.

Unidentified contaminants

Atmospheres that are known or suspected to contain a chemical that hasn't been identified are very difficult to assess. It should not be necessary to make a normal entry into

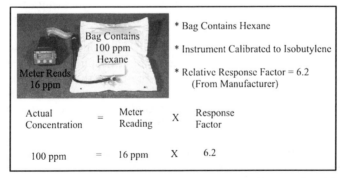

Figure 4.12 The relative response to 100 ppm of hexane by the instrument when it is calibrated to isobutylene causes it to read only 16 ppm. By applying the response factor to the reading, one can estimate the actual concentration.

such a space without getting more information; however, it may be necessary to attempt to rescue someone who has fallen victim to an unknown hazard. Even so, some information can be learned.

For instance, if a CGI gives no reading in the space that has no obvious dust present, it is relatively safe to rule out a flammable atmosphere. Oxygen content also can be measured in any atmosphere. If abnormal oxygen and flammability are ruled out, that leaves only the possibility of a toxic atmosphere. It may still be possible to enter such a space with the highest level of respiratory protection and adequate skin protection. Such an entry would be made only if it were necessary to make a rescue.

Mixtures

When an atmosphere contains a mixture of chemicals that cannot be measured separately, accurately assessing the hazard of entering the space is difficult. Once again, measuring the oxygen atmosphere and the flammability provides some information. Measuring any known chemicals with chemical-specific instruments or detector tubes will also be helpful. One is still left with addressing the toxicity of the remaining mixture.

One approach to this problem would be to use a survey instrument such as a PID to assess a total reading for the mixture. If the chemicals that make up the mixture are known and all can be measured by a PID, the user can identify the chemical with the highest response factor and multiply the reading by that factor. The result then could be compared to the lowest exposure limit of the chemicals present. If the result were lower than the exposure limit, it would be reasonable to consider that entry into the space could be made safely.

An industrial hygienist or other appropriately trained safety professional should use air sampling with laboratory analysis to characterize the space before an entry is made. Once the content and makeup of the atmosphere is known, appropriate air monitoring procedures can be selected. Action levels be established based on the mixture.

Summary

OSHA's PRCS standard requires employers to perform air monitoring to assess a space known or suspected to contain a hazardous atmosphere. Various types of equipment and techniques are available to make such assessments. Since many chemicals are otherwise undetectable, failure to properly perform air monitoring before and during entry can lead to the injury or even death of entrants. With appropriate air monitoring, entrants can enter spaces and work with confidence.

5

The Entry Permit

A written permit is required for every entry into a confined space that has been determined by the employer to be a permit-required space. According to the PRCS standard, an employer who decides that employees will enter permit spaces must develop and implement a compliant written-permit space program. A written permit is required for every entry into the identified spaces.

A wastewater treatment plant's two operators, a father and his son, were draining a digester into tank trucks. The bottom sludge was not draining well, so the two men climbed onto the concrete floating cover in the 30-foot-deep digester and used garden hoses to spray water onto the sludge. The pumper truck driver reversed his pump to blow air up through the sludge to loosen it. The operators lowered a lightbulb on an extension cord through a manhole on the floating cover. The bulb broke and ignited combustible gases in the digester, blowing the cover up and over the men. The cover dropped back onto the workers, killing both. The employer had no written procedures or training in hazard recognition. There was no permit program or written-entry permit.

The mistakes leading to the death of the two wastewater treatment plant workers were probably due to ignorance on the part of two people just trying to get the job done. An OSHA-compliant entry permit system forces managers and entrants to consider all the potential hazards presented by a confined space, and to document procedures for eliminating the hazards and protecting entrants. Protective measures in this case would have included air monitoring and intrinsically safe lighting, among others. With the permit program requirements, OSHA puts the burden on employers to identify dangerous conditions and ensure safe entry. A permit must be filled out and signed for every entry into a permit-required confined space.

Summary of the Permit Program

The written-permit program must be available for inspection by employees and their authorized representatives. It must include the elements discussed in Chap. 2 and summarized in the following list:

- An explanation of measures to prevent unauthorized entry

- Explanations of hazard identification and evaluation methods

- Identification of safe entry provisions and methods to achieve them

- Description of methods for evaluating conditions before entry and during entry

- Provision of attendant(s)

- Identification of personnel roles and duties and description of their training

- Description of rescue procedures

- Outline of entry permit preparation, issue, use, and cancellation
- Means to coordinate entry of multiple employers
- Explanation of procedures for concluding an entry
- Outline of how the permit space program will be reviewed

One of the required elements of the program is a permit form that is prepared and signed before each entry into a permit-required confined space. The permit system prescribes procedures for drawing up the written-entry permit.

The Permit System

The permit system is the employer's written procedure for preparing and issuing permits for entry and for returning the permit space to service following termination of entry. It is a procedure to document the completion of the procedures and practices necessary for safe permit space entry operations, and must be completed before entry is authorized. Procedures include, but are not limited to, the following:

- Specifying acceptable entry conditions
- Providing each authorized entrant or that employee's authorized representative with the opportunity to observe monitoring or testing of permit spaces
- Isolating the permit space
- Purging, inerting, flushing, or ventilating the permit space as necessary to eliminate or control atmospheric hazards
- Providing pedestrian, vehicle, or other barriers as necessary to protect entrants from external hazards
- Verifying that conditions in the permit space are acceptable for entry throughout the duration of an authorized entry

The permit system spells out who will sign the permit and how it will be made available to authorized entrants at the time of entry. Options for the latter include posting the permit at the entry portal, but other equally effective means are allowed as long as entrants can confirm that preentry preparations have been completed.

The system ensures that the duration of the permit does not exceed the time required to complete the job identified on the permit. In other words, an employer cannot use the same entry permit day to day for multiple jobs in a confined space.

The permit system allows the entry supervisor to terminate entry and cancel the permit when operations covered by the permit have been completed or a condition that is not allowed under the entry permit arises in or near the space. It includes provisions for retaining each canceled entry permit for at least one year, and calls for any problems encountered during an entry operation to be noted on the pertinent permit.

The Entry Permit

Although OSHA mandates the use of written permits and lists the required components, the agency does not require the use of a particular format. The permit documents the employer's compliance with the permit program and permit system requirements. It authorizes entry into the permit space.

A permit is unique to the entrant, the job, and the job duration, and cannot be used for another job, although it may cover multiple entrants if all are listed on the permit or in a referenced attachment. It is not valid beyond the duration written on its face for the specified entrant(s) and job(s). Several permit spaces may be listed on the same permit, but they must be associated with entrant(s), job, and duration. For each space, the required data from initial and periodic tests must be listed on the permit. The interpretation that an entry permit may serve for several spaces was offered by OSHA's compliance section in answer to a question regarding permits to be issued for a sewer line with multiple manholes.

Written information on the permit

The entry permit must identify the space to be entered, the purpose of the entry, and the date and authorized duration of the permit. Authorized entrants are listed by name or by some other means that will enable the attendant to quickly and accurately determine which authorized entrants are inside the permit space. This requirement may be met by inserting on the entry permit a reference such as a roster or tracking system.

The name of the individual currently serving as entry supervisor is listed on the permit. There is a space for the signature or initials of the entry supervisor who originally authorized entry. These two supervisors are not necessarily the same person.

All the hazards of the permit space to be entered are written onto the permit. In an industry where the spaces are entered often and are well known, forms may be copied showing a list of hazards that are always assumed present, with spaces for filling in data gained from atmospheric testing. The permit lists the measures used to isolate the space and eliminate or control these hazards before entry and describes acceptable entry conditions; for example, it lists the atmospheric concentration of oxygen or chemical vapors that are deemed acceptable. Initial and periodic test results must be entered onto the permit form, accompanied by the names or initials of the testers and an indication of when the tests were done.

The permit lists the rescue and emergency services that will be summoned and the means for summoning them. It lists the equipment to be used (telephone, radio, etc.) and the numbers to call for rescue. It is no longer acceptable simply to write "Call 911" on the permit. Advance plans, agreements, and practices must be arranged with the designated rescue service, and a call should be made to them prior to each entry to be sure they are available. See Chap. 2 for regulatory restrictions on rescue.

Communications between authorized entrants and attendants obviously are very important. Communications procedures and signals are spelled out on the entry permit. Communications equipment that will be used is listed, as is equipment for atmospheric testing, alarms, and rescue. Personal protective equipment is specified on the permit.

Given the circumstances of the particular confined space, any other information necessary to ensure employee safety must be written on the entry permit. Additional permits that have been issued to authorize work, such as permits for hot work, are also included.

Information specific to the workplace. A good entry permit reflects the needs of the workplace and the spaces it covers. The employer can write up a standard permit for duplicate or similar spaces. A well prepared form will eliminate the need to draw up separate permits for different types of entries. Some examples will illustrate differences in permit needs between employers.

Pulp and paper mill. A company producing pulp and paper has several kinds of permit-required confined spaces, but some are similar in the hazards they present and all are regularly entered for maintenance as part of the overall operation of the mill. The spaces are all vessels in which materials are stored or processed. Tables 5.1 through 5.3 show portions of one paper company's confined space entry permit and checklist.

Tank cleaning contractor. This small company has four full-time employees who contract with a variety of industries for tank cleaning. They do repeat business with some companies, but must always be on the lookout for unique hazards. Safety directors (who are entrants or attendants as need dictates) request a permit form and briefing from

TABLE 5.1   **Potential Existing Hazards (Pulp-and-Paper Mill)**

Hazard	Yes	No	Hazard	Yes	No
Hazardous vapors or fumes			Radiation sources		
Combustible gases			Temperature extremes		
Lack of oxygen			Electric shock		
Engulfment			Injury from mechanical equipment		
Burns (thermal or chemical)			Falls		
Hazardous liquids			Overhead work		

TABLE 5.2   Production Supervisor's Checklist (Pulp-and-Paper Mill)

	Yes	Not required
Confined space drained and cleaned		
Confined space flushed with water		
Confined space purged with steam		
Confined space thoroughly purged with air and ventilated		

TABLE 5.3   Testing Checklist and Log (Pulp-and-Paper Mill)

Indicate test required (circle)	OSHA levels safe for entry	Initial test level	Retest as needed					Safe for entry without respirator	
			1	2	3	4	5	Yes	No
Combustible gas	10% LEL								
Oxygen level	19.5–23.5%								
Hydrogen sulfide	10 ppm								
Methyl mercaptan	0.5 ppm								
Carbon monoxide	35 ppm								
Sulfur dioxide	2 ppm								
Chlorine	0.5 ppm								
Chlorine dioxide	0.1 ppm								
Ammonia	25 ppm								
Other _____	PEL _____								

contracting employers and supplement the employer's permit with their own. Although OSHA requires the contracting employer to brief the contractor about all the hazards of the space, and to ensure that the contractor's employees are trained, equipped, and informed, this doesn't always happen. The tank cleaning contractor can use the permit developed by the owner of the space, or can use a general permit form he developed with spaces for types of hazards and tests required but no hazard assessment requirements already filled in.

Designing a permit.   The first step in designing a permit to suit a workplace may be to look at sample permits drawn up by others. OSHA provides two sample permits in Appendix D to the PRCS standard (Figs. 5.1 and 5.2). Build on what others have written, copying the components that suit your facility and adding others you need. Another general use permit is shown in Fig. 5.3.

Organization of the permit.   The format should facilitate easy use. Organize the permit clearly, with actions listed sequentially in the order in which they should be taken. Use a type font that is easy to read, and at least 12-point in size. Have a clean master copy available so that third- and fourth-generation copies are not used, as these get more blurry and less legible each time they are copied.

If your workplace includes a variety of permit-required confined spaces, and therefore several permit forms, arrange the common elements and the order of elements in the same way on each. Clearly state on each page of a multicopy form who should receive the page following conclusion of the entry.

Single or multiple user(s).   Some companies designate one person to fill out the entire permit, and others prefer that individuals with different areas of expertise and responsibility complete the different sections. An industrial hygienist, for example, may be responsible for conducting air monitoring and would fill in the atmospheric data sections and sign or initial them. A process supervisor, pipe fitter, or maintenance foreman might be assigned to fill in and initial the permit sections describing the accomplishment of energy lockout. The attendant may have a separate checklist section for noting that access is restricted, communications are working, and a rescue team is on standby. A permit designed for multiple users should group each user's sections together for ease of use.

Sample Permit I (Adapted from OSHA 1910.146 App. D)

Date and Time Issued: _____ Date and Time Expires: _____ Job site/Space I.D.: _____
Job Supervisor:_____ Equipment to be worked on: _____ Work to be performed: ____
Stand-by personnel: _____ _____ _____
1. Atmospheric Checks: Time   _____
        Oxygen   _____%
        Explosive _____% L.F.L.
        Toxic   _____PPM
2. Tester's signature: _____
3. Source isolation (No Entry):   N/A   Yes   No
Pumps or lines blinded, disconnected, or blocked
4. Ventilation Modification: N/A   Yes   No
Mechanical
Natural Ventilation only
5. Atmospheric check after isolation and Ventilation:
    Oxygen _____%   >   19.5 %
    Explosive _____% L.F.L   <   10   %
    Toxic _____PPM   <   10   PPM H(2)S
    Time _____
Tester's signature: _____
6. Communication procedures: _____
7. Rescue procedures: _____
_____
8. Entry, standby, and back up persons: Yes   No
Successfully completed required training?
Is it current?
9. Equipment: N/A   Yes   No
Direct reading gas monitor-tested
Safety harnesses and lifelines for entry and standby persons
Hoisting equipment
Powered communications
SCBAs for entry and standby persons
Protective Clothing
All electric equipment listed Class I, Division I, Group D
    and Nonsparking tools
10. Periodic atmospheric tests:
    Oxygen   ___% Time ____ Oxygen   ___% Time ____
    Oxygen   ___% Time ____ Oxygen   ___% Time ____
    Explosive ___% Time ____ Explosive ___% Time ____
    Explosive ___% Time ____ Explosive ___% Time ____
    Toxic   ___% Time ____ Toxic   ___% Time ____
    Toxic   ___% Time ____ Toxic   ___% Time ____

We have reviewed the work authorized by this permit and the information contained herein. Written
instructions and safety procedures have been received and are understood. Entry cannot be approved if any
squares are marked in the "No" column. This permit is not valid unless all appropriate items are completed.

Permit Prepared By: (Supervisor)_____
Approved By: (Unit Supervisor)_____
Reviewed By (Cs Operations Personnel) : _____

_____   _____
   (printed name)   (signature)

Figure 5.1 Sample permits provided by OSHA can serve as the basis for a site-specific permit.

User qualifications. Permits vary in the amount of detail they contain. The industrial hygienist is qualified also to judge whether the proper personal protective equipment (PPE) (respirator and chemical protective clothing) are being used, so the PPE section of the permit may just ask "Correct PPE selected?" On the other hand, if the PPE is chosen by a supervisor or entrant not trained in selection, the permit should be more specific in listing exactly which items are to be worn.

Hot-work permit. When grinding, welding, brazing, or torch cutting will be done inside a confined space, a special permit is required (Fig. 5.4). This can be a separate permit or an attachment to the basic entry permit. The hot-work permit includes the basic permit sections—date, job description, duration, hazard assessment and control, equipment

SAMPLE PERMIT II (Adapted from OSHA 1910.146 D)

PERMIT VALID FOR 8 HOURS ONLY. ALL COPIES OF PERMIT WILL REMAIN AT JOB SITE
UNTIL JOB IS COMPLETED

DATE:_____ SITE LOCATION and DESCRIPTION _____
PURPOSE OF ENTRY _____
SUPERVISOR(S) in charge of crews   Type of Crew   Phone #
_____
_____

COMMUNICATION PROCEDURES _____
RESCUE PROCEDURES (PHONE NUMBERS AT BOTTOM) _____
_____
    * BOLD DENOTES MINIMUM REQUIREMENTS TO BE COMPLETED AND REVIEWED PRIOR
                        TO ENTRY*

REQUIREMENTS COMPLETED   DATE   TIME
Lock Out/De-energize/Try-out
Line(s) Broken-Capped-Blanked
Purge Flush and Vent
Ventilation
Secure Area (Post and Flag)
Breathing Apparatus
Resuscitator - Inhalator
Standby Safety Personnel
Full Body Harness w/"D" ring
Emergency Escape Retrieval Equip
Lifelines
Fire Extinguishers
Lighting (Explosive Proof)
Protective Clothing
Respirator(s) (Air Purifying)
Burning and Welding Permit

        RECORD CONTINUOUS MONITORING RESULTS EVERY 2 HOURS

TEST(S) TO BE TAKEN   Entry Level   1   2   3   4   5
PERCENT OF OXYGEN   19.5% to 23.5%
LOWER FLAMMABLE LIMIT   Under 10%
CARBON MONOXIDE   +35 PPM
Aromatic Hydrocarbon   + 1 PPM * 5PPM
Hydrogen Cyanide (Skin)   * 4PPM
Hydrogen Sulfide   +10 PPM *15PPM
Sulfur Dioxide   + 2 PPM * 5PPM
Ammonia   *35PPM
Short-term exposure limit: Employee can work in the area up to 15 minutes.
+ 8 hr. Time Weighted Avg.: Employee can work in area 8 hrs (longer with appropriate respiratory
protection).
REMARKS:_____
GAS TESTER NAME   INSTRUMENT(S)   MODEL   SERIAL &/OR UNIT #

_____   _____   _____   _____

_____   _____   _____   _____

SAFETY STANDBY PERSON(S)   ENTRANT   ENTRANT
_____   _____   _____
SUPERVISOR AUTHORIZING - ALL CONDITIONS SATISFIED_____

Figure 5.2 Companies can adapt one of the OSHA sample permits to their own confined space requirements.

requirements, and working procedures—and describes additional precautions to be tak-
en. These may include stripping linings or coatings from the surface near the hot work,
ventilation to control vapors or dusts resulting from the hot work, depressurization of piping systems containing materials that may become hazardous if heated, and methods to
protect any equipment inside the space that can be harmed by the hot work. The 29 CFR
1910.251-257 standards on welding, cutting, and brazing should be consulted. Remember
that 1910.252 mandates certain shutdown procedures for hot work in confined spaces. Fire
watch is required.

CONFINED SPACE ENTRY PERMIT

A. General: Date: _____ Building: _____ Location: _____

Work description: _____

Purpose of entry: _____ _____

Hazards of space: _____

Permit duration: Start time: _____ Close time: _____ Extended/entry spvr.: _____/_____

B. Preparation/Isolation: (Minimum requirements of owning department prior to permit authorization and confined space entry; enter: **Yes** = completed, **No** = incomplete - to be reviewed/initialed by entry supervisor or **NA** = not applicable.)

Space clean for entry (flushed/purged/vented) _____ Hazardous energy controlled / LOTO _____

Space isolated (lines separated/capped/blanked) _____ Preentry ventilation required _____

Engulfment hazards eliminated or controlled _____ Internal mechanical hazards eliminated or controlled _____

Thermal hazards eliminated or controlled _____ Structural and/or overhead hazards eliminated or controlled _____

C. Equipment: (Minimum requirements to be reviewed by entry supervisor prior to permit authorization and confined space entry; enter: **Yes** = completed or **NA** = not applicable, and **List** specific PPE/RPE)

Fire Extinguisher _____ Ingress/egress _____ Warning sign posted _____ Barriers/shields _____

Continuous ventilation provided _____ Communications _____ Emergency equip. _____ Lighting _____

Entrant PPE/RPE _____

Full body harness w/D ring _____ Emergency escape retrieval system _____ Lifelines _____

Rescuer PPE/RPE _____

D. Atmospheric Evaluation: (Initial and periodic - minimum 2 hour intervals)
Time/Init's.   % Oxygen   % LEL/LFL   Visible Dust   CO ppm   _____   _____

_____	_____	_____	_____	_____	_____	_____
_____	_____	_____	_____	_____	_____	_____
_____	_____	_____	_____	_____	_____	_____
_____	_____	_____	_____	_____	_____	_____
_____	_____	_____	_____	_____	_____	_____

Instruments used:
_____

E. Authorization: Name   Date   Time   Signature   Closed   Initials

Permit Approver _____ _____ _____ _____ _____ _____

Entry Supervisor _____ _____ _____ _____ _____ _____

Contractor Safety _____ _____ _____ _____
Coordinator   (preentry authorization required if contractors are entrants)

(Maintain Personnel-Time log on the back side of the white copy of this form during confined space entry operations.)
(Post white copy on clipboard near entry portal and yellow copy in owning department control room or office.)

Figure 5.3 This sample permit extensively covers a variety of confined space concerns. (*Printed with permission of Mine Safety and Health Appliances.*)

CONFINED SPACE ENTRY PERMIT

F. Acceptable Entry Conditions:
1. Atmospheric oxygen concentration between 20.5% and 22.5%
2. Flammable gas, vapor, or mist less than 1 % of its lower flammable limit (LFL). Any positive reading on a combustible gas instrument will require additional control measures before entry will be authorized.
3. No visible airborne combustible dust.
4. Atmospheric concentration less than 50% of any regulated substance's dose or permissible exposure limit. Respiratory protective equipment will be required if additional cleaning / purging / ventilation methods do not reduce the concentration to acceptable levels.
5. Interior surfaces cleaned properly, free from chemical contamination (liquids & solids). If chemical contamination can not be adequately eliminated, additional / suitable PPE will be required for entry.
6. All applicable requirements of the Hazardous Energy Control / Lockout / Tagout procedure have been completed.
7. All necessary controls to address engulfment hazards have been implemented.
8. No electrical hazards from improperly used or maintained lights, tools, equipment connected to or to be used within the permit space.
9. No thermal hazards due to inadequate cooling of process equipment / piping / vessels.
10. No structural hazards due to potential failure of vessels, tanks, roof or roof supports, etc.
11. No overhead hazards that could result in materials, tools, equipment falling into the permit space.
12. Proper illumination of the work area inside and outside the permit space.
13. Good housekeeping practices maintained inside and outside the permit space (slip, trip, and fall hazards protected, flammable and combustible materials stored properly, chemical spills cleaned up immediately, egress routes maintained open).

List any additional *Prohibited* conditions: _____

_____

_____

G. Rescue & Retrieval Plan: (To be completed by entry supervisor prior to permit authorization.)

ERT CSR Preplan Completed: Yes   No   NA   Communication methods tested: Yes   No   NA

ERT Rescue Personnel:_____

H. Personnel-Time Log: Name	Initials	Entry	Exit	Entry	Exit	Entry	Exit
Entrant: _____	_____	_____	_____	_____	_____	_____	_____
Entrant: _____	_____	_____	_____	_____	_____	_____	_____
Entrant: _____	_____	_____	_____	_____	_____	_____	_____
Entrant: _____	_____	_____	_____	_____	_____	_____	_____
Entrant: _____	_____	_____	_____	_____	_____	_____	_____
		On duty	Off	On duty	Off	On duty	Off
Attendant: _____	_____	_____	_____	_____	_____	_____	_____
Attd. relief: _____	_____	_____	_____	_____	_____	_____	_____
Attendant: _____	_____	_____	_____	_____	_____	_____	_____
Attd. relief: _____	_____	_____	_____	_____	_____	_____	_____

I. Comments and Recommendations: (To be completed by entry supervisor for program improvements.)

_____

_____

_____

_____

Audit Log / EHS&S Dept. use only.   Reviewed by: _____   Date: _____

IN CASE OF AN EMERGENCY NOTIFY THE ENTRY SUPERVISOR OR CALL ERT/SECURITY

Figure 5.3 (*Continued*)

Hot Work Permit

Valid on this day, up to day shift of the following day   Permit is void as follows:

Date_____   1. When conditions change making continuation hazardous

Work to begin at _____AM PM   2. When starting work delayed or stopped for ____hr

Work to begin at _____AM/PM   3. Permit expires at _____AM/PM

SECTION 1

Permission is granted to _____to use_____to enter_____

Description of work planned and specific location:

SECTION 2 ITEM	YES	NO
Proposed work checked with supervisor and/or operator in charge		
Pressure in equipment and pipeline		
Equipment properly drained and purged or cleaned		
Equipment properly blinded or blanked		
Equipment properly tagged and/or locked out		
Precautions taken against release of gas or oil in area		
All combustible or flammable liquids removed or protected		
Underground drawings checked before excavation		
Sewer openings covered		
Facilities available for control and disposal of hazardous materials		
Special warning/caution signs posted		
Overhead work has been barricaded		
Gauge glass columns drained and closed or protected		
Welding machine safely located, safely grounded, and sparks controlled		
Standby fire equipment needed (list):		
Ventilation equipment installed		
Proper means of access or egress is available		
Personal protective equipment is needed and will be worn (check boxes) Safety glasses__ Face shield__ Safety goggles__ Boots__ Helmet__ Gloves__ Dust respirator__ Cartridge respirator__ Compressed air supply respirator__ Fresh air supply respirator__		
Protective suit		
Safety harness/lifelines		
Other personal protective equipment (list):		
Rescue personnel designated (list):		
Craftsmen are trained for hazard they may encounter		
Additional precautions listed in section 3 will be followed		
Signature		

SECTION 3 VAPOR TESTS AND SAFETY REPORT __Hydrocarbons __CO __Oxygen __Other (list)

Type						
Time						
% or PPM						
Initials of tester						

Vapor tests to be repeated every_____Hrs Reinspection every____Hrs

Additional precautions required:

SECTION 4 APPROVAL SIGNATURES

Safety Checked_____   Work Authorized_____

Figure 5.4   Hot work (welding, brazing, cutting) inside a confined space requires a special kind of permit. This sample can be adapted according to individual needs.

Suitability to the workplace. The most accessible permit forms are those that are specially designed to suit the confined spaces in the workplace where they will be used and the jobs that will be performed during entry. Permits are complicated enough even when they are pared to the minimum information needed for safe entry, and do not need to be made increasingly complex and lengthy by the inclusion of unnecessary elements. The more burdensome the permit is to fill out, the greater the temptation to shortcut the process and the easier it is to overlook important sections.

Posting the permit

Post the completed permit on the space near the entry, or otherwise make it available to all concerned. If you choose not to post it on the space, you must ensure that everyone with a legitimate interest knows where the permit is and has access to it. If a contractor's employees will enter the space and comply with the space owner's permit, these workers must also have access and all rights to hazard and testing information. If the contract employees write their own permit, the company must ensure that the permit is OSHA-compliant and indicative of all hazards, as discussed in Chap. 2.

Summary

A written permit is required for every entry. All the components listed by OSHA, describing all the hazards, testing, limitations, controls, and conditions, are shown on the permit. The written permit is the controlling and informational document for entries into confined spaces for work and for rescue. OSHA does not mandate the format of the permit, allowing confined space owners to draw up a permit appropriate to each space, but does mandate that the permit be complete and available.

Protective Measures

6

Controlling Confined Space Hazards

Two maintenance workers at a tannery were assigned to repair a crack in the top of an overflow standpipe within a concrete manhole that was part of the facility's effluent system. The manhole was 4 ft in diameter and 10 ft deep, with access available through a 2-ft-square opening at the top using steel rungs set into the sides of the manhole. Prior to entry, no confined space safety procedures were implemented, no atmospheric monitoring was conducted, and no ventilation of the space was performed. Hardhats and safety boots were the only personal protective equipment used by the workers.

Evidence indicates that one worker entered the manhole to begin repairs, while the coworker remained at the entryway to observe the operation and hand down tools and equipment as needed. The initial entrant apparently was overcome by hydrogen sulfide gas, and fell into approximately 3 ft of wastewater in the bottom of the manhole. Presumably the coworker entered to attempt rescue of the initial entrant and was also overcome by hydrogen sulfide and fell into the wastewater in the bottom of the space.

Some time later, the maintenance foreman discovered the incident and entered the manhole to attempt rescue. The foreman became dizzy, exited the space, and fell unconscious onto the ground outside the space. On regaining consciousness, the foreman notified plant personnel of the emergency, and local EMS and fire department personnel were alerted. While awaiting the arrival of emergency personnel from offsite, four more plant personnel entered the manhole in unsuccessful rescue attempts and were exposed to hydrogen sulfide.

After arriving on the scene, local fire department personnel donned self-contained breathing apparatus, entered the manhole, attached ropes to the victims, and removed them from the space. Local EMS personnel arrived on the scene and determined that both the initial entrant and his coworker had been dead for some time. As a result of exposure to hydrogen sulfide, the foreman was hospitalized for 2 days, while the other four employees who attempted rescue were treated and released the same day. Six days after the incident, a state OSHA compliance officer measured a hydrogen sulfide concentration of 200 ppm just inside the manhole. That concentration is twice the current IDLH level for hydrogen sulfide published by NIOSH.

The Role of Hazard Control in Confined Space Operations

Confined space incidents such as the one described above often can be prevented by identifying and eliminating the hazards present before entry is made. For example, atmospheric hazards frequently can be abated through effective ventilation of the space. Other incidents occur as a result of accidental activation of energy sources, startup of machinery, or release of material into an occupied space. Those types of accidents typically can be prevented by isolating the space from the hazard sources before entry is made.

When preventive measures fail and rescue operations are required, hazard control procedures are critical to the safety of the rescuers. They also significantly enhance the probability that the victim or patient inside the space will survive the incident.

In this chapter we will explore procedures for controlling confined space hazards. These procedures include those commonly used to abate atmospheric hazards and isolate the space from sources of energy or material, as well as other procedures and considerations for hazard control.

Controlling Atmospheric Hazards

Many confined space incidents are caused by oxygen-deficient, oxygen-enriched, flammable, or toxic atmospheres. Eliminating or abating such atmospheric hazards is a vital step in making spaces safe for entry, whether for routine work assignments or rescue.

A number of procedures may be utilized to control atmospheric hazards. For example, some spaces can be flushed to remove residual product that is the source of vapors contaminating the atmosphere. In some cases a suitable sorbent material may be used to eliminate hazardous vapors, such as when activated charcoal is used to remove an organic contaminant. In other cases a space may be inerted, or filled with an inert gas, in order to purge the space. For example, nitrogen may be used to displace a highly flammable atmosphere, thereby replacing it with an extremely oxygen-deficient one. The most common method, and in most cases the most effective method for controlling atmospheric hazards is through ventilation of the space.

Basic concepts of purging and ventilation

Procedures used in purging and ventilation have the effect of pushing or pulling contaminated or "bad" air out of a space and introducing uncontaminated or "good" air into the space. The initial process of clearing a space of bad air before entry is referred to as "purging" the space. The term "ventilation" indicates the process of continuously providing a good atmosphere in the space throughout the duration of entry. Either process can be driven by differential air pressures and/or differences in vapor densities between contaminated and uncontaminated air.

Keep in mind that this text attempts to provide a simplified coverage of a complex topic. More detailed coverage is available through references such as *Fundamentals of Industrial Hygiene* (Plog et al. 1996) and *Industrial Ventilation, A Manual of Recommended Practice* (ACGIH 1995). In ventilating confined spaces, the surest option is to consult certified industrial hygienists or other trained safety professionals with expertise in ventilation.

No one ventilation procedure is appropriate for all confined spaces. Selection of proper techniques, equipment, and precautions requires sound professional judgment and a careful assessment of the confined space to be ventilated. Important factors to consider in selecting ventilation techniques include

- Size, shape, location, and internal configuration of the space to be entered

- Number, size, location, and configuration of available openings

- Source and characteristics of the contaminant, such as degree of flammability, vapor pressure, vapor or gas density, and toxicity

- Environmental conditions such as natural air currents, temperature, and thermal gradients due to solar gain

- Equipment and time available to effect the ventilation process

The reason for entering the space is an important initial consideration. In making entry for rescue, time is a critical factor. If purging is required before a rescue entry, it must be carried out rapidly and for the shortest duration that will allow rescuers to enter safely with respiratory protective equipment. Permitted work entries and body recovery operations allow more time for preparation of the space prior to entry. Personnel may purge and ventilate the space thoroughly to make entry safe without respiratory protection and make working conditions more comfortable.

The source of contamination is also an important consideration. For example, some spaces may contain only residual gases that will not be replaced after the space is purged. In such a case, ventilation following initial purging may not be required to maintain acceptable entry conditions. In another example, residual product with a high vapor pressure may act as a constant source of flammable and/or toxic vapors within a space. In such a case, unless the hazard can be otherwise mitigated by measures such as by flushing out the residual product, continuous ventilation will most probably be required throughout the entry to maintain acceptable conditions. In some operations, the actual work activities to be done within a space may be the source of the air contaminants involved. Examples of such operations include welding, abrasive blasting, use of cleaning solvents, and painting or application of coatings. These operations may require the use of local exhaust ventilation, as described below.

As a general rule, continuous ventilation is recommended throughout all entries. This is due to the added safety margin and enhanced worker comfort it may provide. OSHA requires that forced-air ventilation be operated continuously throughout all entries made under the "alternate procedures" provision of the PRCS standard.

Strategies and techniques of purging and ventilation

Ventilation strategies and techniques may be classified, compared, and contrasted in several ways, as discussed below. These include

- Natural ventilation versus mechanical ventilation

- Positive-pressure versus negative-pressure ventilation

- General or dilution ventilation versus local exhaust ventilation

Natural ventilation

The process of natural ventilation can be driven by, or affected by, different forces. In some cases natural air currents or wind alone may effect ventilation by forcing good air into a space and forcing bad air out. In other cases differing densities between contaminated air within a space and uncontaminated air outside the space may cause the bad air to flow out of the space and be replaced with, or diluted by, good air from outside. Also, thermal updrafts caused by solar heating of a container may affect the process. In many examples of natural ventilation, all three mechanisms interact during the ventilation process.

Natural ventilation for lighter-than-air contaminants. As a simple example, assume that we have a vertical cylindrical vessel containing residual methane gas and equipped with bolted manways at the top and bottom of the vessel (see Fig. 6.1). Since the relative gas density of methane is around 0.8, we expect the methane to be more concentrated in the upper part of the vessel.

If we open the manway on top of the vessel, the methane will have a natural tendency to rise upward through the opening; however, this will not be very efficient because both outbound and inbound air must share the same pathway. On the other hand, if we also open the bottom manway, replacement air will flow in through the bottom of the vessel to replace the contaminated air flowing out though the top (see Fig. 6.2). The resulting "chimney effect" will provide more effective ventilation than in the previous example. In this type of scenario, the lighter the contaminant gas or vapor (i.e., the lower the relative gas density or vapor density), the faster the ventilation process should proceed.

If the vessel is exposed to solar radiation, the process may proceed significantly faster because of the thermal updrafts produced as the vessel and its contents are heated.

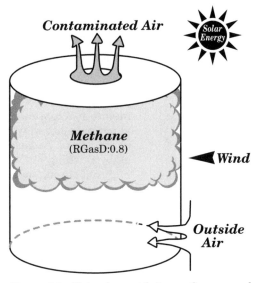

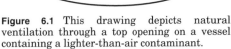

Figure 6.1 This drawing depicts natural ventilation through a top opening on a vessel containing a lighter-than-air contaminant.

Figure 6.2 Natural ventilation of a vessel containing a lighter-than-air contaminant may be more effective when both top and bottom openings are used.

Natural air currents may also affect the process. For example if the wind direction is such that it forces outside air into the bottom manway, then the process may be significantly enhanced (see Fig. 6.2).

Natural ventilation for heavier-than-air contaminants. As another simple example, assume that we are dealing with the same vessel, except that it contains residual propane (see Fig. 6.3). Since the relative gas density of propane is around 1.6, we expect it to be more concentrated in the lower part of the vessel. If both the top and bottom manways are opened and the outside air is still, we expect the contaminated air to flow out through the bottom opening, while the replacement air comes in through the top opening. In this type of scenario, the greater the density of the contaminant gas or vapor, the faster the ventilation process should proceed. Of course, this process could also be affected by environmental factors such as wind- or solar-induced thermal gradients.

One problem with this scenario is that the vicinity of the bottom manway may not be a safe location to release the propane, especially since it is highly flammable and will tend to accumulate and concentrate in low-elevation areas. As we will see later, mechanical ventilation would offer safer options in this example. One option might be to simply reverse the process by using a blower to force air into the bottom manway in order to push the contaminated air out the top of the vessel, thus allowing an opportunity for the vapors to disperse. An even better option might be to leave the bottom manway closed, and use negative-pressure ventilation (NPV) to draw off the flammable gas from the lower part of the vessel and discharge the gas at a safe location (see Fig. 6.9).

Limitations of natural ventilation. Although natural ventilation is used effectively in some instances, it has significant limitations. In the absence of favorable wind currents, it may not be an effective technique for contaminants with vapor densities near that of air (i.e., 1). It may not be effective for contaminants that are highly toxic in low concentrations. Furthermore, the effects of environmental factors such as wind, temperature, and solar energy on the process may be hard to predict. Natural ventilation is more time-consuming and less dependable than mechanical ventilation.

The extra time required for natural ventilation may not be a critical consideration for ventilating prior to routine permitted entries into confined spaces. In contrast, natural

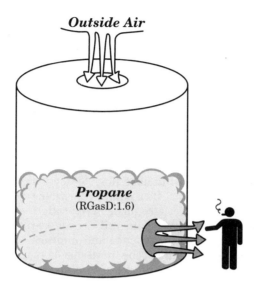

Figure 6.3 Natural ventilation of a heavier-than-air contaminant through an opening at the bottom of a vessel can cause hazards to personnel.

ventilation is not a viable option for ventilating in preparation for confined space rescue because of the time-critical nature of rescue operations. Potential rescuers should be acquainted with the basic principles of natural ventilation because the same principles also come into play when conducting mechanical ventilation. The inappropriate selection or use of natural ventilation techniques may be the reason why a rescue is required in some situations.

Mechanical ventilation

Mechanical ventilation is driven by differences in air pressure between contaminated air within a space and uncontaminated air from outside. Eductors, fans, or some other type of machinery are required, hence the term "mechanical."

Equipment used in mechanical ventilation. Confined space ventilation is a specialized type of operation that requires specialized equipment. Various types of air moving devices are commonly utilized for mechanical ventilation. Some of these devices are described below.

Venturi eductors. *Venturi eductors,* or air horns, are used to purge and ventilate some confined spaces (see Fig. 6.4). These devices are powered by compressed air or steam and operate on the venturi principle. The air or steam is released into the eductor through a nozzle near the air inlet. As the air or steam flows through the nozzle at high velocity, it induces ambient air into the inlet, forces it along the tube, and discharges it through the outlet horn. Venturi eductors can be used for both positive and negative ventilation techniques. Some eductors are designed to allow ducts to be attached to them, which can offer certain advantages, as discussed below.

Compared to fans, venturi eductors are lighter, more compact, cheaper, and able to move a more heavily contaminated airstream without damage. On the other hand, eductors are not able to move large volumes of air and require a significant supply of compressed air or steam to operate, and they may introduce hazards to the situation. The high-velocity air movement may generate large amounts of static electricity that may serve as an ignition source for flammables. For this reason air horns should always be electrically bonded, using a suitable bonding wire, to the space being ventilated. Any large debris present inside the horn when the air or steam supply is activated could fly out and strike anyone in the area with considerable force. They are quite noisy in operation.

Fans.   Fans or vent blowers are commonly used for mechanical ventilation of confined spaces. They include both axial-flow and centrifugal-flow fans. Both types of fans are available with a variety of power sources including electricity, gasoline, or natural gas and pneumatic or hydraulic power sources.

Axial-flow fans are designed to move air parallel to the axis of rotation of the blades (see Fig. 6.5). The operating principle is basically like that of a window fan or box fan you may use to cool off at home. In contrast, centrifugal-flow fans move air perpendicular to the axis of rotation of the blades, and are also known as *radial-flow-type* fans for this reason (see Fig. 6.6). Centrifugal-flow fans are sometimes referred to as "squirrel cage" blowers.

Either type of fan may be designed to have flexible ducts or tubing attached to both the air intake and discharge locations. If so equipped, either type can be used in either the positive-pressure or negative-pressure mode of operation, as discussed below. The use of ducts can provide other advantages as well. By separating the inbound and outbound airstreams, they may reduce turbulence and improve efficiency of ventilation, especially when only one opening must be used for both inbound and outbound airstreams. The use of tubing allows the intake or discharge points in a ventilation system to be placed for best advantage.

Centrifugal-flow fans (Fig. 6.6) operate well under high airflow resistance. This means that less reduction in airflow occurs in moving air through a given sequence of tubing with a centrifugal-flow fan than with an equivalent axial-flow fan. Centrifugal-flow fans tend to be heavier, bulkier, and more expensive than equivalent axial-flow fans. Axial-flow fans are effective for moving high volumes of air under relatively low airflow resistance, such as when minimal or no vent tubing is attached to the fan. Axial-flow fans generally are not recommended for negative-pressure ventilation (as described below) when flammables are present because the fan motor is in the direct path of the airflow and can act as an ignition source.

Figure 6.4 The Petro-Vent® by Air Systems International is a venturi eductor fitted with an adapter that allows it to thread directly to the 4-in vent pipe found on many storage tanks. (*Photograph   courtesy   of   Air   Systems International.*)

Figure 6.5 Axial-flow fans move air parallel to the axis of rotation of the blades. (*Photograph courtesy of Air Systems International.*)

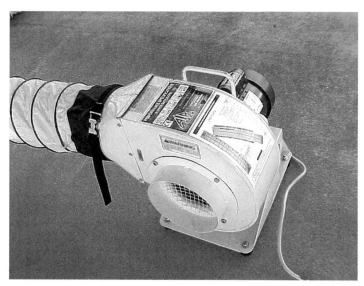

Figure 6.6 Centrifugal-flow fans move air perpendicular to the axis of rotation of the blades. (*Photograph courtesy of Air Systems International.*)

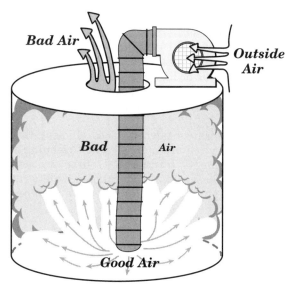

Figure 6.7 Positive-pressure ventilation pushes outside air into the space.

Positive-pressure ventilation. *Positive-pressure ventilation* (PPV) is accomplished by forcing outside air under pressure into a space (see Fig. 6.7). PPV is also known as *forced-air* or *supply ventilation*.

PPV is considerably more efficient than negative-pressure ventilation, because air is easier to push than to pull. One general rule related to ventilation is that a given force will push a given volume of air about 30 times further than the same force will pull the same volume of air (see Fig. 6.14). The efficiency of PPV can result in a strong ventilation action that may eliminate "dead spots" that otherwise might not ventilate well within a space.

When used alone, PPV operates by forcing good air into a space, thereby forcing the bad air out through whatever openings are available. The incoming air has the effect of diluting the contaminant concentration within the space, a process referred to as *dilution ventilation* or *general ventilation* (see Fig. 6.7). General ventilation works well when contaminants are evenly dispersed throughout the atmosphere within the space. It is the method of choice for abating atmospheres that are oxygen-deficient or oxygen-enriched, contain flammables at concentrations well below the LEL, or contain low concentrations of contaminants with a low toxicity.

In some instances PPV may make conditions more dangerous instead of safer for personnel involved in confined space operations. As an example, assume that a space is heavily contaminated with residual product that is emitting highly flammable and toxic vapors (see Fig. 6.8). As a result, the atmosphere is flammable-rich, because it contains vapors well above the UEL of the substance. PPV alone in this case may accomplish nothing more than "leaning out" the flammable vapors to a concentration below the UEL, thus making ignition possible. The incoming airstream may become saturated with vapors that are very quickly replaced by vapors generated from the residual product. The airstream may then disperse the vapors throughout the space, thereby increasing the contaminant concentration throughout most of the space. The airstream may also carry high concentrations of vapors outside the space. Since PPV alone doesn't provide good control of the airflow exiting the space, the escaping vapors may pose a significant hazard to personnel working outside the space (see Fig. 6.8).

Negative-pressure ventilation.   Negative-pressure ventilation (NPV) is accomplished by pulling contaminated air from inside the space and discharging it to the outside (see Fig. 6.9). NPV is also known as *exhaust ventilation*. When NPV is used alone, contaminated air pulled out of the space is replaced by outside air that is pulled into the space through whatever openings are available. As a result, contaminants in the space are diluted to lower concentrations in a process referred to as *general-exhaust ventilation*.

NPV works well when atmospheric contaminants are concentrated in a certain part of a space rather than evenly dispersed throughout. NPV may provide a safer option than PPV for abating flammable or highly toxic atmospheres. By placing the end of the negative-pressure duct in the part of the space where contaminants are more concentrated, it may be possible to selectively remove them and release them at a safe location outside the space.

As previously noted, NPV is considerably less efficient than PPV. Fans or other equipment used for NPV may require more frequent cleaning and maintenance, especially if used in dirty environments. Intrinsically safe or explosionproof blowers must be used for NPV when flammables are involved. Intrinsic safety ratings were discussed in Chap. 4. Such blowers are significantly more expensive than regular blowers.

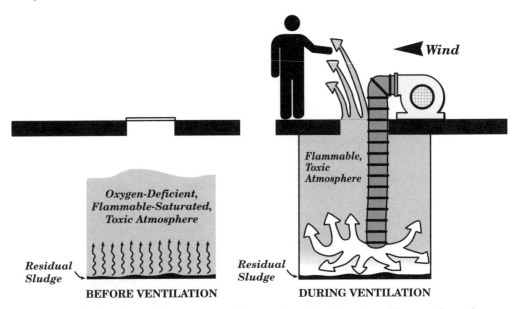

Figure 6.8   Positive-pressure ventilation can make operations more hazardous if improperly used.

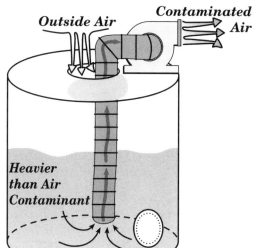

Figure 6.9 Negative-pressure ventilation pulls contaminated air out of the space.

Local exhaust ventilation. *Local exhaust ventilation* is a variation of NPV that can be used when there is a specific point of origin of an air contaminant within a confined space. In local exhaust ventilation, the negative pressure is applied directly at the point of origin of contaminants (see Fig. 6.10), and the contaminants are pulled out of the space in the fashion of a vacuum cleaner. Examples of the points of origin of contaminants might include locations of welding operations, abrasive blasting, use of cleaning solvents, or leaking pipes, valves, or fittings.

In order for local exhaust ventilation to operate sufficiently, the effective fan capacity must be adequate at the point where contaminants are to be pulled into the exhaust duct, as discussed below. The end of the exhaust tubing should be fitted with an appropriately designed intake hood, which should be placed as close to the point of origin of the contaminants as possible.

Combination positive-pressure/negative-pressure ventilation. If more than one device is available for ventilation, it may be possible to ventilate a space using a combination of PPV and NPV. This approach simultaneously uses PPV to push air into the space and NPV to pull air from the space. It is also known as "push/pull" ventilation. In some cases it represents a "best of both worlds" approach intended to provide enhanced efficiency and safety.

Figure 6.11 shows a combination configuration for a heavier-than-air contaminant in which NPV is used to pull contaminated air from the bottom of the container, while PPV is used to push outside air into the top of the container. For a lighter-than-air contaminant, the configuration could be reversed to pull contaminated air from the upper part of the container while pushing clean air into the bottom.

Considerations for conducting confined space ventilation

Before attempting to ventilate a confined space, size up the space and consider a number of factors in order to decide on the appropriate equipment and procedures. These factors are important considerations when planning for permitted entries or conducting preemergency planning for confined space rescue.

Fan capacity required. The capacity of a fan is rated by the manufacturer in terms of how many cubic feet of air the fan can move per minute [cfm (ft^3/m)]. The capacity of a fan has a direct bearing on how quickly purging of a space can be accomplished with it. When using a fan fitted with ducts or hoses, the actual cfm moved within the space will be less than the cfm rating of the fan due to friction loss as air is moved through tubing, around bends or elbows, and through fittings. Fan manufacturers sometimes attach

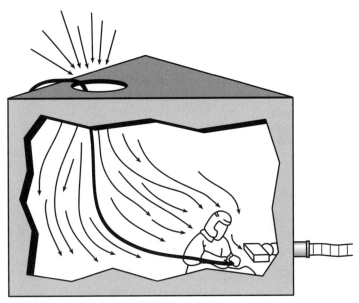

Figure 6.10 Local exhaust ventilation removes contaminants at or near their location of origin.

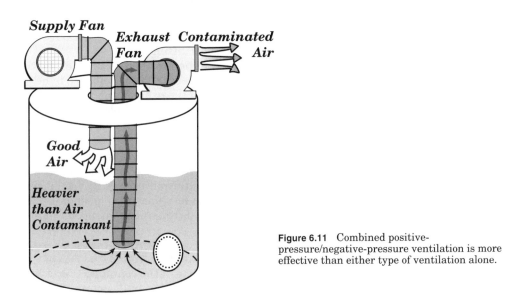

Figure 6.11 Combined positive-pressure/negative-pressure ventilation is more effective than either type of ventilation alone.

decals to their equipment stating the rated capacity in free air (i.e., with no duct attached), the effective capacity at the end of a section of duct, and the effective capacity at the end of a section of duct with one bend and two bends in it (see Fig. 6.12).

The reason why ventilation is being conducted will impact the fan capacity required. If fans are being selected to equip a rescue team, then speed and efficiency in purging spaces will be given top priority. The team may therefore select the most powerful fans that are practical to use, given size and weight restrictions. Portable fans are commonly available with 3000 cfm and greater free-air capacity. On the other hand, if ventilation equipment is intended for use in permitted work entries only, then other factors may be considered. In such cases, smaller, lighter, and cheaper equipment might be selected even though it will take longer to do the job.

Fan capacity requirements for general ventilation. An important step in planning for ventilation operations is to size up the spaces to be ventilated. Considerations for deter-

Figure 6.12 Fan manufacturers sometimes attach decals to fans indicating free-air capacity and effective capacity with various duct configurations. (*Photograph courtesy of Air Systems International.*)

mining the blower capacity required to ventilate a space include the volume of the space and the time available to purge the space prior to entry.

The volume of a space can be estimated roughly from the space's dimensions using simple geometric formulas, as shown in Fig. 6.13. If in doubt, be liberal in estimating space size, since it is better to err in the direction of caution.

We can relate volume, fan capacity, and time using the following equation:

$$\frac{V}{Q} = T$$

where V = volume of space, ft^3
$\quad\;\; Q$ = effective fan capacity, ft^3/min
$\quad\;\; T$ = time required for one air exchange, min

As an example, assume that we wish to obtain a fan that will be used to purge a space that is contaminated with organic vapors. The space is 20 ft long, 10 ft wide, and 10 ft high, indicating a volume of 2000 ft^3. At this point, we must determine how many air exchanges we wish to perform, and how quickly we would like to perform those exchanges.

Recommendations vary on the number of air exchanges required to purge a space. CMC (1996) noted that five exchanges commonly are used for situations involving flammables or toxics in industrial settings. Another source noted that 10 to 15 air exchanges usually are required to purge a space, assuming that there are no active sources of air contaminants within the space (Sargent 2000). Still another source used seven complete air exchanges in developing a chart used to estimate purge times from space volume and blower capacity (Pelsue 1993). The disparity between these recommendations reinforces the importance of always using air monitoring equipment to verify the effectiveness of ventilation, rather than merely assuming that a space is safe for entry on the basis of purge time.

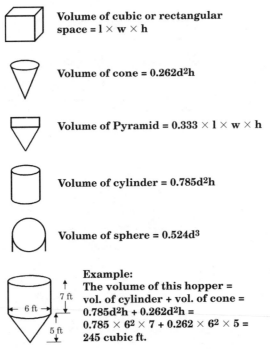

Volume of cubic or rectangular
space = l × w × h

Volume of cone = 0.262d²h

Volume of Pyramid = 0.333 × l × w × h

Volume of cylinder = 0.785d²h

Volume of sphere = 0.524d³

Example:
The volume of this hopper =
vol. of cylinder + vol. of cone =
0.785d²h + 0.262d²h =
0.785 × 6² × 7 + 0.262 × 6² × 5 =
245 cubic ft.

Figure 6.13 Simple geometric formulas can be used to make a rough estimate of the volume of many spaces from their dimensions.

For our example, assume that we desire to perform 10 air exchanges before testing the atmosphere prior to entry. Assume that we wish to be able to complete those 10 exchanges in 20 min; thus, 2 min would be required for each air exchange. We can rearrange the above equation as follows:

$$\frac{V}{T} = Q$$

where V = volume of space, ft³
T = time desired for one air exchange, min
Q = effective fan capacity, ft³/min (cfm)

Using this equation, we can determine that we will need ventilation equipment providing 1000 ft³/min (cfm) effective capacity in order to ventilate the 2000 ft³ space at the desired rate of 10 air exchanges in 20 min. This assumes uniform mixing of the incoming air throughout the space, which is probably not the case. Some lesser number of actual air exchanges will probably occur.

Factors such as the shape, internal configurations, and available openings of spaces must also be considered. Spaces with internal obstructions such as baffles will be more difficult and time-consuming to ventilate than spaces that have an unobstructed interior configuration. These factors also determine the length of duct and the number of bends and fittings required to deliver the airstream, which in turn have a direct bearing on the fan capacity required. For instance, assume that the layout of our example space is such that we will need to use 25 ft of flexible duct with one 90° bend in order to deliver airflow to the space for PPV. We will use an 8-in-diameter duct. We will need a vent fan with a free air capacity of at least 1500 ft³/min in order to provide the 1000 ft³/min effective capacity we need at the end of the duct (see Fig. 6.12).

Another significant factor is the "throw" or "reach" of the airflow, meaning the distance beyond the end of the duct that the airflow can reach with sufficient velocity for effective ventilation. In PPV once the airstream exits the end of the duct, it begins to disperse and lose velocity with increasing distance beyond the end of the duct (see

Fig. 6.14). A minimum air velocity of 200 ft/min is recommended for mixing and moving air contaminants (Sargent 2000).

In PPV the airstream will retain only 10 percent of the air velocity it had at the duct face at a distance beyond the duct face equal to 30 times the diameter of the duct (see Fig. 6.14). For our example, assume that the airstream needs to penetrate up to 20 ft beyond the end of the duct. The volumetric flow rate at the end of the duct is represented by the following equation:

$$Q = A \times v$$

where Q = effective fan capacity, ft³/min (cfm)
 A = cross-sectional area of duct, ft² ($A = 0.7854\, d^2$, with d in feet)
 v = velocity of airstream, ft/min

We can rearrange this equation to calculate velocity of the airstream on the basis of the duct diameter and the effective capacity of the blower, as follows:

$$v = \frac{Q}{A}$$

For our example, assume that the blower we are considering using has an effective capacity of 1000 cfm with an 8-in-diameter (0.35 ft²) duct, the air velocity at the duct face would be 2857 linear feet per minute. At a distance of 20 ft (30 times the duct diameter), one-tenth of this velocity should remain. The blower in question should provide around 286 ft/min velocity at the far reaches of the space, which should be adequate for our example. In addition to effective fan capacity, factors such as expense, weight, size, power source, maintenance requirements, and versatility must also be considered.

Purge time required. How long should a space be ventilated prior to entry? The only safe answer to this question is "As long as it takes to make the space safe for entry, as indicated by air monitoring readings." We can estimate times required to purge a space based on factors such as the volume of the space and effective blower capacity. We must remember that such estimates are no substitute for air monitoring prior to entry, because any number of factors could intervene to make them inaccurate. We can use these estimates

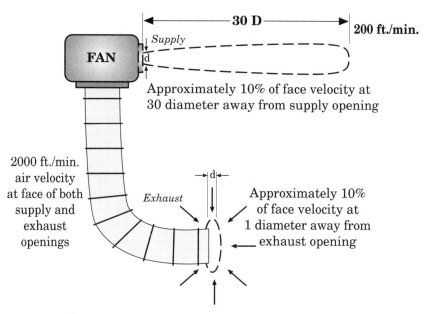

Figure 6.14 Supply airflow is significantly more efficient than exhaust airflow. [Adapted from ACGIH (1995) Fig. 1-9.]

of likely intervals required for purging to guide us in determining how long ventilation should be conducted before checking the space with air monitoring equipment before making permitted entries. For rescue entries the best procedure is to begin ventilating, then check the space continuously or at frequent intervals so that we know how soon it is safe to enter. We can estimate purge time with the following equation:

$$T_P = N \times \frac{V}{Q}$$

where T_P = time required to purge space, min
$\quad N$ = number air exchanges desired
$\quad V$ = volume of space, ft^3
$\quad Q$ = effective fan capacity, ft^3 min (cfm)

Manufacturers of ventilation equipment sometimes make charts available that can be used to estimate purge time based on effective fan capacity and the size of the space to be ventilated (see Fig. 6.15).

Adequate and complete purging and ventilation can be difficult to achieve. In some cases, such as when airflow is directed into the center of a space, the central area of a space may purge quickly, while contaminants in peripheral areas slowly diffuse into the central area to be purged. In other cases, peripheral or isolated areas may not be purged at all. This points out the importance of having personnel with the appropriate expertise involved in ventilation operations. It also points out the critical importance of monitoring the atmosphere of the entire space thoroughly before drawing any conclusions about whether it is adequately purged or ventilated.

Fan capacity and other considerations for local exhaust ventilation. Local exhaust ventilation involves using NPV with the end of the duct located to pull contaminants into the ventilation system at or near their point of origin. For this procedure in confined space work, the end of the duct is usually fitted with a capture hood. Local exhaust ventilation is a very effective ventilation tactic for situations in which contaminants have a very localized point of origin, such as a welding arc or a leaking valve. Using local exhaust, we may be able to remove contaminants that would be much more difficult to control through general ventilation once they contaminate the atmosphere of the space.

While local exhaust ventilation can be an efficient tactic, the local exhaust process is not very efficient in terms of airflow. This is because pulling air is not nearly as efficient as pushing air, as noted earlier. For example, at a distance of only one hood diameter from the face of a capture hood, the velocity of the airstream being pulled into the hood is only one-tenth of the velocity at the face of the hood (see Fig. 6.14). For a local exhaust system to work, the airstream in the vicinity of the point of origin of the contaminants must have sufficient velocity to overcome natural air currents and pull the contaminants into the system. This is referred to as the "capture velocity" of the contaminants. Capture velocity varies with the physical state and rate of generation of contaminants and the amount of turbulence in the air (see Table 6.1).

In determining the feasibility of using local exhaust, important factors are (1) the effective blower capacity of the fans available and (2) how close to the point of origin of contaminants the capture hood can be placed. We can calculate the velocity of air at a given distance from the hood using the following formula:

$$v = \frac{0.1\,Q}{X^2 + 0.1\,A}$$

where v = air velocity at a distance X from the hood, ft/min
$\quad Q$ = effective blower capacity at the hood face, ft^3/min (cfm)
$\quad X$ = distance out from hood to capture location, ft
$\quad A$ = area of hood opening, ft^2 (for circular openings, $A = 0.7854\,d^2$, with d in feet)

This formula assumes that X is less than 1.5 hood diameters and that the hood used has a circular or square opening.

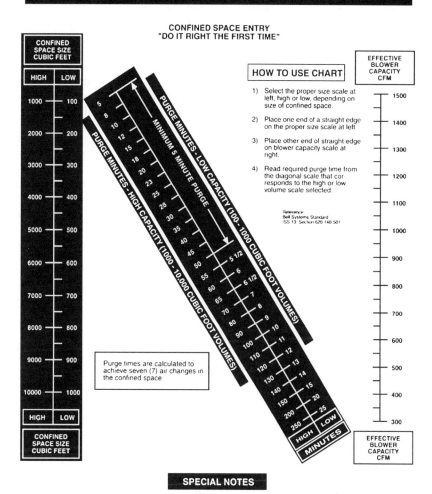

ESTIMATING APPROXIMATE PURGE TIMES

CONFINED SPACE ENTRY
"DO IT RIGHT THE FIRST TIME"

HOW TO USE CHART

1) Select the proper size scale at left, high or low, depending on size of confined space.

2) Place one end of a straight edge on the proper size scale at left

3) Place other end of straight edge on blower capacity scale at right.

4) Read required purge time from the diagonal scale that corresponds to the high or low volume scale selected.

Reference
Bell Systems Standard
ISS 10 Section 620.140.501

Purge times are calculated to achieve seven (7) air changes in the confined space.

SPECIAL NOTES

1) Air quality of the confined space should be tested prior to ventilation.

2) Ventilate confined space for the minimum times as determined in the above chart and then retest air quality.

3) If toxic (combustible) gases or low oxygen is encountered, increase purge times by 50%.

4) If 2 blowers are used, add the two capacities, then proceed with the "How to use chart" above.

5) Effective blower capacity is measured with one or two 90° bends in 8" diameter 25 ft. blower hose.

Figure 6.15 Fan manufacturers often provide charts used to estimate purge time based on fan capacity and volume of space. (*Photograph courtesy of Air Systems International.*)

TABLE 6.1   Range of Contaminant Capture Velocities

Contaminant type	Capture velocity, ft/min
Solvent vapors	50–100
Welding fumes or spray-paint mists	100–200
Grinding or abrasive blasting dusts	500–2000

SOURCE: Adapted from ACGIH (1995).

As an example, assume that we are wondering if it will be feasible to use a fan providing 1200 ft^3/min (cfm) flow into an 8-inch (0.67-ft)-diameter hood (or 0.35 ft^2 in area) to provide local exhaust for welding fumes. The hood will be placed at a distance of one foot from the location where the welding fumes will be generated. Using the formula above, we estimate that the air should be moving at about 116 linear feet per minute at the point of origin of the contaminants. This is in the low end of the capture velocity range for welding fumes shown in Table 6.1. While this may be marginally effective, we would have a greater margin of confidence if we could use a more powerful blower or place the hood closer to the source of the contaminants.

When using local exhaust ventilation in confined space work, a capture hood should always be used. The use of a properly designed hood significantly increases the efficiency of local ventilation. The hood should be placed as close as possible to the source of contamination. The velocity of the air in front of the hood varies inversely with the square of the distance from the hood. Doubling the distance from the hood to the source will reduce the air velocity at the source by a factor of 4. After contaminants are captured and pulled into the hood, the rate of airflow within the duct must be at a sufficient velocity to prevent them from settling out. Examples of recommended transport or duct velocities are shown in Table 6.2. In some cases, it is necessary to use air cleaners to remove contaminants from the airstream before they reach the fan to prevent possible damage or rapid wear to the fan.

Ventilation is a complex topic. This applies especially to local exhaust ventilation. The assistance of personnel such as certified industrial hygienists or other trained safety professionals with expertise in ventilation is invaluable in confined space ventilation.

Pitfalls and problems of ventilation

A number of mistakes or problems commonly occur during ventilation operations. In some cases these problems merely reduce the effectiveness of ventilation, making the process more time-consuming. In other cases they may pose a serious danger to those involved in confined space operations.

Short-circuiting and dead-air spots. Short-circuiting occurs when the fresh air being delivered into a space follows a relatively short pathway back out of the space, leaving a significant portion of the space unventilated (see Fig. 6.16). This often occurs because supply and exhaust locations are too close together. For example, the problem in Fig. 6.16 might be corrected by adding a duct to allow the supply air to be released at the far end of the space from the fan. A better alternative might be to simply open the hatch at the far end of the tank and close the one directly over the supply location.

Corners and other peripheral areas within a space may harbor contaminants even after the more central areas of the space have been purged. PPV with a strong effective blower capacity helps scour out dead-air spots that otherwise might not ventilate well. Repositioning the end of the duct at intervals during ventilation may also help ventilate dead-air spots in some spaces. This reinforces the importance of performing air monitoring throughout the entire space before drawing any conclusions regarding the effectiveness of ventilation.

Recirculation and contaminated supply air. Recirculation occurs when contaminated air exiting a space is picked up by ventilation equipment and returned to the space (see Fig. 6.17). In PPV operations this can occur when blowers are positioned too close to the location where contaminated air is exiting the space, or downwind from that location. To prevent this, the blower should be located as far as possible upwind from the location where contaminated air is exiting the space. In some cases it may also be necessary to add tubing to the intake of the blower and position the end of the tubing to pull the supply air in from a clean location.

TABLE 6.2   Range of Contaminant Duct Velocities

Contaminant type	Duct velocity, ft/min
Vapors, gases, or smokes	1000–1200
Welding fumes	1400–2000
Grinding dust	3500–4000
Abrasive blasting dust	4000–4500

SOURCE: Adapted from ACGIH (1995).

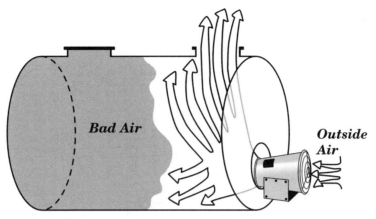

Figure 6.16 Short-circuiting occurs when supply air takes a short pathway back out of the space, leaving most of the space unpurged or unventilated.

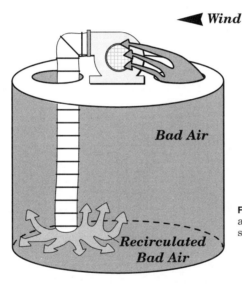

Figure 6.17 Recirculation occurs when contaminated air exiting the space is picked up and returned to the space by ventilation equipment.

In all instances care must be taken to ensure that the air going into the space is of good quality. Contaminants from industrial processes may be present in the ambient air. During street operations, carbon monoxide from automobile exhausts may be present. When gasoline-powered fans are used, the fan itself can be a source of high levels of carbon monoxide in the air supply (see Fig. 6.18). To prevent this, some manufacturers provide flexible tubing that adapts to the engine exhaust to route the exhaust emissions to a safe location (see Fig. 6.19). Other options include positioning the fan so that the engine exhaust is downwind of the air intake (assuming that a brisk, steady wind is blowing), or attaching a duct to the intake and positioning the open end well upwind of the engine.

Hazards introduced by the ventilation process. The ventilation process may create hazardous conditions inside and outside the space being ventilated. The contaminants released may be hazardous to support personnel outside the space or other workers in the area, especially if the space is inside a building or some other enclosure. In such cases it may be desirable to use NPV to remove the contaminated air and release it at a safe location. We may violate air pollution control regulations if we simply release the contaminants into the atmosphere. The bad air may need to be removed from the space using NPV and exhausted through a purification system.

Ventilation ducts may obstruct the entry and exit locations when ventilation is required during entries. One specialized piece of equipment intended to alleviate this is

Figure 6.18 Gasoline-powered engines used to drive fans, as well as automobile exhaust emissions, may be a source of carbon monoxide introduced into spaces during ventilation operations.

Figure 6.19 The Saddle Vent® by Air Systems International is one example of equipment used to conduct ventilation air into a space without obstructing the entry/exit point. (*Photograph courtesy of Air Systems International.*)

the Saddle Vent by Air Systems International (see Fig. 6.19). Another approach is the use of ventilation tubing that is made completely of flexible plastic, designed to allow an entrant to pass unhindered.

Static electricity generated by moving air can serve as a source of ignition for flammable atmospheres within a space, especially when venturi eductors are used. Grounding and bonding of ventilation equipment is required in some cases.

Spaces that cannot be effectively ventilated. Some spaces may be impossible to ventilate adequately no matter how many fans or which ventilation procedures are used. This is due to the size, configuration, level of contamination, or other factors present.

For example, it may be impossible to completely eliminate atmospheric hazards from a work area within a sanitary sewer system because of this area's interconnection with the rest of the system. If the actual work area is isolated from the rest of the system (such as by using inflatable bladder bags at either end of the work area), then effective ventilation may be possible (see Fig. 6.28).

As another example, assume that a vessel must be entered that is heavily contaminated with residual product and contains a highly flammable, highly toxic atmosphere. In many situations such as this, unless the vessel can be flushed to remove the product, the best ventilation procedures available are able to pull the concentration of flammable vapors only below 10% of the LEL, while a significant toxicity hazard remains. Workers may be required to use the highest available level of respiratory protection for the entry despite the use of continuous ventilation.

In a variation on the previous example, assume that the same conditions exist, except that a flammable and toxic atmosphere remains in spite of the best available ventilation technology. In such a case it may be necessary to fill the space with an inert gas, thereby creating an extremely oxygen-deficient atmosphere to prevent potential ignition. Workers might then be able to enter safely, provided appropriate respiratory protective equipment was used.

Isolation Procedures

A worker was assigned to change a faulty atomizer within a blender drum. Prior to entry, no confined space entry permit was issued. The atomizers were deenergized, but were not locked out, and the source of electrical energy for the blender drum was not deenergized. As the worker was working inside the space, the blender drum was activated. The worker fell between the rotating drum and the outer shell of the blender onto a rake conveyor about 14 ft below. The worker was transported about 100 yd through the conveyor system, then discharged through a flume into a double-screw auger. The coroner determined that the worker died as a result of the initial fall.

The role of isolation in preventing confined space incidents

Other confined space incidents have occurred when an energy source was encountered during an entry, material was released into a space during an entry, or machinery was activated within an occupied space. Accidents such as those can be prevented by isolating permit spaces prior to entry. According to OSHA's PRCS standard, isolation procedures are those by which a permit space is removed from service and completely protected against the release of energy or material into the space. In addition to the PRCS standard, another significant regulation addressing isolation is OSHA's *Control of Hazardous Energy standard* (29 CFR 1910.147). The provisions of both standards were described in Chap. 2.

Lockout/tagout and other isolation procedures

Lockout/tagout (LO/TO) procedures are usually regarded as being applied to electrical equipment. Other types of equipment, such as valves in piping systems, can also be locked and tagged out. In addition to LO/TO, a variety of other control procedures may be utilized.

Basic lockout/tagout procedures for electric circuits and equipment. Hazards of electric circuits or equipment usually can be controlled by locating the main switch at the energy source, placing it in the OFF position, and installing a lock and tag (see Fig. 6.20). The lock keeps the switch in the OFF position, and the tag identifies who performed the lockout procedure and warns others that the system is out of service and should not be reactivated. It is important that this be done at the energy source instead of merely at a control switch between the source and the work location. There may be more than one control switch, and in some systems control switches can be bypassed accidentally or intentionally.

For operations involving more than one entrant, a group lockout device and tag are installed so that each group member has a place to install a lock (see Fig. 6.20). Ideally each entrant should take the only key to his or her lock into the space to ensure that the lock cannot be removed until after the space is vacated. Other equally effective means of key control, such as the use of a lockout box, are also allowable. Lockout boxes may be used on large jobs to maintain control over a large number of keys. The bottom line is that the procedure must ensure that only after all entrants have exited the space can the electrical source be unlocked and reactivated. Other precautions, such as disconnecting electrical leads, may be used to isolate electrical hazards in some cases.

Basic lockout/tagout procedures for valves in pipes, tubing, or duct systems. Lockout/tagout devices can be applied to valves in pipe, tubing, or duct systems. Isolating these systems may require the installation of hoods on gate valves, ball valve lockout devices, chaining valve wheels, or the removal of valve handles (see Fig. 6.21).

Other isolation procedures for pipe, tubing, or duct systems. It is important to note that merely closing and locking valves alone is not considered sufficient to isolate a space. Additional precautions, as described below, are required in case a valve leaks in the closed position.

Blanking. Blanking or blinding is performed by unbolting a set of flanges within the system, installing a solid plate between the flanges, and bolting the flanges back together (see Fig. 6.22). The plate, known as a "blank," "blind flange," or "skillet," must be able to withstand the maximum pressure on the system with no leakage past the location where it is installed.

Line breaking. As the term implies, *line breaking* is accomplished by unfastening connections between sections of piping, tubing, or ducts and moving the adjacent sections out of alignment (see Fig. 6.23). The technique is also referred to as *misalignment*.

Figure 6.20 Lockout/tagout devices are one means of isolating a space from a source of electrical energy.

Figure 6.21 Lockout/tagout devices for valves include hoods for gate valve wheels and ball valve lockout devices.

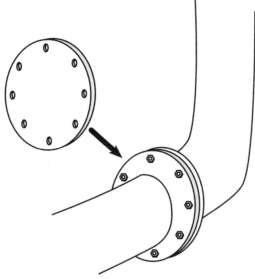

Figure 6.22 Blanking or blinding a piping system is performed by installing a solid plate that completely covers the bore.

Double block and bleed.   Double block and bleed is accomplished by using lockout/tagout procedures to lock two in-line valves in the closed position, and locking a drain or vent valve between the two closed valves in the open position (see Fig. 6.24).

Hazards to personnel performing isolation procedures.   Personnel involved in isolation procedures may be seriously injured if procedures are not carried out properly. For example, unfastening pipe connections on a system that is still under pressure could result in a release of materials that are toxic, corrosive, hot or cold, under high pressure, or otherwise dangerous.

Tagout alone as an isolation procedure

When used alone, the term "tagout" refers to placing a switch or valve in the safe position, then attaching a tag to warn others that the device is out of service and should not be reactivated. Tagout alone doesn't provide failsafe isolation. It is far too easy for an unauthorized worker simply to remove the tag and reactivate the device while a space is occupied. For this reason lockout should always be used in conjunction with tagout procedures. OSHA's *Control of Hazardous Energy* standard (29 CFR 1910.147) allows the use of tagout alone in situations in which the employer demonstrates that tagout alone is sufficiently protective and for devices that cannot be locked out.

Other forms of isolation

In addition to dealing with situations involving electrical sources, valves, and piping or duct systems, a variety of additional energy sources may impinge on a confined space. Some of these sources may not be obvious. Energy may be stored in hydraulic or pneumatic systems, springs under pressure, counterweighted flywheels, and a variety of other ways. Isolating spaces from the energy stored in these types of systems may require a wide array of different procedures.

In some cases chocks or cribbing (e.g., hardwood blocks of various sizes) must be placed to wedge or block the machinery to prevent movement and stabilize it. "Full cycle" machines must be stabilized before the energy is disconnected. Otherwise they will complete their cycle of operation when they are deenergized. In some cases procedures such as removing drive belts or drive chains from equipment or clamping forward and reverse sides of conveyor belts together are required to control energy hazards.

Figure 6.23 Line breaking can be used to isolate a space from contents of pipes, tubes, or ducts.

Figure 6.24 A double block and bleed is performed by locking two in-line valves in the closed position and locking a vent valve located between the two closed valves in the open position.

Problems in achieving a true zero-energy state

In some situations, isolating a permit space from energy and/or material sources may be a relatively simple and easy process; however, in dealing with large and complex systems, it may be very difficult and complicated to achieve. Complex systems may have multiple sources of energy that affect a single space. Isolation procedures involving such systems may be further complicated by energy stored in springs, flywheels, fluids under pressure, suspended loads, or other sources that are not obvious. The ultimate goal of all isolation procedures is to achieve a true zero-energy state prior to entry.

The complexities involved in performing isolation procedures in some industrial settings can be almost overwhelming. Innovative procedures, including the use of zero-energy worksheets, are required to carry out isolation procedures in some situations. Activate all system controls to test the effectiveness of isolation procedures before making a permitted entry.

Special considerations for isolation procedures during rescue operations

Isolating spaces before making entry for rescue can be very challenging, especially for outsiders not familiar with the process and equipment involved. This requires that fire service personnel interface with facility personnel to control the hazards prior to entry. Preemergency planning of such measures is invaluable when the clock is running and adrenaline is flowing.

Facility personnel in preparing for a permitted entry may have already done isolation procedures. In that case, rescuers must determine that the procedures in place are adequate. After doing so, rescue service members install their own LO/TO devices over those initially in place (see Fig. 6.25). If such devices are not available, then someone

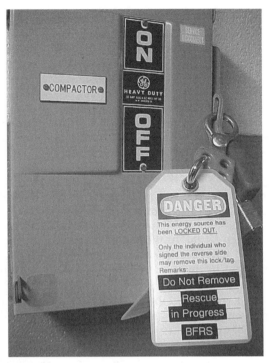

Figure 6.25 The rescuer's lock and tag should "overlock" those initially in place.

must be posted with express instructions to ensure that the isolation measures in place are not tampered with.

Other Hazard Control Procedures

Personnel involved in confined space operations may be exposed to a variety of hazards in addition to atmospheric hazards and hazards related to energy or material impinging on the space. These hazards may threaten not only the entrants but also attendants or other support personnel working outside the space.

Controlling hazards of the work area and incident scene

The incident scene itself may be a hazardous location from which to stage a permitted entry or a rescue operation. In some cases merely opening the space could expose personnel to significant hazards. To address this the PRCS standard requires that conditions making removal of an entrance cover unsafe be eliminated before the cover is removed.

As another example, manhole operations are frequently carried out on city streets where personnel may be threatened by vehicular traffic, automobile exhaust emissions, and contaminants emitted during ventilation operations. Similar operations carried out in the industrial sector may involve hazards due to normal workplace operations, such as equipment traffic, materials-handling operations, and hazardous materials in the area of operations.

Vertically oriented settings such as manholes create the opportunity for personnel to fall into openings. Loose tools or equipment can be kicked or dropped into open spaces and strike people inside.

Unauthorized personnel may interfere with or endanger operations. This is especially true for rescue operations, when distraught coworkers or family members may distract rescuers or even try to commandeer control of the operation.

Using barricades and barriers

The PRCS standard requires barriers to be provided to bar pedestrians, vehicles, and other types of traffic from the vicinity of the entryway. When entrance covers are removed from a space, the opening must be immediately guarded using railings, temporary covers, or other types of temporary barriers to prevent accidental falls through the opening and to protect entrants from foreign objects entering the space. Examples of such devices include shields and railings for manhole operations (see Fig. 6.26). The outer perimeter of the area of operations surrounding the entryway should be delineated using traffic cones, flagging tape, ropes, saw horses, warning signs, or lights.

Zoning the work area or incident scene

Zoning is a helpful concept in controlling hazards around the entryway of a confined space. This is a familiar concept to anyone with training in hazardous-materials emergency response. The concept involves dividing the area of operations into geographic zones based on factors such as the hazards present, the levels of protective equipment required, and the types of activities carried out in different areas. The location of a confined space operation might be divided into the following zones, as shown in Fig. 6.27.

Hot zone.   The *hot zone* is considered the area within the permit space. This is the location where the actual permitted work activities are carried out. It is generally the most hazardous location at the scene.

Warm zone.   The *warm zone* is the area outside the space occupied by the attendant and other support personnel. Support operations such as ventilation are carried out in this area. Some hazards to personnel may exist there, such as exposure to hazardous atmospheres from the space or from the ventilation process.

Cold zone or support zone.   The area outside the warm zone but inside the perimeter is considered the *cold zone.* Equipment and supplies are staged and support personnel are active there. Rescuer personnel are on standby there during certain operations.

Figure 6.26 Shields and railings should be used around open entryways.

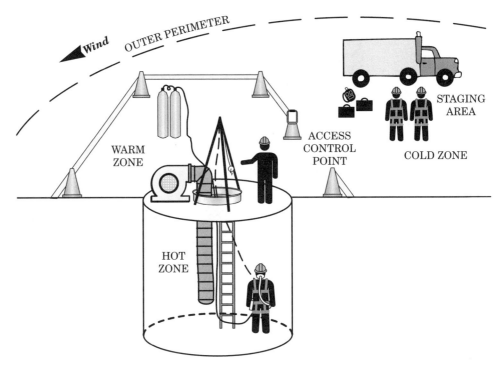

Figure 6.27 Zoning is an effective way to control activities at the scene of confined space operations.

Isolating a particular location from the rest of the space

In some cases it is desirable to isolate or divide the part of a confined space where operations are being carried out from the system as a whole. Consider permitted work in a sanitary sewer system. Hazards commonly associated with that type of space include toxic and oxygen-deficient atmospheres and engulfment. In some cases it may be impossible to adequately ventilate a work area within the sanitary sewer because of its interconnection with the system as a whole. One alternative is to isolate the actual work area from the rest of the system. Inflatable plugs can be installed on either end of a section of the sewer where work is to be conducted (see Fig. 6.28). In addition to providing protection from engulfment and other hazards of the effluent, this approach may also make adequate ventilation of the work area possible.

Hot-work permitting

Welding, cutting, or brazing frequently are required inside confined spaces. Such operations not only introduce a ready source of ignition but also may introduce harmful air contaminants such as welding fumes. These operations require the use of welding machines or dangerous gases such as oxygen and acetylene inside the space. Because of these special hazards, operations such as welding, cutting, and brazing within confined spaces require a special hot-work permit in addition to the regular entry permit (see Chap. 2).

Controlling ignition sources

As discussed previously, one of the primary hazards of many confined space entries is the potential for flammable atmospheres. The only safe way to deal with flammability hazards is to eliminate them by flushing, inerting, purging, or ventilating the space to eliminate the hazard altogether. No one should ever knowingly work in a flammable environment.

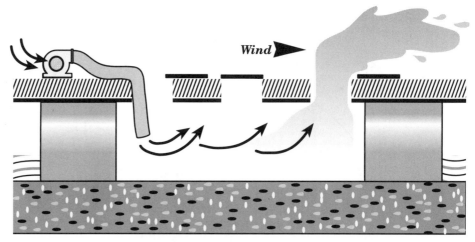

Figure 6.28 Special measures, such as the use of inflatable bags in a sewer system, are sometimes used to isolate a work area within a larger confined space.

Measures intended to prevent ignition of flammable atmospheres do not make it safe to conduct operations in such atmospheres. Anyone who has ever been shocked by the discharge of a static electric charge picked up when shuffling the feet across a rug or sliding across a car seat realizes this. Measures to prevent ignition of flammables may provide an extra margin of safety in the event that hazard control measures fail or an entrant unknowingly encounters flammable conditions. Commonly used measures include the use of intrinsically safe electrical equipment and nonsparking tools. Ignition prevention precautions are described in Chap. 14.

Controlling electrical hazards

Workers in confined spaces may be exposed to electrical hazards other than those discussed under isolation procedures. For instance, electrically powered tools, equipment, and lighting systems can produce serious electrical shocks.

As a minimum, all electrical equipment used in confined space operations should meet standard safety requirements. All equipment used inside spaces should be fitted with ground-fault circuit interrupters or operated on low-voltage direct current (dc). Low-voltage dc is recommended because of the difficulty that may be encountered in attempting to establish grounding protection for normal power in confined spaces.

Controlling general hazards of the working environment

A variety of hazardous conditions may be encountered in the general environment where confined space operations are carried out. For example, the environment in and around confined spaces may be very hot or very cold. These areas may be dark and wet, with slippery walking and working surfaces. Vertical spaces, elevated entryways, and elevation differences within spaces may pose fall hazards to personnel.

All confined space work areas, as well as surrounding areas of operation outside the space, must have adequate lighting to allow operations to be carried out safely. Wet, muddy, or otherwise slippery work surfaces should be eliminated, if at all possible, before operations begin. All stairways, ladders, and catwalks should comply with applicable standards to prevent falls. Measures should be taken to protect personnel from potential falls in accordance with relevant fall protection standards (see Chaps. 13 and 14).

Some confined spaces are very hot and humid, posing a very real hazard of heat-induced illness to entrants. Other spaces are very cold and damp, threatening entrants with hypothermia and other cold-related injuries. Very few confined spaces are comfortable working environments in their normal states. General guidelines for avoiding heat- or cold-related injuries were provided in Chap. 3. In addition to controlling typical atmospheric hazards, the ventilation process can favorably affect temperature conditions within a space. For example, cooler, dryer outside air may be pumped into a hot space, such as an aboveground tank in direct sunlight. The air forced into cold spaces can be warmed using special heaters designed for use with ventilation equipment.

Summary

Many of the potential hazards of confined space operations can be eliminated or controlled prior to entry. Operations such as ventilation, isolation, and a variety of other measures can be used to make entry much safer. Without measures such as these, workers' personal protective equipment and safe work practices become the only factors keeping them from being injured during permitted entries. Work is much safer when hazard control procedures are used to reduce or completely eliminate the hazards before entry. When rescue is required, the same measures can provide a safe area of operations for emergency personnel, and enhance the survivability of the patient as well.

7

Respiratory Protection

Confined spaces are often tight, enclosed areas with poor natural ventilation. Such places can trap hazardous substances and allow them to build up into hazardous atmospheres. OSHA regulations, as well as good safety and health practice, require that, whenever feasible, airborne hazards be controlled by engineering controls. Such controls include ventilation, isolation, substitution of hazardous substances, and other means of protecting the environment around the entrant.

The enclosed, isolated nature of confined spaces often makes ventilation a good choice for control of atmospheric hazards. Ventilation was discussed in Chap. 6 in some detail. Ventilation serves to dilute or remove contaminants from the air. It is easier to control the atmosphere of a space that is isolated from the surrounding areas. Ventilation can successfully render the atmospheres of many spaces safe for entry.

Many spaces have features that make ventilation difficult or ineffective. A space with internal walls, baffles, or other obstructions to airflow may create "dead spaces" or other unventilated areas. Spaces may have sludge or other persistent sources of hazardous substances that may overpower ventilation systems. Sometimes the work being done creates or liberates additional hazards. Rescue of a victim in a space may entail entering a hazardous atmosphere where ventilation was inadequate. In some cases engineering controls are not sufficient to control the atmospheric hazards to a sufficient degree to protect entrants.

In such cases, OSHA allows the use of personal respiratory protective equipment by entrants (Fig. 7.1). Such equipment provides breathable air to the entrants while they are in the space. Because the entrant is working in an area known or suspected to contain a hazardous atmosphere, respiratory protective equipment must be used only with appropriate planning and under special control. This chapter describes the safe use of respiratory protective equipment for confined space entry and rescue.

Relevant Standards

Chapter 2 summarizes a number of regulatory and consensus standards that provide guidance on the safe use of respiratory protective equipment. Several legally enforceable OSHA regulations place requirements on employers that use respiratory protection. NFPA standards provide guidelines for the establishment of respiratory protection programs in fire departments. ANSI has a regulation covering the use of respiratory protective equipment. Significant requirements of these standards are described below.

OSHA's *Respiratory Protection Standard*

Employers must comply with all OSHA regulations that apply to their workplaces. The primary regulation controlling the use of respirators is the *Respiratory Protection Standard* (29 CFR 1910.134), which was substantially revised in 1998. Employers with respiratory protection programs developed prior to the revision must be sure that the programs comply with the current standard.

Figure 7.1 OSHA allows the use of respirators to protect confined space entrants when other controls alone are not adequate.

Substance-specific standards found in Subpart Z of Part 1910 include special respirator requirements for the workplaces that they cover. OSHA revised the respirator provisions of these standards with the revision of 29 CFR 1910.134. Now the substance-specific standards have respirator provisions only if they are unique to protection from that substance. Otherwise employers are to use 29 CFR 1910.134 as the standard for compliance.

The standard covers all aspects of the safe use of respirators during confined space entry and rescue. Some of the major requirements are discussed in the next section. Other requirements are addressed later in the appropriate sections describing respiratory protection.

Written program. The standard requires employers to develop a written respiratory protection program with worksite-specific procedures. The program is to be administered by a suitably trained program administrator and updated as needed to reflect changes in the workplace. The program must include the following provisions:

- Selection procedures
- Medical evaluation
- Fit testing procedures
- Procedures for proper use in routine and emergency situations
- Procedures for cleaning, inspecting, and maintaining respirators
- Procedures to ensure adequate air quality, quantity, and flow for atmosphere-supplying respirators
- Training employees in the hazards to which they are exposed
- Training employees in the proper use of respirators
- Program evaluation

OSHA provides a sample program in the *Small Entity Compliance Guide* that can be downloaded from the OSHA Website or obtained from any OSHA office. Commercial vendors also offer software that can be used to develop a program. In either case, the employer must be certain to adapt the generic program to be worksite-specific by addressing the actual hazards and procedures of the company or facility.

Respirator selection. The employer must evaluate the respiratory hazards in the workplace and identify the relevant workplace and user factors. Respirator selection is based on these factors. Evaluation of the respiratory hazards includes a reasonable estimate of the employee exposure levels and an identification of the contaminant's physical state. Any employer who cannot identify or reasonably estimate the employee's exposure must consider the atmosphere immediately dangerous to life and health (IDLH). Such atmospheres require the highest level of respiratory protection as discussed later in the chapter.

Employers must select respirators that are certified by NIOSH. The respirators must be used in compliance with the conditions of their certification. Employers must be certain that respirators are not modified or used in any way that would void the NIOSH certification.

OSHA specifies the types of respirators to be selected for both IDLH and non-IDLH atmospheres. These and other issues related to selecting respirators are discussed in the respirator selection section later in this chapter.

Medical evaluation. Employees who wear respirators must be physically able to do so without being harmed by the burden that the respirators place on them. OSHA requires the employer to arrange for a medical evaluation of the employee by a physician or other licensed healthcare professional (PLHCP). The evaluation is to be performed before the employee is fit tested or required to use the respirator in the workplace. The purpose is simply to evaluate the employee's ability to use the respirator.

OSHA allows the PLHCP to use a questionnaire to determine which employees need to receive a physical exam and which employees have no indications of a potential for problems using respirators. The questionnaire to be used is mandated by OSHA and must be provided to the workers confidentially in a way that they can understand. Based on responses to the first part of the questionnaire, the PLHCP either determines the worker fit for using a respirator or recommends follow-up evaluation. The PLHCP provides the employer with a written statement indicating one of the following conclusions:

- The worker is able to wear a respirator.

- The worker is able to wear a respirator with limitations.

- The worker is not physically able to wear a respirator.

The specific test results and findings of the exam are considered privileged and confidential and are not provided to the employer.

There is no requirement for annual or periodic reevaluation of the worker. Reevaluation is called for under any of the following conditions: (1) an employee reports signs or symptoms related to the ability to wear a respirator; (2) the PLHCP, administrator, or supervisor determines it is necessary; (3) information from the respiratory protection program indicates a need for reevaluation; or (4) a change in workplace conditions substantially increases the physiological burden placed on the employee.

Use of respirators. The regulation addresses certain practical aspects of the use of respirators in a section by that title.

Facepiece seal protection. The employer must prohibit the use of respirators by employees with facial hair or other conditions that interfere with the face-to-facepiece seal. Where the employee requires corrective eyeglasses, they must be worn in a way that does not interfere with the facepiece seal. This may require a special lens holder available from the facepiece manufacturer that mounts inside the facepiece without temple bars that protrude out of the facepiece. OSHA allows the use of contact lenses while wearing a respirator.

Procedures for IDLH atmospheres. The respiratory protection standard requires that the employer post at least one employee outside the IDLH atmosphere if employees enter such an area. This section does not specifically reference the confined space entry standard (29 CFR 1910.146); however, the requirements are consistent with the duties of the attendant required for permit-required confined space entry. The section also addresses procedures for interior structural firefighting that are beyond the scope of this book.

Training. The standard requires effective training for each employee who is required to wear a respirator. In sections numbered as below, each employee should demonstrate knowledge of at least the following:

1. Why the respirator is necessary and how improper fit, usage, or maintenance can compromise the protective effect of the respirator

2. What the limitations and capabilities of the respirator are

3. How to use the respirator effectively in emergency situations, including situations in which the respirators malfunction

4. How to inspect, put on and remove, use, and check the seals of the respirator

5. What the procedures are for maintenance and storage of the respirator

6. How to recognize medical signs and symptoms that may limit or prevent the effective use of respirators

7. The general requirements of the section

The training is to be provided prior to requiring the employee to use a respirator and in a manner that is understandable to the employee. Retraining is to be conducted annually and whenever there is evidence that it is needed to ensure the safe use of the respirator by the employee.

NFPA standards

NFPA standards are consensus standards that establish a level of competency for firefighters and fire departments. Two standards have sections that specifically pertain to respiratory protection (Fig. 7.2): NFPA 1404 *Standard for a Fire Department Self-Contained Breathing Apparatus Program* (NFPA 1996) and NFPA 1500 *Standard on Fire Department Occupational Safety and Health Program* (NFPA 1997).

NFPA 1404. This standard (NFPA 1996) outlines the required elements of an effective self-contained breathing apparatus (SCBA) program. The SCBA is the primary form of respiratory protective equipment used by firefighters. The standard addresses these areas in detail:

- Provision of SCBAs

- Use at emergency scenes

- Training

- Inspection

- Maintenance

- Breathing air program

- Program evaluation

The requirements of the standard apply to all environments in which firefighters use SCBAs. Although there are no specific requirements for entry into confined spaces, fire

Figure 7.2 NFPA standards 1404 and 1500 outline
the requirements for the proper use of respiratory
protective equipment by firefighters.

departments can use the standard to develop a program that will ensure that their respirators are adequate, in good shape, and used properly.

NFPA 1500. This standard (NFPA 1997) covers comprehensive safety and health programs, which includes the use of respiratory protection. While general in nature, the standard provides some specific guidance affecting SCBA usage. Detailed discussion of the
standard is beyond the scope of this book, but some of the key points are discussed below.

NFPA does not allow firefighters to compromise the integrity of the SCBA while operating in a hazardous atmosphere. This disallows the practice during rescue operations
of "buddy breathing," the practice of a rescuer and victim sharing a single SCBA by
alternately breathing air through the facepiece. Buddy breathing allows brief exposure
of the firefighters to the hazardous atmosphere that may result in their being overcome.
This is discussed further in the section on SCBAs in confined spaces.

The standard specifically prohibits the use of an SCBA by firefighters with beards or
facial hair at the point where the facepiece is designed to seal with the face. Head coverings or other equipment that passes through the facepiece seal area are also prohibited. These prohibitions apply regardless of what fit test measurements can be obtained.
Fit testing is discussed later in the chapter.

Other general provisions of the NFPA standards may apply but are beyond the scope
of this book. Fire departments should verify that their programs meet all the requirements of these standards.

ANSI standard

ANSI standards serve as consensus standards for the private sector. ANSI Z88.2,
Practices for Respiratory Protection, describes elements of an effective respirator program

for private-sector employers. The standard is more detailed in nature than the OSHA regulation that covers these employers.

Basic Operation of Respirators

Respirators are classified in several ways. Some of the classifications overlap, but each classification describes a distinctive feature of respirators. Three primary classification systems are facepiece design, source of breathable air, and mode of operation.

Facepiece design

The *facepiece* provides the barrier between the environment and the user's respiratory system. Facepieces range from the mouthpiece with nose-clip design for escape respirators to hood-style respirators. The degree of hazard, conditions of the workspace, and user comfort are all factors considered in selecting a facepiece style. Each facepiece style offers both advantages and disadvantages.

Loose-fitting facepieces such as hoods and helmets drape over the head and sometimes the shoulders of the user. The facepiece does not form a tight seal but creates an enclosure around the user's head. Copious amounts of fresh air are blown across the user's face to dilute or diffuse any contaminants that migrate into the breathing area. Hoods and helmets most frequently are used to protect against particulate contaminants, since these migrate more slowly than gases or vapors. Because these facepieces do not fit tightly across the face, they are often more comfortable for the user than tight-fitting facepieces. Because they do not seal tightly, contaminants can migrate into the breathing area and expose the user.

Tight-fitting facepieces form a tight seal with the user's face. Because of this seal, all air entering the facepiece can be controlled. In order to provide full protection, the seal must be complete, with no points of leakage. Fit testing determines the quality of the seal. Tight-fitting facepieces are available in either half-mask or full-facepiece designs (Fig. 7.3).

Half-mask respirators form a seal around the user's nose and mouth only. They include filtering facepieces, dust masks, and elastomeric facepieces. The body of a filtering facepiece or dust mask is constructed of a filtering medium. The elastomeric facepiece has an inlet through which clean air enters. Half-mask respirators are less expensive than full-facepiece respirators and allow the user to wear eyeglasses. But because they seal against the curved surface of the nose, they generally provide a lesser-quality fit and should not be used in higher-concentration atmospheres.

Figure 7.3 Employers can choose either full-facepiece or half-mask respirators for the protection of their employees.

Full-facepiece respirators cover the entire face, including the eyes. The elastomeric facepiece has a clear visor for visibility. Since this facepiece seals to the smoother surface of the forehead, it offers a better fit and a higher degree of protection than does a half-mask facepiece. It also protects the eyes from splashing and exposure. Users who wear eyeglasses must use spectacle kits to mount lenses inside the facepiece.

Source of breathable air

Respirators also are classified by how they provide breathable air to the user. Respirators provide air either by purifying existing air or through supplying a complete atmosphere.

Air-purifying respirators. *Air-purifying respirators* (APRs) remove contaminants from the air inside the space by means of a filtering medium. They do not replace or alter the oxygen concentration of the air. APRs typically use filtering or adsorption media contained in either cartridges (smaller capacity) or canisters (larger capacity) that are attached to the facepiece inlet (Fig. 7.4). Cartridges typically attach directly to the facepiece inlet and are smaller and lighter than canisters. Canisters may attach to the chin of the facepiece or may be mounted on the user's back or chest and attached to the facepiece inlet by a breathing tube. Canisters contain more media, but are heavier and more uncomfortable to use.

The medium must be selected on the basis of chemical and physical properties of the contaminant. Air contaminants take the form of either particulates or gases and vapors. Mechanisms for filtering particulates differ from the sorbing methods used for removing gases and vapors.

Particulate APRs. Particulate respirators filter dusts, fibers, fumes, and mists from the air by means of interception or electrostatic capture. *Interception* involves trapping a particle onto the surface of the filter as air passes through it. In *electrostatic capture,* the particle has a charge that is attracted to the opposite charge in the filter fibers.

In 42 CFR Part 84, NIOSH classifies particulate respirators according to two criteria: (1) the impact of oil mist in the air on the effectiveness of the filter media and (2) filter collection efficiency. Oil particles, such as mists of lubricants or cutting fluids, can coat filter fibers as the air is breathed through the filter. This coating can reduce the electrostatic attraction of the particles to the filter, thus affecting some filters' efficiency. Other filters, which rely less on electrostatic attraction and more on interception, are less affected by the presence of oil mist. The three categories of resistance to filter efficiency degradation by oil are labeled N (*not* resistant to oil), R (*resistant* to oil), and P

Figure 7.4 Air-purifying respirators may be equipped with purifying media contained in either cartridges or canisters.

(oil*proof*). N-series filters should not be used when oil mist may be present. R- or P-series filters can be used if oil mist is present, although R-series filters should not be used for more than one workshift.

NIOSH recognizes three levels of filter collection efficiency: 95, 99, and 99.97% (rounded up to 100% for labeling). Respirator manufacturers label their particulate cartridges according to both criteria. An N95 cartridge will collect 95% of the particles out of air that has no oil mist present. A P100 cartridge will collect 99.97% of particles out of the workplace air regardless of the presence of oil mist.

Gas and vapor respirators. APR media remove gases and vapors from the air by sorption or catalysis. These methods involve interaction between the gas and vapor molecules and the granular solid materials. Sorption involves the attraction of molecules onto the surface of the granular, solid sorbent (adsorption) or into the molecular spaces of the sorbent (absorption). Catalysis uses a catalyst to initiate a reaction between the contaminant and another chemical. For example, hopcalite initiates the reaction between carbon monoxide and oxygen to form the less toxic carbon dioxide.

Unlike filtering methods, which work well regardless of the chemical composition of the particulate, methods for removing gases and vapors depend on properties that vary with different chemicals. Cartridges can be designed to remove individual chemicals, such as mercury or ammonia; or groups of chemicals, such as organic vapors or acid gases. In order to select the proper cartridge, the user must know the identity of the contaminant.

A gas-and-vapor respirator effectively removes nearly all the contaminant from the air until its medium becomes saturated. At that point, nearly all of the contaminant passes through the cartridge and enters the facepiece to expose the user. The time needed for the contaminant to saturate the medium, called the service life of the cartridge, depends on factors such as

- Amount of sorbent material in cartridge

- Concentration of chemical in the air

- Humidity

- Temperature

- Breathing rate of the user

The OSHA standard requires the employer to calculate a schedule for changing cartridges based on the service life of the cartridge. Respirator manufacturers can provide computer software to aid in the calculation of such a schedule on the basis of their cartridge and the conditions inside the space.

Atmosphere-supplying respirators. Atmosphere-supplying respirators provide a complete atmosphere from outside the confined space. They do not purify or otherwise use the air inside the space, so they can be used in situations where APRs cannot. The air must be of at least grade D quality as specified by the Compressed Gas Association Commodity Specification for Air, G-7.1-1989. The two categories of atmosphere-supplying respirators are the supplied-air respirator (SAR) and the self-contained breathing apparatus (SCBA).

Supplied-air respirators. SARs (also called *airline respirators*) provide breathable air to a user in a confined space directly from a source outside the space, such as a compressor with purification unit or from large air cylinders (Fig. 7.5). The air is delivered to the user through a hose or airline by way of a regulator that controls the airflow. This system provides a practically limitless work period if used with a compressor.

Because of the resistance to air, airline length is limited to 300 ft. SARs must meet NIOSH criteria for the minimum and maximum airflow rates into the facepiece for all hose lengths.

Figure 7.5 Supplied-air (airline) respirators provide breathable air to the entrant from a source outside the space.

According to the OSHA respirator standard, SARs can be used for entry into an IDLH atmosphere only if they are equipped with a self-contained auxiliary air supply for escape. These auxiliary air cylinders are often classified according to the breathing time they provide, typically 5-, 10-, or 15-min escape bottles. It is important to remember that individual breathing and air consumption rates vary widely, so these times should not be taken literally. If the primary breathing air supply fails, entrants must exit the space immediately.

Hybrid SAR/APR units. Respirator manufacturers have added air-purifying elements to the facepiece of their SAR units to form a hybrid respirator that offers a choice of operating modes. The exhalation valve of the facepiece is switched between the positive-pressure and negative-pressure modes. When operated as an SAR, air is supplied under positive pressure into the facepiece for breathing. Should the air supply fail, the user is protected by the air-purifying elements as he makes his escape out of the contaminated environment. These units can be used only where APRs can be used, as described later in this chapter. They cannot be used in IDLH conditions.

Self-contained breathing apparatus. SCBAs provide all the breathing air from a source carried into the space by the entrant (Fig. 7.6). This typically involves a compressed-gas cylinder worn in a harness on the entrant's back. A regulator is mounted either on the facepiece or on a harness strap with a breathing tube connecting with the facepiece. The regulator controls the flow of air into the facepiece as described later in this chapter. An audible and/or vibrating alarm warns the user when the amount of air pressure drops below 25% of the working pressure.

Figure 7.6 The self-contained breathing apparatus provides the highest level of respiratory protection and allows the entrant to move around without airline restrictions.

Certain safety requirements apply to the use of SCBAs in any environment, including confined space entry. NIOSH (1987) includes the following requirements for all certified SCBAs:

- Pressure gauges visible to the wearer which indicate the quantity of gas remaining in the cylinder

- Alarms that signal when only 20 to 25% of service time remains

- Bypass valves that allow the user to get breathing air in the event of regulator failure

- Breathing air fittings that are incompatible with other gas fittings

SCBAs are classified as closed-circuit or open-circuit according to how they handle the air exhaled by the user.

Closed-circuit SCBAs (or "rebreathers") capture and reuse the breathing gas exhaled by the user (Fig. 7.7). Exhaled air contains some oxygen (15 to 17%) plus carbon dioxide added by the user. Closed-circuit SCBAs recirculate the exhaled air through a "scrubber" to remove the carbon dioxide and then add oxygen from a compressed-gas cylinder to render the air breathable. Since almost all of the breathing air volume is conserved, rebreathers can provide greatly extended work periods, sometimes as long as 4 hours. The carbon dioxide scrubbing process gives off heat that can add to heat stress problems inside a confined space.

Open-circuit SCBAs direct all the exhaled air out of the facepiece through an exhalation valve. Open-circuit SCBA cylinders are constructed of steel or aluminum wrapped in fiberglass, with working pressures of 2200 or 4500 lb/in^2 (psi). Since all the air is

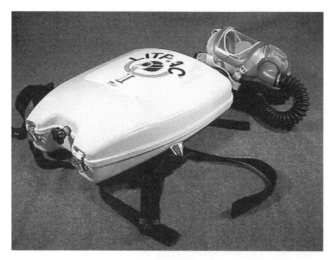

Figure 7.7 Closed-circuit SCBAs provide extended entry time because the wearer rebreathes the exhaled air after it is scrubbed and oxygen is replaced.

exhausted to the environment, they have rated service lives of 30 or 60 min, much lower than that of closed-circuit SCBAs. Cylinders with a higher working pressure provide more breathing air from the same-size cylinder. The cylinder diameter and weight can be reduced by using a high-pressure cylinder rated for 30 min.

Escape SCBAs. Some SCBA units are designed only for escape from atmospheres that suddenly become hazardous. These units are compact and can be donned quickly (Fig. 7.8). Escape-only SCBA units (ESCBAs) have a very limited air supply, usually 5 to 15 min. Many ESCBAs have a clear plastic hood with an elastic collar that the user dons during an emergency, and breathing air is fed into the hood in a continuous flow. Other units use a standard full facepiece and are equipped with a pressure-demand regulator. These units are not suitable for entry into a hazardous atmosphere.

Hybrid SAR/SCBA units. Several respirator manufacturers market a combined or hybrid SAR/SCBA respirator (Fig. 7.9). These respirators are essentially SCBAs with a port on the regulator for attaching an airline. This allows the user to enter a hazard area while breathing from the SCBA cylinder and then connect to an airline for the work operation. When used in this way, no more than 20% of the air supply should be used during entry.

Another application of this device is to use the airline for breathing during normal entry work and save the SCBA cylinder as an emergency escape air supply. This might be appropriate when the configuration of a space may make exiting the space slow. In the event of airline failure, the SCBA cylinder would provide extended escape time. In either case, the airline extends the work period indefinitely while the SCBA provides the safety of an air supply always with the user.

Quick-Fill® systems. One manufacturer, Mine Safety Appliance Company (MSA), has marketed a Quick-Fill® system for using a special quick-connect high-pressure hose to refill SCBA cylinders while the user continues to wear and operate the SCBA. The high-pressure hose coming from a cylinder bank or other breathing air source attaches to a special adapter installed on the high-pressure side of the regulator (Fig. 7.9). The cylinder is recharged without requiring the wearer to remove the unit or stop using the SCBA. Rescuers can use this system to quickly and safely connect a trapped victim to a spare SCBA cylinder for breathing air during rescue or extraction. According to MSA, NIOSH has extended certification to the SCBAs equipped with the Quick-Fill® system.

A transfill hose allows a rescuer equipped with the Quick-Fill® system to share air from his cylinder with another similarly equipped user who is trapped or otherwise unable to get to a source of clean air. In this case, the donor and recipient cylinders

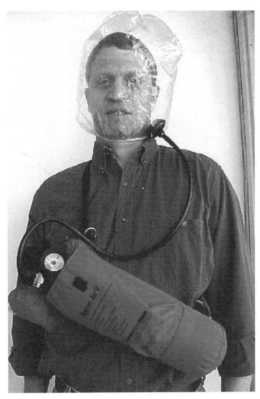

Figure 7.8 Escape SCBAs provide breathable air for entrants to escape the space in an emergency.

QUICK-FILL
CONNECTOR

AIRLINE
CONNECTOR

Figure 7.9 This regulator is equipped with both an airline connection and a Quick-fill® connector.

equalize pressure between them. This is potentially very dangerous if the rescuer shares so much of his breathing air that what remains is inadequate to ensure safe exit from the space. A transfill operation should be performed only when a rapid exit of the rescuer can be assured.

Mode of operation

On the basis of the pressure generated within the facepiece during use, respirators are classified as operating in either a negative-pressure or a positive-pressure mode of operation.

Negative-pressure respirators. With a negative-pressure respirator the user inhales, pulling a volume of air out of the facepiece and creating a negative pressure or vacuum. In response, air enters the facepiece through the inlet. If the respirator is an APR, the air comes from the environment through the purifying element. If an atmosphere-supplying respirator is equipped with a regulator operating in the demand mode, the vacuum opens a valve that allows air in from the cylinder or airline. Once the air is replaced, the pressure equalizes and airflow into the facepiece stops.

Any leak or break in the facepiece seal offers an easier path for air to enter. A negative-pressure respirator with a poorly fitting facepiece will allow a potentially significant exposure to the user. Negative-pressure respirators have a lower protection factor (discussed later in this chapter) than do positive-pressure respirators.

Positive-pressure respirators. Positive-pressure respirators maintain a slight positive pressure within the facepiece at all times. If there is a leak in the facepiece seal, air should blow out the leak and prevent contaminated outside air from entering. Respirators that operate in the positive-pressure mode provide more protection than equivalent respirators operating in the negative-pressure mode. Entry into IDLH conditions requires the use of either SCBAs or SARs that operate in the positive pressure mode.

Powered air-purifying respirators (PAPRs) are APRs that use a fan or pump to draw air through the filtering medium and into the facepiece. This creates a positive pressure within the facepiece that forces air out of any leaks in the seal. Since the user does not have to draw a negative pressure to move air into the facepiece, PAPRs place less strain on the user's breathing.

Two positive-pressure designs are currently used:

- Continuous-flow respirators continuously provide breathable air into the facepiece under pressure.

- Pressure-demand respirators are designed to maintain a slight positive pressure within the facepiece. Inhalation reduces that positive pressure and opens the regulator to allow replacement air to enter. These units use less air because the breathing air flows only during inhalation.

One concern regarding the use of positive-pressure respirators is that heavy breathing while performing strenuous tasks may cause the entrants to temporarily "overbreathe" the positive pressure within the facepiece. If there is a poor facepiece seal, this could lead to brief exposures to toxic contaminated air. Even when using positive pressure respirators, entrants must use a properly fitted facepiece.

Selection and Use Considerations for Confined Space Entry

Each type of respiratory protective equipment has certain advantages and limitations. Entry supervisors must select respirators that provide adequate protection against all the airborne hazards that have been identified in the space. As a minimum, the assigned protection factor (APF) of the respirator (as described below) must be adequate. The specific conditions in the space and the specific task(s) to be performed must also be considered. All the factors must be considered to ensure that the entrants are properly protected. The following discussion considers the major advantages and disadvantages of each type of respirator.

Selection considerations for APRs

Air purifying respirators are an option for protecting entrants while they work inside confined spaces. Because atmospheres inside confined spaces may accumulate high concentrations of chemicals, care must be used in selecting and using APRs.

Advantages of APRs.   The advantages of APRs include their light weight, ease of use and maintenance, low price, and minimal restriction of the user's movement. Most employers will find it desirable to use APRs whenever they provide adequate protection to the entrants.

Disadvantages of APRs.   Despite their advantages, there are a number of limitations to using APRs. The limitations are primarily related to the conditions of the space in which they may be used. The limitations include the following:

- All potential atmospheric hazards in the space must be identified prior to entry in order to select the proper purifying medium and ensure that APRs will offer adequate protection.

- APRs cannot be used in oxygen deficient atmospheres (atmospheres containing less than 19.5% oxygen).

- APRs cannot be used if the identified contaminants are highly toxic in small concentrations (such as hydrogen cyanide).

- APRs cannot be used if the contaminant's concentration is higher than its IDLH level.

- A leak in the facepiece seal may allow a significant amount of contaminant to enter, since most APRs work in the negative-pressure mode.

- The service life of the cartridges or canisters (discussed below) must be calculated so they can be changed before breakthrough occurs.

- Breathing through negative-pressure APRs is difficult; PAPRs can be used if needed.

- Conditions in the space, such as high temperature or humidity, can reduce the effectiveness of the purifying medium.

Other considerations for using APRs

Cartridge service life.   APRs are considered safe to use only when the user has some way of knowing when the end of the canister or cartridge service life has been reached. Ideally, the canister or cartridge would have an *end-of-service-life indicator* (ESLI). This would be a system or device on the cartridge that would change color or otherwise warn the user when the end of adequate protection by the respirator was approaching. Very few such ESLIs exist on the market.

Good warning properties allow the APR user to smell, taste, or experience irritation in the event of breakthrough or leakage at contaminant concentrations below appropriate exposure limits. Generally APRs should be used only for contaminants having good warning properties. Users should leave the hazard area immediately at the first sensation of a warning property and evaluate the cause. It may be cartridge saturation or a poorly fitting facepiece.

While warning properties are useful, OSHA requires that employers develop cartridge change schedules for gas or vapor cartridges and canisters with no ESLI. The schedule is based on

- Estimated concentration of chemical present

- Exposure limit of chemical

- Cartridge capacity

- Environmental conditions such as temperature and humidity

- Estimated user breathing rate

The calculation of such a schedule involves rather complex formulas. Respirator manufacturers offer computer programs that can calculate the schedule on the basis of information about their products and workplace information supplied by the user. Since some of the information is unique to the manufacturer's product, the calculator program should be obtained from each manufacturer and used only for their products. Particulate cartridges do not require such a schedule since increased filter loading leads to increased filter efficiency.

Purification elements should not be used if the manufacturer's expiration date has passed. They should be used immediately once the package seal is broken and discarded after completion of the work assignment, at the end of service life, or when breakthrough begins to occur, whichever comes first.

Service limits. APRs cannot be used if the concentration of air contaminants in the space exceeds the service limit concentration of the cartridges or canisters used. A *service limit* is the maximum use concentration that NIOSH certification allows for the respirator. Depending on the specific contaminants involved, service limit concentrations vary from 10 ppm (0.001%) to 1000 ppm (0.1%) for cartridges and from 0.5% (5000 ppm) to 3% (30,000 ppm) for canisters. Information about any applicable service limit concentration is provided by the respirator cartridge manufacturer in the instructions.

The role of APRs in confined space entry. For routine work operations in confined spaces, APRs offer a number of advantages, as described above. If it can be determined prior to entry that APRs will offer adequate protection, then APRs should be used. This allows entrants to operate with minimal mobility restrictions, as opposed to using atmosphere-supplying equipment.

In some cases, entry into a confined space with direct-reading air surveillance equipment may be required in order to fully characterize the breathing hazards in all areas within the space. In such cases, either SAR or SCBA should be utilized. All precautions for entry into IDLH atmospheres should be taken until IDLH conditions are proved absent. Respiratory protection may then be downgraded for routine operations within the space.

APRs should not be used for rescue operations in confined spaces for which respiratory protection is required. Given the large number of restrictions on the use of APRs, the process of determining that APRs will provide adequate protection is complicated and time-consuming. Atmospheric hazard levels sometimes increase during the course of rescue efforts. The use of a SAR or SCBA saves valuable rescue time while providing maximum protection to rescue personnel.

Selection Considerations for Supplied-Air Respirators

Supplied air respirators improve the protection provided to entrants by supplying a complete breathable atmosphere rather than cleaning the existing air. The following factors should be considered before choosing SARs.

Advantages of SARs. SARs offer several advantages for work in confined spaces where atmosphere-supplying respirators are required. They provide a much longer work time by using large cylinder banks or compressors as breathing air sources. They are less cumbersome and provide a lower profile to the entrant, allowing him to fit through tighter spaces. They are also lighter to wear and place less physical demands on the entrant.

Disadvantages of SARs. Problems with the use of SARs in confined space entry primarily involve the airline. The airline limits the distance entrants can travel from the source of air to 300 ft and forces the entrant to exit the space by the same path he entered. The airline can become tangled (Fig. 7.10), cut by physical hazards, or contaminated and permeated by chemical hazards while in the space. SAR systems are expensive to purchase and require extensive maintenance. They can be used in IDLH

Figure 7.10 The SAR's airline can become tangled, restricting the entrant's mobility.

environments only if they are equipped with an escape air supply, such as a 5-min escape air cylinder.

The role of SARs in confined space entry. SARs offer definite advantages over SCBAs for confined space entries in which APRs cannot be used. These advantages include

- Greatly extended air supply

- Greatly reduced apparatus weight

- Minimal restriction of ability to pass through small passages or entryways

In some cases airline-related problems (such as the potential for entanglement of the air hose) outweigh these advantages. Careful consideration must be given to the increased stress levels placed on entry personnel by extended duration of entry made possible by SARs.

SARs equipped with escape self-contained breathing apparatus (SAR/ESCBA) may prove useful for confined space rescue. This equipment provides the extended entry time of the SAR while ensuring that the rescuer can exit safely should an airline-related problem develop. The following conditions should be met before SAR/ESCBAs are selected for use during rescue efforts:

- The layout of the space is known prior to entry.

- Rapid escape from the space is possible.

- The rescuers' source of breathing air is independent of that used by the authorized attendants.

■ The volume of the rescuers' air supply should be at least twice the estimated total volume required to complete the rescue operation.

Selection considerations for SCBAs

The self-contained breathing apparatus typically is reserved for use in emergency activities, but it may also hold advantages for some confined space entries. The following considerations may help determine whether SCBA use is appropriate for an entry.

Advantages of SCBAs.   The primary advantage of the SCBA is that it provides breathable air from a source that is carried and controlled by the entrant. This provides the benefit of atmosphere-supplying respiratory protection without the airline restrictions. SCBAs offer the highest level of respiratory protection available to confined space entrants.

Disadvantages of SCBAs.   Disadvantages of SCBAs are related primarily to the air cylinder, which is both heavy and cumbersome. It limits the entrant's passage through tight openings and spaces (Fig. 7.11) and entry time is limited by the capacity of the air cylinder. SCBAs are expensive and require detailed maintenance procedures.

The role of SCBAs in confined space entry.   Because of mobility restrictions and the limited duration of the air supply, SCBAs are considered less desirable than SARs for many routine confined space entries. In some confined work areas for which APRs cannot be used, specific conditions may prevent the use of SARs, leaving SCBAs as the only remaining option. Only those SCBAs that operate in the positive pressure mode should be used for confined space entry.

Because of the high protection factor and lack of airline-related mobility restrictions, positive-pressure SCBA has traditionally been the mainstay of respiratory protection for emergency response personnel. For confined space rescue operations, SARs (see text above) may be a better choice.

Special Considerations for Using SCBAs and SARs during Confined Space Emergencies

A worker deep inside the hold of a ship was breathing air supplied through an airline by an electricity-powered compressor. The attendant observed a power interruption and, when he failed in his attempt to contact the worker, called the fire department. Two fire-

Figure 7.11 The bulky SCBA air cylinder makes entry into tight spaces difficult.

fighter rescuers began a difficult on-air entry through a tube, pushing their airpacks in front of them. As the first rescuer reached the end of the tube, his airpack dropped into the hold and ripped the mask from his face. He was in distress immediately, as the hold contained gasoline vapors. The second rescuer shared his mask and air with his partner in turns, but was forced to leave the tube. He had just reached safety when he, too, was overcome. This man lived, but the worker and the first rescuer did not.

Emergency breathing procedures

In emergency conditions during confined space entry, emergency breathing procedures may be required in order to use limited air supplies more efficiently. This is especially true in situations in which personnel become trapped or lost during confined space operations and are awaiting rescue with a limited air supply. These procedures require specialized training and practice, such as are gained in SCBA specialist classes, also known as "smoke diver schools."

Controlled breathing. *Controlled breathing* means simply limiting the frequency of breaths taken. Exertion and anxiety may cause rapid breathing, which depletes the available air supply very quickly. Remaining calm in an emergency is much easier if controlled breathing practice in nonemergency situations has taken place. Not only is less air used, but the enforced calm allows more efficient work and better communications. Once learned, controlled breathing techniques should be used at all times during respirator use.

Skip breathing. In contrast to controlled breathing, skip breathing is strictly an emergency technique used only to extend a low air supply. To use this technique, the breather inhales normally, holds the breath for the length of time normally required for exhalation, and then inhales again before exhaling. This specialized technique requires special training to perform well.

Emergency use of SCBA bypass valve. During normal SCBA operation, the mainline valve is fully open while the bypass valve is fully closed. In a situation where the regulator malfunctions, opening the bypass valve and closing the mainline valve provides a flow of air directly from the cylinder into the facepiece. Manufacturers of SCBAs typically instruct users of their products to open the bypass valve and leave it open for a continuous flow of air while exiting the contaminated area. An alternative is to open the bypass valve when a breath is needed and close it between breaths so as not to waste air that otherwise would be lost in a continuous flow through the exhalation valve. This procedure compromises the positive pressure of the SCBA and should be used only when it is required to ensure that adequate air will be provided during the escape.

Using a cylinder transfill or Quick-Fill® system. The Quick-Fill® and transfill options described earlier can be used to address breathing air issues of SCBAs during emergencies. The transfill option allows the transfer of air from a rescuer's cylinder to a victim's or other rescuer's cylinder, reducing the air supply of the donor rescuer. This should be done only in emergencies and only when the rescuer has sufficient air supply for both parties to exit the space safely.

The Quick-Fill® system can be used to provide breathable air to a trapped victim, provided the victim is using an SCBA equipped with the Quick-Fill® adapter. In that case, rescuers can bring spare cylinders or extend a special high-pressure hose into the space to recharge the victim's cylinder. This can provide time for rescuers to remove the victim safely.

Changing cylinders in toxic atmospheres. If it becomes necessary to replace an empty SCBA cylinder in a toxic atmosphere, two methods are possible. Both are done without removing the facepiece. If a Quick-Fill® system is in place, this is the safest method. The valve on the airline does not open until plugged into the receptacle so no contaminated air enters. The only other option is to hold one's breath during the exchange.

Buddy breathing.   A controversial procedure related to the use of SCBAs in confined spaces is "buddy breathing." Two scenarios when this procedure might be used are when an entrant (1) depletes his own air supply or (2) becomes trapped in an atmosphere where atmosphere-supplying respiratory protection is needed. A rescuer removes his facepiece and applies it to the face of the victim who breathes several breaths. The rescuer places the mask back on his own face to breathe a few breaths, and the sequence is repeated.

This practice assumes that some good air is better than all bad air. Rescuers believe that they will be able to hold their breath while the victim is breathing, or that small amounts of toxic atmosphere will not overcome them. Both of these beliefs have proved fatal to both victims and rescuers, as demonstrated in the scenario described above. In addition, these practices usually result in faster consumption of the available air supply than do other options such as getting assistance or removing the victim to a safe atmosphere.

NFPA 1500 specifically disallows the practice of buddy breathing and provides good guidance in steps to avoid the need for it. Proper planning through an effective respirator program, good training in SCBA use, and constant attention by a user to the air consumption prevent most needs for buddy breathing. Where victims are trapped, alerting other rescuers and getting assistance are the best procedures to help the victim. Remember the old motto, "Be part of the solution, not part of the problem."

Reduced profile maneuvers

One of the major limitations on the use of SCBAs within confined spaces is the inability of the entrant to pass through small entries or passageways while wearing an SCBA. In the past this limitation has been overcome through practices referred to as "reduced profile entry procedures" by entrants such as fire service personnel during rescue missions. These procedures require that the SCBA harness be removed from the entrant's back while the facepiece remains in place. The SCBA harness assembly is passed through the entryway just ahead or behind the entrant. The SCBA is donned after entry.

Strictly speaking, these procedures are a violation of OSHA's respirator standard, since the unit is used in a way other than that approved by NIOSH. Obviously the procedure used by the rescuers in the attempted ship-hold rescue described earlier is not recommended. It is not safe, and can usually be avoided by proper emergency planning;

Figure 7.12   A short length of webbing or rope attached between the harness and SCBA can prevent the facepiece from being pulled off the face if the unit is dropped after being doffed for reduced profile entry.

(a)   (b)

Figure 7.13 (*a*) In the first step of the two-person pass, the first rescuer doffs his airpack and climbs through the entryway while the second rescuer holds the pack. The second rescuer then passes the airpack through the entryway to the first rescuer. The first rescuer then dons the pack. (*Note:* The rescuers are moving from left to right through a plywood portal simulator in this training session.) (*b*) In the second step of the two-person pass, the second rescuer doffs his airpack and passes it through the entryway to the first rescuer. The first rescuer then holds the airpack while the second rescuer climbs through the entryway. The second rescuer then dons the pack.

however, reducing the rescuer's profile may be the only way he can reach a victim in a confined space emergency. If a reduced profile maneuver is attempted, there are several things rescuers can do to make it safer.

Tether line. Using a class III harness (described in Chap. 11), don the harness prior to donning the SCBA. Attach a short length of webbing or rope from the harness to the SCBA cylinder frame (see Fig. 7.12). The tether line must be short enough to prevent the facepiece from being pulled off the face even if the unit drops unexpectedly. The unit can be retrieved with the tether line.

Two-person pass, If the entry portal is small and cannot be entered wearing an SCBA, but the plane of entry is the only tight spot, rescuers can work in pairs, as shown in Fig. 7.13. The first rescuer doffs his pack but leaves the facepiece on and continues breathing air from the SCBA. He then climbs through the portal while the second rescuer holds his airpack. The second rescuer passes the airpack though the portal (see Fig. 7.13a) and the first rescuer dons it inside the confined space. The second rescuer doffs his airpack and passes it through the portal to the first rescuer. The first rescuer then holds the second rescuer's airpack while he climbs through the portal (see Fig. 7.13b). The second rescuer then dons his airpack inside the space.

Fit Testing and Assigned Protection Factors

A properly selected respirator will not provide adequate protection if a leaking facepiece allows contaminated air to enter. A tight-fitting facepiece must form a complete, unbroken seal with the face in order to prevent the outside air from bypassing its protective barrier. Unfortunately, even relatively large leaks can go undetected by the user. Respirator users must use an approved protocol to test the quality of the facepiece fit if they are to be certain they are getting the best protection.

Terminology

Several terms must be clarified in order to discuss fit testing and the quality of respiratory protection.

The *fit factor* (FF) is the quantitative measurement of the fit of a particular respirator to a particular individual. Quantitative fit test methods, described below, provide this direct measurement. This measurement assumes proper function of any purification elements and assesses only the facepiece-to-face seal.

The *protection factor* (PF) describes the overall performance of the respirator in terms of a ratio of the concentration of an agent outside the facepiece to the concentration inside the facepiece. The factor accounts for the quality of the fit as well as the penetration through purifying elements and other factors.

The *assigned protection factor* (APF) is defined as a measure of the minimum anticipated workplace level of respiratory protection provided by a properly functioning respirator or class of respirators to properly fitted and trained users. APFs are typically estimates, not absolute values based on extensive research. They are useful for selecting a type of respirator for a given hazard as long as some safety margin is included. For instance, half-mask air-purifying respirators have an APF of 10. A user could expect that if he uses a half-mask APR with an organic vapor cartridge in an atmosphere containing 45 ppm of phenol, he should not actually breathe more than about 4.5 ppm (45 ÷ 10) of phenol. While this would be below the OSHA PEL of 5 ppm, the user may not actually experience that exact level of protection and so should consider a full facepiece respirator with an APF of 50 to provide for extra safety.

The *workplace protection factor* (WPF) is a protection factor resulting from the simultaneous measurement of air contaminant inside and outside the facepiece of a user in an actual workplace. Because such measurements are difficult to make, few WPFs are available for respirators. When they are available, NIOSH adjusts the APFs for a class of respirators accordingly.

The *maximum-use concentration* (MUC) is the highest concentration of a particular chemical in which a class of respirator can be used. The APF of a respirator is multiplied by the exposure limit for the chemical. In the example of phenol above, the MUC for full facepiece respirators would be 250 ppm (50 × 5 ppm).

OSHA regulation

In the revision to the respiratory protection standard (29 CFR 1910.134), OSHA added very specific requirements for respirator fit testing by employers. These are outlined in Paragraph (f) and Appendix A of the standard. In the standard, OSHA specifies the kinds of fit tests allowed, the procedures for conducting them, and how the results of the fit tests must be used.

Employers must fit-test employees prior to the initial use of a respirator and at least annually thereafter. In this way the user is assured of a proper fit from the beginning. The annual retest should catch changes in the user that affect the quality of the fit. If the user, supervisor, or respirator program administrator detects a condition that may change the fit between fit tests, a retest should be performed earlier. Such conditions include significant weight change, facial scarring, and other similar changes.

Employers may use only the fit-test protocols approved by OSHA and described in Appendix A of the standard. The descriptions of the protocols are complete, and commercially available fit-test kits meet these requirements. The discussion of these methods is based on the requirements in the standard.

Fit testing of tight-fitting atmosphere-supplying respirators must be performed in the negative-pressure mode, regardless of the mode of operation (negative or positive pressure) that is used for respiratory protection. This may require the use of an adapter that allows the use of APR cartridges with the SCBA or SAR facepiece. Respirator manufacturers can provide such adapters for their equipment.

Respirator fit checks

Each time a worker puts on a respirator, he must perform two simple fit checks (Fig. 7.14). The positive and negative fit checks do not qualify as respirator fit tests that are described below. They do help the user ensure that the mask is properly positioned and that there are no major leaks. OSHA describes these tests in Appendix B-1 of the standard and requires that employees perform them each time they don a respirator.

The positive-pressure check is performed by lightly covering the exhalation valve with the palm of the hand and exhaling into the facepiece. The user should feel the facepiece expand away from his face and remain that way for several seconds. If there is a leak, air will leak out and the facepiece will collapse back to its original position. The user may be able to feel the airflow on the skin and thus locate the leak.

Figure 7.14 Negative- and positive-pressure fit checks assure that the facepiece has been properly donned.

To perform a negative-pressure check, the user lightly covers the cartridges, canister inlet, or other respirator inlet with the palm of the hand. He inhales slightly and holds the breath. The facepiece should collapse against the face and hold the position for several seconds. A leak will allow air to enter and the facepiece will relax away from the face.

In addition to these tests, users can employ isoamyl acetate (or banana oil) ampoules to check the respirator fit. This is a quick check and not a qualitative fit test that is described below. The user must be wearing an APR with organic vapor cartridges. The user breaks a small ampoule of the sweet-smelling vapor and waves it slowly around the facepiece. If the vapor is detected, the facepiece seal has a leak, and it should be adjusted; however, a saturated cartridge may allow the vapor to pass through and falsely indicate a leak. This check should be performed with fresh cartridges.

Approved fit test protocols

OSHA has approved both qualitative and quantitative fit-test protocols for testing the fit of any tight-fitting facepiece. Loose-fitting facepiece respirators do not require fit testing. The specific procedures of each protocol are spelled out in Appendix A of the standard. Each protocol includes a challenge agent and a set of exercises that is to be performed by the user during the test to challenge the fit.

Qualitative methods can be used to fit-test all positive-pressure respirators (PAPRs and atmosphere-supplying) and any negative-pressure APR that must achieve a fit factor of 100 or less. The required fit factor of 100 or less derives from multiplying a protection factor of 10 by a safety factor of 10 to account for the difference between fit-testing results and the actual protection achieved during work. In other words, a negative-pressure APR can be qualitatively fit-tested if it will be used only in atmospheres where the level of hazardous contaminant is 10 times or less than the exposure limit for the contaminant. Quantitative methods must be used when a negative-pressure respirator will be used in atmospheres that contain a hazardous contaminant at a level greater than 10 times the exposure limit.

Qualitative methods.   Qualitative methods involve the exposure of a person wearing a respirator to a challenge agent that can be tasted or smelled, or that can elicit a response from the user (Fig. 7.15). If the user does not detect or respond to the agent, an acceptable fit has been achieved. This is referred to as a "pass/fail test."

The approved qualitative methods use one of these challenge agents:

- Saccharin mist

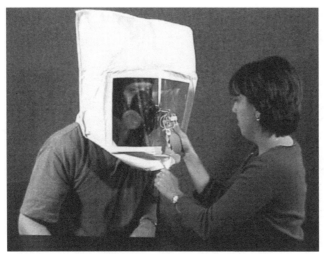

Figure 7.15 Qualitative fit tests use a challenge agent to check the fit of the facepiece to the face.

- Isoamyl acetate (banana oil)
- Bitrex (denatonium benzoate)
- Irritant smoke

The saccharin mist and banana oil tests rely on the user's detection of sweet-tasting or sweet-smelling agents. The Bitrex and irritant smoke agents evoke an involuntary response by the user. These agents frequently are used to prevent deception by individuals who may be trying to qualify for a job that requires respirator usage.

All of these methods include a sensitivity test as part of the protocol. The user enters an enclosure or dons a hood (except for irritant smoke tests) without wearing the respirator and the agent is introduced into the space until he detects it. The sensitivity test determines that the user can detect the agent and at about what level. Next the user dons the appropriate respirator and enters the enclosure or dons the hood again. A hundredfold stronger concentration of the agent is introduced into the space and the user performs a prescribed series of exercises to challenge the fit. If the user detects or responds to the agent, the fit is considered inadequate and the respirator is adjusted or exchanged. If the user does not detect the agent, the fit is considered adequate.

Quantitative methods.   Quantitative methods measure the quality of the fit and provide a numerical fit factor. Unlike the APF, the fit factor is a direct assessment of the fit of a particular facepiece model and size to the individual user. The tests can be performed on the user's personal mask or on a representative mask of the same brand, model, and size (Fig. 7.16).

According to OSHA regulations, when quantitative methods are used, half-mask respirators must achieve a measured fit factor of at least 100 and full-facepiece respirators must achieve a fit factor of at least 500. NFPA 1404 requires a minimum fit factor of 1000 for negative-pressure (demand) SCBAs, but does not specify a satisfactory fit factor for positive-pressure SCBAs.

OSHA has approved three quantitative methods:

- Generated aerosol
- Condensation nuclei counter
- Controlled negative pressure

The required procedures are described in Appendix A and are based on the instructions from the test equipment manufacturers.

Figure 7.16 Quantitative fit tests provide a numerical fit factor that describes the quality of the fit of a respirator facepiece.

Generated aerosol. This method entails the generation of a known concentration of an aerosol such as corn oil mist inside a test chamber. The user enters the chamber wearing a facepiece fitted with a port that allows sampling of the air inside the mask. A real-time aerosol monitor measures the concentrations of the aerosol inside the enclosure and inside the facepiece. The *fit factor* is the ratio of the aerosol concentration of the enclosure to the concentration inside the facepiece.

Condensation nuclei counter. This technology often is referred to by the brand name of the tester equipment, PORTACOUNT®, by TSI. It uses ambient dust as the challenge agent and alternately samples air from outside and inside a facepiece equipped with P100 particulate filters and a sampling probe. The sampled air passes through an alcohol-rich atmosphere and then through a condenser where alcohol vapors condense around the dust particles. The airstream then passes through a laser beam where the dust/alcohol nucleus diffracts the beam and creates a flash of light. A photocounter detects the flash, and each particle is counted. Here again the ratio of the concentration of aerosol outside the facepiece to the concentration inside the facepiece determines the fit factor.

Controlled negative pressure. The brand name for this technology is Fit Tester 3000® by Occupational Health Dynamics. The tester uses normal air as its challenge agent. Special adapters replace the cartridges to block the respirator inlets. During the fit test the user draws a normal breath and holds it. Tubing attached to the adapters then allows the tester to draw a target negative pressure inside the facepiece and try to maintain it during the 8-second test. If leakage in the seal allows a volume of air to enter the facepiece, the tester draws an equivalent volume of air out to maintain the target negative pressure. The additional pump flow rate needed to maintain the pressure represents the leakage flow rate that is compared to the breathing flow rate to establish an equivalent fit factor.

Care and Maintenance of Respirators

The proper care of respirators is vital to their continued effectiveness. The manufacturer's instructions provide the proper procedures and practices for care of the equipment. In addition, the OSHA standard provides general requirements that employers must follow as part of their respiratory protection program. NFPA 1404 gives extensive guidance on the inspection and maintenance of SCBAs used by fire departments.

Inspection

The user must inspect the respirator before each use and during cleaning according to the manufacturer's instructions. Employers must inspect respirators maintained for emergency use at least monthly as well as before and after each use. Emergency escape-only respirators must be inspected before being carried into the space for use.

The inspection must include a check of the proper respirator function, tightness of any connections, and the condition of various parts. General procedures that apply to all types of respirators include

- Check material condition of harness, facial seal, and breathing tube (if so equipped) for pliability, signs of deterioration, discoloration, and damage.
- Check faceshields and lenses for cracks, crazing, and fogginess.
- Check inhalation and exhaust valves for proper operation.

Additional guidelines for inspecting air-purifying respirators focus on the purifying elements. Examine cartridges or canisters to ensure that

- They are the proper type for the intended use.
- The expiration date has not passed.
- They have not been opened or used previously.

When inspecting SCBAs, the user should

- Check the air supply.
- Check all connections for tightness.
- Check for proper setting and operation of regulators and valves (according to manufacturers' recommendations).
- Check for leaks in the lines and fittings.
- Check operation of alarms.
- Check harnesses and straps.

Inspecting supplied-air respirators should include the following steps:

- Inspect the airline prior to each use, checking for cracks, kinks, cuts, frays, and weak areas.
- Check for proper setting and operation of regulators and valves (according to manufacturers' recommendations).
- Check escape air supply (if applicable).
- Check all connections for tightness.

Respirator users should be thoroughly trained in the proper inspection procedures for the respirators they will be using. It is prudent for the inspection to follow a checklist to facilitate inspection and serve as documentation that the inspection occurred.

Repairs

Respirators failing an inspection or otherwise found to be defective must be removed from service and adjusted, repaired, or discarded. Repairs and adjustments can be performed only by appropriately trained individuals who use only the respirator manufacturer's NIOSH-approved parts. Regulators, valves, gauges, and alarms must be adjusted or repaired by the manufacturer or a technician trained by the manufacturer.

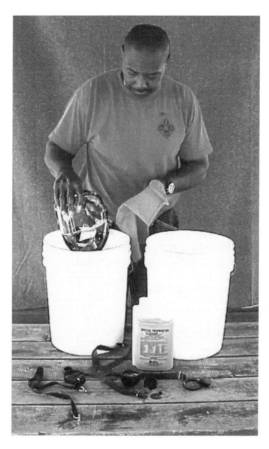

Figure 7.17 Respirator facepieces should be cleaned with an appropriate solution recommended by the manufacturer.

Cleaning

Users should clean respirators after each use to keep them sanitary (Fig. 7.17). OSHA requires that respirators used by more than one individual be cleaned before use by a different individual. Emergency respirators are to be cleaned and disinfected after each use, as are respirators used for fit testing and training.

Employers must follow the manufacturer's instructions on cleaning. OSHA includes instructions in Appendix B-2 of the standard that include the following steps:

1. Disassemble and inspect the facepiece and other components.

2. Wash components in warm water with a mild detergent or other cleaner recommended by the manufacturer.

3. Rinse the components in clean, warm water.

4. Disinfect the components with a laundry bleach solution in water (one milliliter bleach to one liter water), an iodine solution (50 ppm iodine in water), or other commercially available disinfectant approved by the manufacturer.

5. Thoroughly rinse the components to remove all detergent and disinfectant.

6. Hand-dry with a lint-free cloth or allow to air-dry.

7. Reassemble the components and facepiece.

8. Test the respirator function.

Storage

Respirators must be stored in a manner that will protect them from damaging agents such as dust, chemicals, temperature extremes, sunlight, and moisture. Use storage cases or bags from the manufacturer to prevent deformation of the facepiece gasket. Other storage requirements may be given by the respirator manufacturer.

Summary

Employers must try to control hazardous atmospheres in confined spaces by engineering controls such as ventilation, isolation, removal, or other effective means. These controls are sometimes not enough to ensure protection of the entrants. In these cases, OSHA allows the use of respiratory protective equipment. The safe use of respirators occurs under a respiratory protection program that meets the requirements of OSHA's respiratory protection standard (29 CFR 1910.134). A number of issues regarding the proper selection and use of respirators must be addressed in the respiratory protection program.

8

Chemical Protective Clothing

In some cases confined space entrants may be exposed to chemicals through contact with the skin. While confined spaces should be emptied and cleaned of chemicals to the maximum extent possible before entry, sometimes chemicals remain as residues or in difficult-to-access areas. These chemicals may cause harm directly on contact with the skin or may pass through the skin to harm other parts of the body. Some chemicals very readily absorb through the skin. These chemicals have a "skin" notation with their airborne exposure limits (e.g., PEL) to remind health and safety professionals diligently to protect the skin as well as the inhalation route of entry.

Chemical protective clothing (CPC) provides a barrier between the entrant's body and the chemicals. If properly selected and used, CPC effectively prevents the chemicals from making contact with the skin. CPC users must consider many factors when selecting the appropriate type and amount of protective clothing. This chapter describes the selection and use of chemical protective clothing.

Selection of Chemical Protective Clothing

Chemical protective clothing must be selected properly to provide any protection to the user. In fact, improper CPC can actually be more harmful to the user than no protection at all, if it allows a chemical to reach the skin and then holds it there. The type of clothing and the material used to make it determine how well the entrant is protected from a particular chemical.

Consider this example. An entry supervisor preparing a permit for a job to clean a tank of methyl ethyl ketone (MEK) has a choice of two pairs of rubber gloves that are made of two different materials, neoprene and butyl rubber. If she chooses the butyl rubber gloves, the entrants should be protected for over 4 hours, but MEK would begin to pass through the neoprene gloves in less than 10 min. Proper selection of the glove material obviously is vital to getting the most protection.

Beyond simply choosing the CPC material, it is important to select the proper type of clothing. In the example above, if the supervisor chose the proper gloves but allowed the entrants to wear regular cotton coveralls and tennis shoes, they would not have adequate protection. Protective clothing must cover enough of the body to prevent contact with the skin.

The following factors must be considered when selecting protective clothing:

- Specific chemical contaminants likely to be encountered, since no single protective material is effective against all potential chemical assaults

- The physical state of contaminants (solid, liquid, or gas)

- How well the protective clothing resists chemical attack and physical damage
- The duration of the entry and the exposure time of the clothing
- Potential for direct exposure by splashing or spraying
- The degree of stress (particularly heat stress) placed on the wearer by the protective clothing
- The degree to which the mobility of the entrant will be restricted by the protective clothing selected

The information presented here is intended to serve as an introduction to the basics of CPC selection. Given the complexities involved and the potential consequences of improper selection, the selection should be made only by individuals who are properly trained, knowledgeable, and experienced. The entry supervisor should confirm the selection before signing the permit and allowing entry to the space.

Chemical attacks on CPC

To provide adequate protection to the wearer, the material from which the CPC is constructed must resist attacks from the chemicals in the space. Chemicals may move through or damage the material by means of three mechanisms (see Fig. 8.1). It is important to understand these mechanisms when selecting and using any type of CPC.

Permeation is the movement of chemicals through the CPC material on a molecular level. Individual molecules of contaminant pass between the molecules of the CPC material. This occurs when the physical properties of the chemical and the material allow the chemical from outside the garment to dissolve into the material and then "offgas" on the inside of the garment. It's important to remember that permeation does not necessarily produce any visible effect on the material.

Penetration refers to the bulk movement of chemical through an opening or flaw in the material. Penetration may occur through imperfect seams, small holes, zippers, or any other opening. There is no chemical or physical process other than simple leakage involved with penetration.

Degradation occurs when there is a chemical interaction between the chemical and the material that causes the material to change form or lose physical integrity. Unlike permeation, degradation typically produces a visible or otherwise detectable change in the CPC material.

Permeation and degradation entail a direct interaction between the chemical and the CPC material based on their physical and chemical properties. These properties vary with the different chemicals and materials, so the severity of these effects also varies with different combinations of the two. It is vital for the appropriate material to be purposefully selected for each chemical that is known to be present.

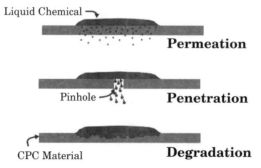

Figure 8.1 Chemicals may attack chemical protective clothing material in three ways.

Measurement of chemical attack

The field of research and development for chemical protective clothing includes manufacturers, academic institutions, and government agencies. A growing number of CPC materials are tested against a vast number of chemicals. These tests follow standard practices laid out by such organizations as the American Society for Testing and Materials and the National Fire Protection Association to ensure consistency and repeatability. The data generated by these tests may be published in journals and magazines or used by manufacturers to develop selection charts that are discussed in a later section of this chapter.

The typical test procedure for permeation tests entails mounting a swatch of material between two glass hemispheres. The test chemical (or challenge agent) is circulated in either liquid or vapor form through one hemisphere. Air or another sampling medium is circulated through the other and returned to a detection device that will measure the chemical. The test begins with the first exposure of the chemical to the material and ends when the chemical is detected on the other side.

The chemical resistance data typically are reported in two terms: breakthrough time and permeation rate. *Breakthrough time* is the time that it takes for the chemical to pass through the test material and be detected on the other side. It is reported in seconds, minutes, or even hours. Generally the longer the breakthrough time the more resistant the material is to the chemical.

Permeation rate is the actual rate of movement of the chemical through the material. It is reported as the amount of chemical that moves through a surface area of material per unit of time. For example, one manufacturer may report that MEK will permeate a particular neoprene glove material at a rate of $7.2 \text{ g mm}^{-2} \text{ min}^{-1}$, while another reports that it permeates butyl rubber at a rate of $1.3 \text{ } \mu\text{g cm}^{-2} \text{ min}^{-1}$. (Notice the difference in units between the data.) The lower the permeation rate, the more resistant the material is to the chemical. Be sure that all of the measurement units match when comparing the rates.

The tests are performed in laboratories under controlled conditions. When the results are reported, the temperature of the testing should be assumed to be about 70°F unless otherwise stated. The results also should indicate the thickness of the material that was tested. The significance of these factors in selection is discussed below in the section on using the selection information.

CPC material selection information

There are many sources of information available to individuals who are responsible for selecting CPC for confined space operations. Once the chemical has been identified, a person can make a selection from the available options by consulting one of the types of references discussed below.

Qualitative and quantitative data.   Selection information typically is provided in either qualitative or quantitative forms (Fig. 8.2). Qualitative recommendations typically rate the material's resistance to chemicals with subjective terms such as "excellent," "good," "fair," "poor," or "not recommended." Such guides are usually tables that list several material options for a number of chemicals or chemical families. Selection can be based on the material with the highest or most favorable rating for the chemical of interest.

Specific criteria for assigning each rating may or may not be provided with the listing. The manufacturers or other providers of this kind of information base the rating on an assessment of permeation or degradation data, even if it is not provided. Users may be somewhat uncomfortable in basing selection decisions on these ratings if the criteria are not provided.

Quantitative selection guides report the actual breakthrough time and, usually, the permeation rate data. The information is given in a table with columns for each CPC material and values listed for each of a number of chemicals. The values listed are usually an average of the results of several tests, although occasionally individual test results are listed.

Many guides now include a degradation rating. Manufacturers assign this qualitative rating principally on the basis of weight gained when the material is immersed in the

		PERMEATION		DEGRADATION
		Normalized Breakthrough	Rate	60 minute soak
CAS #	CHEMICAL TESTED	Time - minutes	µg/cm²/minutes	Rating
67-64-1	Acetone (99.5%)	5.2	>1900	P
79-10-7	Acrylic Acid (99%)	>480	ND	E
1336-21-6	Ammonium Hydroxide (29%)	358.0	0.153	E
71-43-2	Benzene (99.9%)	>480	ND	E
123-86-4	Butyl Acetate (99.5%)	23.9	474.2	P
75-15-0	Carbon Disulfide (99%)	>480	ND	E
79-04-9	Chloroacetyl Chloride (98%)	218	45.3	E
108-90-7	Chlorobenzene (99.5%)	>480	ND	E
67-66-3	Chloroform (99%)	271	6.62	E
56-23-5	Carbon tetrachloride (99.9%)	>480	ND	E
110-82-7	Cyclohexane (99%)	>480	ND	E
68-12-2	Dimethylformamide (99.8%)	13.1	131.2	P
100-41-4	Ethyl Benzene (99%)	>480	ND	E

COBRA™ FLUOROSOLV™ (15 mil Fluoroelastomer/Nitrile)

Key to Permeation Ratings:
Normalized Breakthrough Time -
   the time from initial chemical contact to normalized detection.
Rate µg/cm²/min. -
   Rate in micrograms per centimeter squared per minute: the rate at which the chemical passes through the glove film.
ND - None Detected
Test procedure - ASTM F739

Key to Degradation Ratings:
Performance Rating
Weight Change
   (E) Excellent   0 to 10%
   (G) Good   11 to 20%
   (F) Fair   21 to 30%
   (P) Poor   Over 30%
NT - Not Tested
Test procedure - Modified ASTM D471

Figure 8.2 This CPC selection chart provides both quantitative and qualitative selection data.

chemical. When provided, this rating is often an addition to quantitative information from permeation tests.

Sources of selection information

Manufacturers. CPC manufacturers and vendors provide most of the selection information available to the CPC user. This is to be expected, since they perform most of the research and development work on new clothing materials. The data generated during the research on the product are used to determine and describe its applicability to different workplace settings. The manufacturers also test established products against new and different chemicals to add information to the selection guides.

Manufacturers and vendors provide the selection data in print form or in electronic database form. Printed selection information is provided as a table relating one or more products to a list of chemicals. The chemicals in the list may represent a group of similar chemicals. The recommendations may be either qualitative or quantitative in nature. The tables usually are offered free of charge as a part of marketing information.

Electronic CPC selection guides are available from most manufacturers. The guides allow the user to search a database for recommendations of the manufacturer's products best suited to protect against a particular chemical. The computer program usually includes some additional information about the chemical, such as chemical and physical properties and exposure limits. Most manufacturers provide the programs free on request, though some charge a fee for the software. Many manufacturers make their selection guides available on their Website as well.

Independent sources. Some selection guides have been produced by independent sources that are not associated with CPC manufacturers. Most notable of these are the *Quick Selection Guide to Chemical Protective Clothing,* 3d ed. (Forsberg and Mansdorf 1997) and *Chemical Protective Clothing Performance Index,* 2d ed. (Forsberg and Keith 1999). These guides are compilations and evaluations of CPC test data from sources all over the world. Other guides have been published in the past but have not been updated and are, therefore, of limited use.

Test data from studies on specific materials and chemicals frequently are published in professional journals. These studies typically limit their focus and do not provide comprehensive guides; however, they provide excellent information for the chemicals and materials described.

Some chemical reference documents include recommendations for CPC materials with their information. For example, the *Emergency Action Guides,* published by the Association of American Railroads (AAR 1993), include such recommendations. The guides are emergency response guides and the CPC recommendations are just one piece of the information they provide. Other published chemical references may include CPC selection data as well.

The Material Safety Data Sheet (MSDS) may also recommend CPC material. OSHA requires employers to acquire MSDSs for each hazardous chemical in the workplace. Not all MSDSs provide specific recommendations for CPC. Some simply recommend "rubber gloves," using the term generically to mean *chemical-resistant* and without reference to a specific material. They are a good place to begin the search for selection information, as they are likely already available in the workplace. Since the accuracy of information provided in MSDS varies, recommendations should be confirmed by another source whenever possible.

Government resources. NIOSH has made available the *Recommendations for Chemical Protective Clothing: A Companion to the NIOSH Pocket Guide to Chemical Hazards.* This electronic database provides recommendations for CPC materials for each of the approximately 600 chemicals in the guide. NIOSH states that they acquired the information in the database from a contractor. The database can be accessed from NIOSH's Website.

The Computer-Aided Management of Emergency Operations (CAMEO®) software developed jointly by the EPA and the National Oceanic and Atmospheric Administration (NOAA) includes CPC recommendations. CAMEO® is a system of software applications used widely to plan for and respond to chemical emergencies. It is distributed by the EPA free of charge to anyone responsible for the safe handling of chemicals. While not directly related to confined space work, the chemical information database can be a useful reference resource.

Using CPC material selection information

The information used to make decisions about CPC comes from various sources and takes different forms. The user should consider some of the limitations of the available data when making the selection. Some of these limitations are discussed below, followed by practical guidelines.

Limitations of the information

Whole suit versus swatch. The tests performed to establish permeation data use a small swatch of material without seams or other inconsistencies; however, the clothing made from the material uses larger swaths of material, often with sewn seams, zippers, and other potential breaks. A garment is more likely to have thinner areas or defects just because there is more material. Manufacturers have extensive quality control procedures, but some variation among garments does occur. Some garments have integrated boots, gloves, and visors that may be made of a different material than the suit itself. Information on the permeation through these other materials may not be provided with the selection data.

Since protective clothing often involves seams, fasteners, zippers, and accessories, there are possible opportunities for leakage. Seams are sewn and taped to prevent penetration. Flaps cover zippers to prevent penetration by splashed liquids.

Single-chemical tests.   Manufacturers usually test a single chemical against a single material, although they may test some common mixtures such as gasoline. With tens of thousands of chemical compounds in use, the number of potential mixtures to be tested makes it unlikely that any meaningful testing of mixtures will occur. Chemical interactions could significantly change the permeation of one or both of the chemicals and cause a more rapid breakthrough. Without direct testing of mixtures, it is impossible to know how much the permeation of chemicals in a mixture would be affected.

Laboratory tests.   Laboratory tests of the material occur under controlled environmental conditions. These conditions may be varied between tests, but they are consistent during the tests themselves; however, the conditions of the workplace can be extreme and may change during the use (Fig. 8.3). Using the material in temperatures higher than the test temperature can lead to a significant decrease in the breakthrough time due to an increased permeation rate. In addition, users move around, flexing the material and causing folds. These actions can cause weakening of the material that may allow chemical permeation to occur more quickly.

Vendor-supplied data.   Much of the selection data available for use comes from CPC manufacturers and vendors. The vendors want to sell as much of their product as they can. While this seems like a conflict of interest, these vendors have strong motivation to provide accurate selection information. They would incur substantial liability by making false claims about their product that resulted in injury to a worker; therefore, they follow standard, approved protocols when conducting the permeation tests. Nonetheless, users should take vendors' marketing claims about their products with caution and rely only on technical data or recommendations for selection.

Use of breakthrough times.   When given breakthrough times for a chemical and a material, users sometimes treat these as time limits for using the selected CPC. Considering

Figure 8.3 CPC material testing is performed in the laboratory environment; however, the garments may be used in very different environments.

the limitations described above, this can be dangerous. Breakthrough times are provided only for comparing one material to another to select the most effective protection.

Actual workplace uses and conditions are different than the test conditions. The breakthrough times should be considered "best case" times since actual work conditions are likely to be worse than the test conditions. If there is any real exposure of the chemical to the material in the workplace, CPC should be used for no longer than the breakthrough time. In fact, the breakthrough time should be lessened by a safety factor.

Basic principles for using selection data.   When selecting CPC material, users should remember these basic principles:.

■ No material protects against all chemicals. The identity of the chemical must be known to select the proper material.

■ No material protects forever. It is only a question of time before a chemical will break through the material.

■ The user must select the material that provides the longest protection against the most chemicals in use.

Selection considerations for confined spaces

Selecting the proper CPC material is the first and most important step. Then other factors must be considered. The user must understand the conditions inside the space and the work to be done in order to select the actual type of chemical protective garment from among the acceptable options. Users should consider these factors when considering the options:

■ Entrant's mobility needs

■ Durability of the garment

■ Entrant's need for dexterity and sense of touch

■ Added weight of the garment

■ Heat stress due to insulating effect of clothing

Confined spaces offer unique challenges to the use of CPC. They are often very small and include tight spaces where entrants may be forced to brush against walls and equipment. Any protruding metal or objects can cause punctures, scrapes, or other physical damage to the garment. Protective garments may need to be thicker and more durable than in open spaces where physical damage is less likely.

The enclosed nature as well as the contents of confined spaces may cause an increased likelihood of contamination by the chemical. Airborne concentrations may accumulate as well. The exposure potential to the entrant may be higher because the area is confined, calling for protection of more of the body. The potential for direct contact with a chemical may require thicker-than-usual material to slow the permeation process and provide longer use time.

Spaces with poor ventilation often are hotter and more humid than open work areas. Such conditions increase the heat burden placed on the entrant and may result in heat injury. To prevent such injuries, entrants may need to substantially shorten the work periods between rest breaks or use special cooling garments. Heat stress is discussed in more detail in Chap. 3.

Increased temperature causes chemical molecules to move faster; therefore, higher ambient temperatures in a space may increase the permeation rates of chemicals through the CPC material. Work in hot spaces may require more frequent changing of articles of CPC. The manufacturer should be consulted to determine the effect of high temperatures on the protection provided by the clothing.

Types of Protective Clothing

Chemical protective garments can provide protection to any or all parts of the body. The degree and nature of the hazards in a space and the work to be done determine the degree of protection needed for the entrant. As a rule, the more of the body to be protected, the greater the burden placed on the wearer. A basic guideline for selecting the type of clothing to use is that chemical protection must be adequate but not overly burdensome to the entrant.

Chemical protective garments

Chemical protective suits are designed to protect the body from exposure to damaging chemicals. The suits may completely enclose the wearer or simply shield the body from splash. The properties of the chemical and the type of work to be done will determine which type of suit is required.

Totally encapsulating chemical protective suits.   The totally encapsulating chemical protective (TECP) suits completely enclose the wearer (Fig. 8.4). They are designed to provide a gastight barrier between the wearer and the environment. They have seams that are sewn, glued, and taped to prevent vapor migration. The zippers on such garments are made gastight by pressing material against material rather than simply interlocking metal teeth. The suits have a polycarbonate faceshield for vision outside the suit.

Some TECP suits have attached boots for chemical and physical protection. Others simply enclose the foot in a bootee that provides chemical protection and slips into a substantial outer boot for physical protection. Likewise, substantial chemical-resistant gloves may be attached to the suit, or the suit may have a simple inner glove and a solid, smooth cuff ring over which an outer glove is stretched and sealed.

Since the TECP suit is completely sealed, it requires the use of atmosphere-supplying respirators, either SCBA or SAR. The SAR requires that a pass-through port be installed in the suit to connect the airline to the SAR unit inside. Air exhaled from the respirator facepiece eventually creates a positive pressure inside the suit that is relieved through pressure-relief valves in the suit body. These suits are designed to hold a slight positive internal pressure, so that any leakage will be from the inside out.

TECP suits are constructed either of reusable materials or limited-use (disposable) materials. Reusable suits are substantially more expensive than the disposable initially (perhaps 6 to 8 times more). The savings gained by reusing the suit may actually make them more economical for longer jobs where repeatedly disposing of suits after one use can be expensive. Reuse requires proper decontamination, which is discussed later in this chapter.

TECP suits seal out gases and vapors as well as liquid splashes and particulates. They provide complete protection to all of the body and skin. They are selected when

- Chemicals are highly toxic or damaging to the skin.

- Chemicals are readily absorbed through the skin into the rest of the body.

- Work conditions involve a high degree of splashing by a skin-damaging chemical.

Figure 8.4   TECP suits completely seal the wearer from the environment.

The TECP suit material must be properly selected to prevent the permeation of gases and vapors into the suit.

Nonencapsulating suits. Nonencapsulating chemical protective (NECP) suits, also known as "splash suits," cover most of the body with gaps around the neck, or hood if so equipped. They provide good protection from moderate liquid splash and particulate contaminant, but allow gases and vapors to enter the suit (Fig. 8.5). Splash protection can be enhanced by sealing the cuffs to the boots and gloves, but this does not convert a splash suit into a gastight TECP suit.

NECP suits can also be constructed of reusable or disposable materials. Although NECP suits are not gastight and preventing permeation is less critical than for TECP, the suit material must still be properly selected. This will prevent degradation and slow permeation and provide better protection to the wearer.

Disposable overgarments. It is possible to use lightweight, disposable garments over more expensive reusable garments. The inexpensive disposable takes most of the physical abuse and minimizes the splash while allowing the reusable garment to provide the primary chemical protection. On exiting the space, the disposable garment is discarded and the reusable is decontaminated. An overgarment increases the effectiveness of the chemical protection and extends the useful life of the reusable garment. As long as the disposable garment is not too heavy, it does not add significantly to the heat stress of the wearer. The same concept is regularly used for boots and gloves to enhance the protection of these areas that are likely to receive the most contamination.

Partial body coverings. In addition to suits for protecting the whole body, there are protective clothing items to protect portions of the body. Such articles include aprons, sleeves, leggings, boots, and gloves. These provide localized splash and contact protection where minimal splashing is expected. Proper selection of these articles is important to prevent permeation and degradation by chemical contact. Boots and gloves are discussed further in Chap. 9

Protective clothing for nonchemical hazards

Confined spaces may have hazards other than chemicals for which protective clothing might be selected. Examples of such hazards include fire, temperature extremes, water, and radiation. The protective clothing for such hazards may have specifications beyond the scope of this discussion, but basic descriptions are provided below.

Figure 8.5 Splash suits protect the wearer from liquid splash.

Firefighter's protective clothing. Firefighters wear a number of different types of protective clothing. Basic turnout or bunker gear (Fig. 8.6) is composed of a fire-retardant outer shell and an insulating inner liner. It is intended to provide limited protection from heat, hot water, and some particles and debris. It does not provide protection from gases or vapors. Exposure to hazardous materials will likely result in chemicals permeating into the fabric. Decontamination of firefighter's turnout gear is difficult at best. It is not designed or intended to provide chemical protection.

Chemical protective clothing is very flammable. In hazardous-materials emergency response operations, a flash cover may be used for entry in situations where flammables may be involved. Flash covers are made from a flame-retardant material with an aluminized exterior and worn over CPC garments (see Fig. 8.7). They are intended to provide brief protection to the responder in the event of a flash fire to allow for a rapid escape.

Flash covers are of limited value for confined space entry. They are intended only to provide limited protection to allow a rapid escape from the hazard area; however, a rapid escape from many confined spaces is not feasible. In addition, the enclosed nature of many spaces is likely to intensify the heat and pressure of a fire, which can quickly overwhelm any protection the cover provides. Flash covers add a significant bulk and weight to the wearer. All of this emphasizes the point that personnel should not enter confined spaces with actual or potential flammable atmospheres.

Ambient-temperature extremes. Protective clothing may be needed to provide protection to personnel entering extremely hot or cold environments. These hazards are discussed in Chap. 3. Cooling garments may hold frozen cooling packets close to the body or circulate chilled water or air through tubing over the body. New high-tech garments use a cooling material that is not as cold as ice. All types provide brief cooling for work in hot, confined spaces; however, for prolonged work in a hot space, ventilating the space with cool air may provide better protection from heat injury.

Clothing can provide protection to entrants in cold spaces. Entrants should wear layers of insulating clothing that will wick moisture away from the skin. Outer shells that

Figure 8.6 Firefighter turnout gear insulates the wearer against heat and steam, but offers no chemical protection.

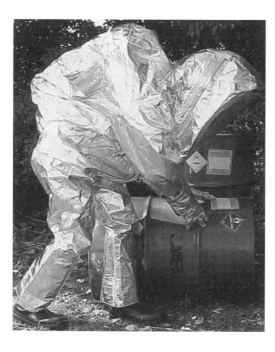

Figure 8.7 Flash cover suits offer limited protection from flash fire.

block the wind and keep moisture out are also helpful. Prolonged contact with cold metal surfaces will result in significant heat loss by conduction, so clothing barriers that prevent such contact should be worn.

Radiation.   Radiation is the spontaneous emission of energy from a source. The energy is given off as either a particle (alpha or beta) or a wave (gamma or X ray). Radiation cannot be detected by any of the human senses. Radioactive materials are tightly controlled by government agencies that require extensive training for anyone who works with those materials. Confined space entrants working with radioactive sources must receive adequate and appropriate training to do their work safely.

Protective clothing worn for radiation work is designed to prevent contamination of the worker by radioactive alpha or beta particles. Clothing provides no protection from gamma radiation. If radiation is known or suspected to be in a space, entry should not be made until the radiation hazard is evaluated by the appropriate radiation expert.

Safe Use of CPC

Once protective clothing has been selected, entrants can ensure that they get the highest protection by using it properly. Personal protective equipment (PPE) programs and levels will be discussed in Chap. 10 as the means of organizing the selection and use of all PPE. The following are some basic actions related to the use of CPC that will enhance its protective capabilities.

Inspection

Entrants must fully inspect CPC before it is used. This applies to all CPC, even new clothing straight out of the package. Manufacturers use quality control procedures to try to catch defective products before they leave the factory; nonetheless, some defective garments do make it to the user and the consequences of using such equipment can be significant. The inspection by the user is the last defense against exposure through defective or improperly selected garments.

Each article of clothing should be inspected according to the following procedures:

- Determine that the proper clothing material and type have been selected for the hazard and the task to be done. Users should check the permit to confirm that the CPC issued to the entrant will protect against the hazard designated on the permit.

- Visually inspect the garment for imperfect seams, thin spots in the material, tears, and broken or defective zippers and closures.

- Hold the garment up to a light to inspect for pinholes or other defects.

- Check the flexibility and texture of the fabric for signs of deterioration during storage.

- Look for staining, changes in flexibility or texture, or other signs of permeation or degradation in reusable items.

For TECP suits, additional inspection procedures apply:

- Check for the presence and proper operation of pressure-relief valves.

- Check that gloves and boots are attached securely.

- Check the faceshield for cracking, scratching, or other signs of wear or potential leakage.

- TECP suits should be pressure-tested according to the manufacturer's guidelines. Two test procedures described in Appendix A of the *OSHA Hazardous Waste Operations and Emergency Response* standard (29 CFR 1910.120) serve as a reference.

Donning CPC

Entrants should don all PPE according to established procedures. Consistency and a reduced likelihood of forgetting something are two benefits of using such procedures. Donning assistants help the entrant put on the CPC and can help check for proper fit. Fold excess material in the legs under into downward-pointing cuffs that prevent the splashed product from pooling in folds. Tape outer gloves or boots to the suit to prevent the splashed product from draining into them.

In-use monitoring

CPC users should always be alert to signs that the garment has been compromised. In particular, they should look for observable changes in the clothing that indicate degradation or permeation, such as

- Discoloration
- Change in texture
- Becoming more or less flexible
- Swelling
- Cracking
- "Waffling" of the material

If any of these indications are observed, the entrant exits the space and removes the clothing immediately. The CPC selection must be reevaluated before any additional entries are made.

In addition to signs of degradation, the entrant should be alert to symptoms of exposure. Before entering, entrants should be fully briefed on the possible symptoms of exposure to the chemicals in the space and told to exit the space immediately on experiencing any of them. Possible symptoms include

- Irritation of skin

- Tingling of skin or extremities

- Difficulty breathing

- Headaches

- Fatigue

- Nausea

Symptoms should be considered warnings of failure of the protective equipment and should be investigated immediately.

CPC change schedules

As discussed above, entrants must remember that CPC does not keep out a chemical forever. Prior to entry, the entry supervisor should establish a schedule for changing CPC to prevent exposure by permeation. The schedule will depend on the potential for physical damage by the work, the breakthrough time of the chemicals, the environmental conditions of the space, and some appropriate safety factor. If more than one chemical is known to be in the space, the change schedule should be based on the chemical with the shortest breakthrough time for the material selected.

Reducing permeation

Entrants should avoid any unnecessary contact with the chemicals. Any contact of the liquid chemical on the CPC material increases the permeation of the chemical. Avoid kneeling or walking in pooled liquid chemical, and stay out of spraying or splashing. Smooth out all folds in the garment that may trap chemicals. Actions taken to avoid exposure result in enhanced chemical protection by the CPC.

Care and Maintenance of CPC

Proper care and maintenance of the CPC will enhance its protection and ensure that it is ready for use. This includes decontamination, maintenance, and storage, all of which require planning.

Decontamination

Chemical protective clothing that is worn into a place where chemicals are present must be either decontaminated or disposed of after use (Fig. 8.8). Decontamination serves two important purposes:

- Removing chemical from the clothing that would otherwise expose either the wearer as he takes off the CPC or future users of reusable garments

- Preventing the spread of the chemical to areas outside the space

Decontamination procedures provide a safe way to remove the chemical from the CPC as a part of the doffing process. Even disposable items must be decontaminated prior to removal to prevent exposure. The decontamination process can be as large or small as it takes to clean the CPC without exposing the wearer or decontamination assistants.

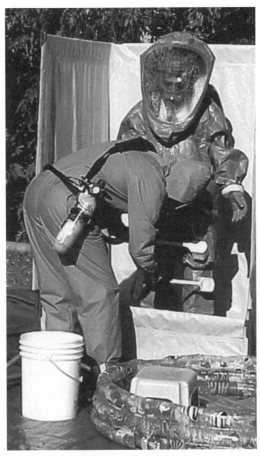

Figure 8.8 Contaminated protective clothing must be decontaminated.

Reuse of CPC

Items of CPC must be completely decontaminated prior to reuse; otherwise, the items cannot be considered safe to use. In some instances, contaminants may permeate the CPC material and be difficult or impossible to remove. Such contaminants may continue to diffuse through the CPC material toward the inner surface during storage, posing the threat of direct skin contact to the next wearer. If permeation is suspected, hang the article of CPC in a warm, well-ventilated place to allow the item to release the permeated contaminant.

Be very careful when reusing CPC to be sure that previous use has not made the garment dangerous to wear. Remember that permeation and degradation might occur without any visible indications. Any chemical that permeates into the fabric of the clothing shortens the breakthrough time for future use.

Maintenance

There is little maintenance involved with CPC except for reusable TECP suits. These garments may require the repair or replacement of components such as pressure-relief valves, boots, gloves, or faceshields. These repairs should be performed by the manufacturer or individuals trained by them.

Storage

Store CPC properly to ensure that it is not damaged by dust, moisture, sunlight, temperature extremes, damaging chemicals, or physical abrasion. Follow the manufacturer's instructions. If it is stored for extended periods before use, inspect it regularly.

Store similar types of clothing that are made with different materials in separate areas with proper labeling to avoid confusion. If there is a shelf life for the clothing, clearly mark the shelf life on packages or storage areas.

Summary

Entrants must be protected from contact with the chemicals during a confined space entry and rescue. Chemical protective clothing provides an effective barrier when it is properly selected and used. Given the possible consequences of failure of the clothing, users must take the proper selection and use of CPC seriously.

Personal Safety Equipment

Confined spaces frequently contain safety hazards that are intensified by the tight spaces and dark conditions. Any of the physical hazards described in Chap. 3 can cause injuries to entrants who are not suitably protected. As mentioned throughout this book, OSHA regulations require employers to control workplace hazards by means of engineering controls when feasible. This applies to safety hazards as well; however, many safety hazards cannot be engineered out of the workplace completely. That is why some personal safety equipment is required.

This chapter provides basic information about specific safety hazards and equipment that protects confined space entrants. It is grouped by the types of protective equipment commonly used, including

- Head protection

- Face and eye protection

- Hearing protection

- Hand protection

- Foot protection

Within each section, the basic hazards are reviewed and the characteristics of the available safety equipment are described. The relevant standards, both consensus and regulatory, are summarized. In some cases, the regulatory standard may include by reference a version of the consensus standard that has since been revised. The current version of the consensus standard is discussed.

Head Protection

The tight, close quarters of confined space operations make the possibility of head injuries high. Entrants pass through tight spaces where heads bump. They may be working at different levels within the same space, so dropped objects are more likely to hit someone as they fall. Because by definition confined spaces are not intended for human occupancy, electrical hazards and hot surfaces may be exposed. As workers squeeze through spaces or move through dark areas, their heads may contact these hazards. Head protection is a vital piece of safety equipment for all confined space entrants (Fig. 9.1).

Standards and regulations

Head protection for industrial applications is governed by two principal standards: OSHA's *Head Protection Standard* (29 CFR 1910.135) and ANSI's *American National Standard for Industrial Head Protection* (Z89.1-1997). Firefighter head

Figure 9.1 Head protective helmets protect the head from impact and penetration by falling objects.

protection is governed by NFPA 1971: *Standard on Protective Ensemble for Structural Fire Fighting*. These standards set minimum performance characteristics for the helmets.

OSHA standard. OSHA's head protection standard has two primary requirements:

- Employers must ensure that each affected employee wears a protective helmet when working in areas where there is a potential for injury to the head from falling objects.

- Employers must ensure that a protective helmet designed to reduce electrical shock hazard be worn by each affected employee when near exposed electrical conductors which could contact the head.

To meet these requirements, OSHA requires that employers use helmets that meet the criteria of ANSI Z89.1-1986 for all helmets purchased after July 1994. No other requirements specific to head protection are given. The general requirements of 29 CFR 1910.132, which are described in detail in Chap. 10, apply as well.

ANSI standard. ANSI Z89.1-1997 (ANSI 1997) establishes the minimum performance characteristics of protective helmets. It describes the types and classes of helmets and their required components as well as what accessories are allowed. The standard describes two types of helmets with regard to the type of impact:

- Type I helmets are designed to reduce the force of impact resulting from a blow to the top of the head only.

- Type II helmets reduce the force of impact that may be received off center or to the top of the head.

The standard also classifies helmets according to their protection from electrical contact:

- Class C (conductive) helmets are not intended to provide protection against contact with electrical conductors.

- Class G (general) helmets reduce the danger of contact exposure to low-voltage conductors.

- Class E (electrical) helmets reduce the danger of exposure to high-voltage conductors.

The standard describes specific test procedures for assessing the performance of helmets against the standard. The rigorous testing involves preconditioning the helmets in heat (120°F), cold (0°F), and water for 2 hours prior to testing. The specific tests performed on the helmets are

- Flammability

- Force transmission to the head

- Apex penetration

- Impact energy attenuation

- Off-center penetration

- Chin strap retention (type II helmets only)

- Electrical insulation

Helmets complying with the standard are labeled according to the type and class with which they comply, such as type I, class C or type II, class E.

NFPA requirements. NFPA 1971 provides the same type of performance standard for structural firefighter ensembles, including the helmet. Given the demanding environments in which firefighters work, the testing procedures and performance requirements for firefighter helmets are more rigorous than industrial helmets. Firefighter helmets meeting these rigorous requirements are labeled as NFPA 1971–compliant.

Head protection types

Helmets. Helmets (commonly called "hardhats") provide protection to the head from impact and penetration of flying objects. Their design and construction may vary for different applications (see Fig. 9.2); however, their basic construction includes these components:

- Outer shell

- Adjustable headband

- Suspension system

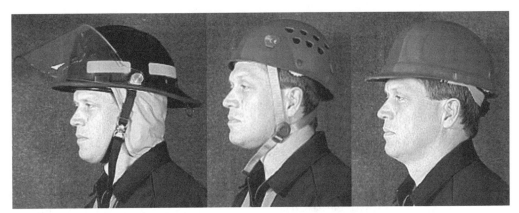

Figure 9.2 Protective helmets provide different coverage depending on their intended use.

The outer shell provides a hard barrier protection against penetration by flying objects. It typically is made of rigid materials such as high-density polyethylene, polycarbonate, and acrylonitrile butadiene styrene (ABS) plastic. These materials may or may not conduct electricity depending on their intended use and classification. Entrants entering spaces where electrical conductors may be exposed must be sure that the hardhat is rated for the appropriate class (G or E).

The distribution of force to the head from impact of objects is achieved through the suspension system. The straps of the suspension system distribute the impact force over a larger area of the head and absorb some of the energy. The suspension system must maintain a minimum clearance specified by the manufacturer between the shell and the head during normal use. Padding may be added to the interior of the shell to enhance impact protection as long as the helmet continues to meet the requirements of ANSI Z89.1.

The headband must be adjustable in size and designed to keep the helmet in the proper position on the wearer's head. Sizing must be consistent with normal hat sizes. The headband may be covered with terrycloth or other material to absorb sweat.

Bump caps.   "Bump caps" are caps or hats that do not meet the ANSI specifications for protection from impact or penetration by flying objects or electrical hazards. They provide protection for the head from bumps when entrants work in tight, low spaces. They should not be selected for work in spaces where entrants are working at different levels, where tools and objects may fall from above, or at other times when substantial head protection is required.

Hoods and soft caps.   Hoods or soft caps made of special fabrics may be used to protect the head and neck from airborne hazards such as sparks, dust and debris, splashes, or temperature extremes. This headwear may be worn by itself or as an accessory to other primary head protection. An example is the hardhat liner worn in cold environments.

Head protection in confined spaces

Work in confined spaces presents many opportunities for head injury because of the tight areas and passageways. Spaces may have low ceilings or other obstructions that may be difficult to see in low lighting. Confined space entrants should always wear head protection.

Some entrants and rescuers find hardhats with brims difficult to wear in tight spaces. Some helmet manufacturers have helmet models with no brim, allowing the head to move around more in a tight space without bumping into walls and ceilings. These helmets still should meet the impact and penetration requirements of the ANSI standard, since the brim is not a required element of the head protective system.

Proper use of head protection

Like other protective equipment, hardhats and other head protection works only if they are selected and worn properly.

Selection. Select head protection on the basis of the possible hazards to the head that may be present in the confined space. Confined space operations may involve people working in close proximity to each other, so the possibility of side impacts from nearby workers is greater. Use ANSI type II helmets that provide limited protection from side impacts as well as blows to the top of the head. Consider the potential exposure to electrical hazards when selecting the shell material.

Size and positioning. Adjust helmet suspension systems to provide a secure fit that is comfortable for the wearer. Position the helmet according to the manufacturer's instructions. Helmets that are improperly positioned may leave areas of the head exposed or may not properly distribute the force of impact.

Some wearers choose to put their helmets on backward to avoid the problem of the brim. If this is done, reverse the suspension system so that the head is properly cradled. Consult the manufacturer to see if this affects the ANSI rating of the helmet.

Cleaning. Clean the helmet periodically as a matter of good hygiene. Wash the shell in a mild detergent and warm water according to the manufacturer's guidelines. Strong cleaners or solvents that remove grease may chemically affect the shell material. If there is any question about the compatibility of a cleaner with the helmet, contact the helmet manufacturer to confirm the cleaner's suitability for use.

Inspection. Inspect head protection before each use to look for signs of wear, aging, or damage. Discard or repair helmets with cracking, discoloration, or texture changes of the shell or torn or broken suspension system components. Replace any helmet that sustains a substantial blow or impact, since imperceptible damage may have been done.

There is no required replacement time for head protection. Rely on inspection to indicate the need for replacement. Hardhats worn regularly in harsh or extreme environments or in direct sunlight may need to be replaced more frequently because of the damage that may be done to the shell.

Head protection needs little repair or maintenance work. Where broken parts are replaced, use approved parts from the manufacturer.

Storage. Store hardhats in clean, dry places. Do not allow them to be subjected to chemical exposure, excess moisture, or physical abrasion. They definitely should not be stored in the rear window of a truck or car or any other location of direct sunlight.

Eye and Face Protection

The face and eyes are among the most sensitive exposed areas of the body. While the skin can tolerate contact with dust, dirt, and debris without significant problem, foreign bodies in the eye can cause damage or can create safety hazards by rendering the entrant unable to see well. Add to that the increased vulnerability to chemical exposure, and one can see that it is important to provide protection of these areas.

Standards

Both OSHA and ANSI have written standards for eye and face protection. OSHA standards describe required employee protection, and reference ANSI standards for the performance of protective equipment.

OSHA. OSHA's *Eye and Face Protection Standard* (29 CFR 1910.133) regulates the protection of the face and eyes. The standard requires the employer to ensure that each affected employee uses face or eye protection when exposed to

- Flying particles
- Molten metal
- Liquid chemicals
- Acids or caustic liquids
- Chemical gases and vapors
- Potentially injurious light radiation

The standard also requires that eye protection include side protection when there is a hazard from flying objects. Any face and eye protection must be marked distinctly to facilitate identification of the manufacturer. OSHA requires that all face and eye protection must comply with the ANSI standard discussed below.

For users who require prescription lenses, the eye and face protection must incorporate the lens in its design or must fit over the prescription glasses without disrupting the proper fit of either the prescription glasses or the protective lenses.

OSHA lists the required level of shading for protection against radiant energy during operations such as welding and cutting. Other requirements for eye protection during welding operations are given in 29 CFR 1910.252, the general requirements section of OSHA's Subpart Q regulations for welding, cutting, and brazing. These include

- Proper selection of eye protection based on the hazard and type of work
- Helmets, shield, and goggles not readily flammable
- Goggles vented to prevent fogging
- Lens shade properly selected with the number marked on the lens

Requirements for other protective equipment and practices related to welding are given in the standard.

ANSI. ANSI Z87.1-1989, *American National Standard Practice for Occupational and Educational Eye and Face Protection* sets the standards for performance of face and eye protection and describes the appropriate testing procedures to evaluate the protective products. The standard provides a helpful description of protective devices, which they classify as

- Spectacles
- Goggles
- Faceshields
- Welding helmets

For each classification, the standard provides testing procedures and the required standard of performance in each of these categories:

- Impact resistance
- Optical (vision) requirements
- Flammability resistance
- Corrosion resistance
- Cleanability

Specific discussion of performance requirements of each of these categories for each classification of protection is beyond the scope of this book. Employers purchasing eye and face protection labeled as compliant with ANSI Z87.1-1989 can have confidence that the equipment will provide good protection and usability in most normal applications in confined space operations.

Two potentially important concerns are not addressed by the standard. One concern is that the standard specifically excludes protection from hazards related to

- X rays
- Gamma rays
- High-energy particulate radiations
- Microwaves
- Radiofrequency radiation
- Lasers
- Masers
- Sports and recreation

Use of protective equipment in these or other applications outside of the standard's scope can result in serious injury.

The other concern not addressed in the standard is electrical hazard. Spectacles, goggles, and faceshields may be constructed of materials that will conduct electrical current if brought in contact with conductors. Currently there is not much mention of this hazard in manufacturer and vendor information about face and eye protection products. If the entry operations may involve potential contact with electrical hazards, employers should contact the manufacturer of the protective equipment for further information.

Types of face and eye protection

There are a wide variety of styles and types of eye and face protection available for use today (see Fig. 9.3). When eye and face hazards are present in the space, supervisors should identify the hazards and then offer the entrants a choice of protection so that the most comfortable protection can be selected. The general types are discussed below.

Spectacles. Spectacles protect primarily the immediate area of the eyes. They are made up of lenses joined by a bridge over the nose and supported by temple bars. The lens may be flat with separate sideshields or curved around the side of the eye. Spectacles are now sometimes styled like popular sunglasses. New materials and designs make spectacles much more comfortable than the old, stiff safety glasses of the past. This is evidenced by the fact that many employees are choosing to wear tinted safety glasses as sunglasses.

Spectacles do not fit tightly against the face, so they offer limited splash protection and no protection from gases and vapors. They are intended primarily for impact protection. ANSI-compliant industrial safety glasses have lenses designed to withstand much greater impact force than common sunglasses or even recreational eye protection. Employers must ensure that only approved eye protection is worn by employees in eye hazard areas.

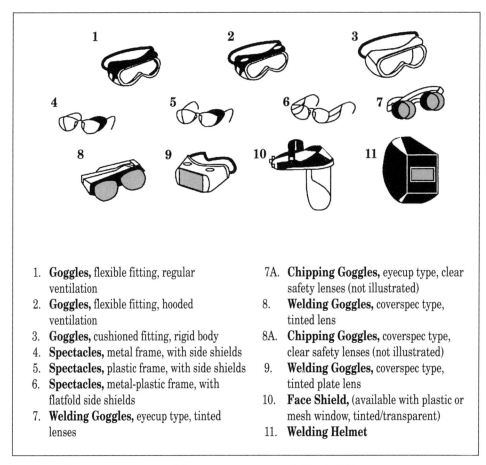

1. **Goggles,** flexible fitting, regular ventilation
2. **Goggles,** flexible fitting, hooded ventilation
3. **Goggles,** cushioned fitting, rigid body
4. **Spectacles,** metal frame, with side shields
5. **Spectacles,** plastic frame, with side shields
6. **Spectacles,** metal-plastic frame, with flatfold side shields
7. **Welding Goggles,** eyecup type, tinted lenses
7A. **Chipping Goggles,** eyecup type, clear safety lenses (not illustrated)
8. **Welding Goggles,** coverspec type, tinted lens
8A. **Chipping Goggles,** coverspec type, clear safety lenses (not illustrated)
9. **Welding Goggles,** coverspec type, tinted plate lens
10. **Face Shield,** (available with plastic or mesh window, tinted/transparent)
11. **Welding Helmet**

Figure 9.3 Various types of eye and face protection are available. (OSHA 2000)

Goggles.   Goggles are protective eyewear that fit against the face. They may be used to cover goggles that have a single lens that covers the face around the eyes or eyecup goggles that have two lenses joined by a bridge that cover each eye separately. Goggles can be worn over prescription glasses to protect them from damage.

Because they fit tightly against the face, goggles provide excellent protection against splash, dust, and debris as well as frontal impact. They provide limited protection from gases and vapors if they are unventilated. Ventilated goggles allow airflow behind the lens to reduce fogging.

Faceshields.   Faceshields provide a transparent vertical barrier in front of and around the wearer's face. The shield may rest on its own suspension system or attach to the brim of a helmet. Shields come in a variety of shapes to accommodate other equipment being worn. Chin and neck protection can be added to the shield as well.

Faceshields protect the face, or portions thereof, against major splashing and some flying objects. Faceshields are not primary protectors, but should always be worn with eye protection such as spectacles or goggles. For additional protection when liquids are being poured or when flying objects may injure the face, faceshields provide wide coverage with minimal discomfort to the wearer.

Welding shields.   Welding shields are specialized faceshields that protect welders from impact and from direct radiant energy from certain welding operations. They may be suspended from a hardhat, supported by their own suspension system, or handheld. According to OSHA's welding standard, the helmet must be positioned to protect the head, neck, and ears from direct radiant energy. They must have filter plates (shaded appropriately) and protective coverplates.

Special features of eye protectors

Many manufacturers now apply special coatings to their lenses that enhance resistance to scratching and fogging. Some coatings absorb almost all of the damaging ultraviolet (UV) light passing through the lens. These coatings improve the useful life of the lenses. Many spectacles have replaceable lenses to reduce cost.

Many spectacles have adjustable components for achieving the best size and fit. These adjustments allow the glasses to sit comfortably in the proper position on the wearer's face. Improved comfort and attractive styling have contributed greatly to workers' acceptance and use of proper eye protection.

Proper use of face and eye protection

Follow these guidelines to enhance the protection of the face and eyes:

- Select the proper type of protector for the hazards present in the space. Use only ANSI-compliant protective devices.
- Make the necessary adjustments according to the manufacturer's instructions to achieve the best and most comfortable fit.
- Always wear the protection in the hazard area.
- Periodically wash the protective devices according to the manufacturer's guidelines.
- Inspect the devices before and after each use.
- Immediately replace any devices that receive a substantial impact or chemical exposure, even if they appear undamaged.

Hearing Protection

Exposure to excessive noise levels causes hearing loss, inability to communicate or hear warnings, and mental stress that can disrupt sleep patterns. *Noise* is created by pressure oscillations (sound waves) transmitted through a medium such as air to a receiver. The sound waves have two key characteristics:

- Pressure level (perceived as volume)
- Frequency (perceived as pitch)

Sound pressure levels are intensified as the sound waves reverberate off rigid surfaces. Since many confined spaces are constructed out of metal walls (tanks, reaction vessels, etc.), noise levels are amplified by the reverberation. Work in confined spaces can be very loud.

Humans receive and process sound through the ear. Hearing loss can result from sudden traumatic injury to the ear or from prolonged exposure to high noise levels that slowly damage the ear. While some hearing loss is temporary, many workers permanently lose their ability to hear because of occupational noise exposure. Reducing or eliminating noise exposure is the surest way to protect an entrant's hearing.

OSHA's *Hearing Conservation Standard*

The noise hazard is described in Chap. 3. A detailed discussion of the requirements of OSHA's *Hearing Conservation Standard* (29 CFR 1910.95) is beyond the scope of this chapter. What follows is a brief discussion of personal hearing protection.

The standard requires employers to control their employees' noise exposures to less than 90 dBA as an 8-hour time-weighted average. The decibel (dB) is the unit of measuring sound pressure, and the A-weighting scale measures sound at all frequencies much in the same way that the human ear hears. The C-weighting scale gives a closer measure of the actual sound pressure level at all frequencies. OSHA's standard is based on A-weighted measurements (dBA), although C-weighted measurements may be needed for application to the noise reduction ratings of hearing protective devices as described below.

The OSHA standard specifically requires employers to control noise exposure by means of engineering controls whenever feasible. Such controls might include isolation of the source or the worker in an enclosure, substitution of noise-producing equipment, mounting vibrating equipment on vibration-reducing pads, and many other potential solutions. The temporary nature of work in confined spaces makes expensive engineering controls impractical; therefore, confined space operations frequently require the use of personal hearing protective devices (Fig. 9.4).

Types of personal hearing protective devices

Earplugs insert into the ear canal and fit tightly against the walls. They are made of sound-absorbing foams or other moldable materials. The wearer rolls the earplug into a small cone, inserts it deep into the ear canal, and lets it expand to fill the canal. If a good fit and insertion are achieved, the plug blocks the passage of the sound waves into the ear.

Semi-inserts or hearing bands have a soft plug-like inserts that block off the external opening of the ear canals. The inserts are held in place by a lightweight headband. These devices do not provide as much noise reduction as inserts, but they are very easy to put on and take off. They are very useful for intermittent exposures to lower noise levels.

Earmuffs have padded cup-shaped enclosures that completely cover the outer ear. The enclosures are lined with sound-absorbing material and have a cushioned seal that fits tightly against the head of the wearer. Since earmuffs rely on a good seal to prevent sound waves from entering the ear, any breaks in the seal, such as from eyeglass temple bars, reduce the effectiveness of the hearing protection. Many hardhats have slots on either side that will allow the attachment of earmuffs to the hat.

Some particularly noisy environments may require the use of a combination of earplugs and earmuffs. This combination enhances the protection, but does not double

Figure 9.4 Hearing protective devices such as these earplugs, semi-inserts, and earmuffs block noise from entering the ear.

the noise reduction as one might think. The effectiveness of such a combination of protective devices should be evaluated by a trained occupational noise professional.

Noise reduction rating

A noise reduction rating (NRR) system has been developed by the EPA to report the noise-reducing capability of hearing protectors. Hearing protector manufacturers determine the level of noise reduction their products provide (expressed in decibels) according to procedures outlined by the EPA system. They are required to mark the NRR value on the packaging of the devices (Fig. 9.5).

Users can subtract the NRR from the sound level in the space to estimate the noise exposure they can expect. The NRR correlates best with C-weighted sound measurements, whereas most sound survey measurements are made with the A-weighting scale. Appendix B of OSHA's *Hearing Conservation Standard* describes the required procedures for using the NRR to determine the adequacy of hearing protection. The procedures take into account the appropriate correction factors when A-weighted or C-weighted measurements are used.

Use of hearing protectors

Select hearing protectors that will provide adequate noise reduction to bring the entrant's noise exposure to less than 90 dBA. This requires that properly trained professionals assess the noise level during confined space operations. Once the level is determined, trained individuals can use the NRR to estimate the noise exposure to the entrant.

Replace earplugs after each use or at least when they become unhygienic or unable to be inserted properly. Entrants must have clean hands when inserting the plugs to prevent contamination that can lead to infection of the ear canal. Insert earplugs according

Figure 9.5 The noise-reduction rating can be used to determine if entrants are properly protected from the noise level in the space.

to the manufacturer's instructions, but not in a way that will cause them to become lodged in or damage the ear.

Clean earmuffs periodically according to the manufacturer's instructions. Replace the sealing cushion whenever it becomes compressed or will no longer form a seal. Earmuffs with nonconductive headbands may be required where potential contact with electrical conductors exists.

Hand Protection

The hands are vulnerable to many different hazards. The OSHA *Hand Protection Standard* (29 CFR 1910.138) requires the employer to select, and requires employees to use, appropriate hand protection when they might be exposed to any of these hazards:

- Chemical exposure
- Severe cuts or lacerations
- Severe abrasions
- Punctures
- Thermal burns
- Cold exposure

The standard's only other requirement is that the employer base the selection on the performance characteristics of the hand protection relative to the tasks to be performed, the conditions present, the duration of use, and the hazards identified.

Types of protective gloves

A number of glove types are available to provide protection from the hazards (Fig. 9.6). While they are discussed separately, some glove designs provide overlapping or multiple protections.

Chemical protective gloves.   Select and use gloves that provide protection of the hands from contamination and exposure to hazardous chemicals according to the guidelines described for chemical protective clothing (CPC) in Chap. 8. As hands are the parts of the body most likely to contact chemicals, it is very important that gloves be properly

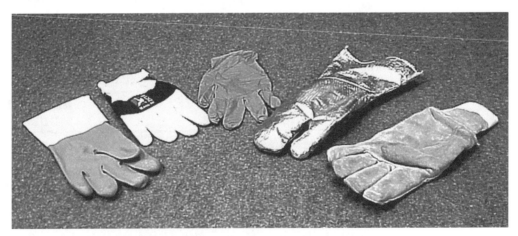

Figure 9.6 Gloves can protect the hand from cuts, burns, abrasion, chemicals, and other hazards.

selected based on the chemicals present and the requirements of the tasks to be performed. Remember that the selection of the proper glove material requires that the identity of all hazardous chemicals in the space be known.

In some situations where multiple chemicals have been identified, chemical protection may be enhanced by layering gloves of two different materials. Each material provides protection from some chemicals, and together the range of chemicals blocked is increased. This works best where the tasks to be performed do not require fine-finger dexterity.

As discussed in Chap. 8, the user must be allowed to change gloves on a schedule that will prevent permeation and exposure. The required frequency is determined by the chemical(s) present, the physical demands of the work, and the likelihood of direct contact with the chemical. See Chap. 8 for more discussion of the selection and use of CPC.

Electrical insulating gloves.   Some gloves place a barrier to prevent the flow of electricity from an electrical conductor through the hand. OSHA's Electrical Protective Equipment Standard (29 CFR 1910.137) provides detailed requirements for the design, testing, and use of gloves and other electrical protective equipment.

Gloves are classified as low-voltage and high-voltage insulating gloves based on the minimum test voltage that they will block. Low-voltage insulating gloves are classified as class 00 for use on circuits up to 500 volts ac and class 0 for use on circuits up to 1000 volts ac. A leather protector glove must be worn over insulating gloves used on voltages higher than 250 volts ac. The protector glove receives the brunt of physical damage, allowing the rubber insulting gloves to remain intact and function as the primary electrical barrier.

Cut-resistant gloves.   Some tasks involve work with knives or sharp metal edges. Gloves made of strong, cut-resistant materials protect the hands. While some gloves are constructed out of stainless-steel mesh, many gloves are made out of strong fabrics such as Du Pont's Kevlar® that are woven with stainless-steel threads. Cut-resistant gloves protect against minor cuts or lacerations, but effective engineering controls such as machine guarding are the best protection against major cuts or amputations.

Temperature-resistant gloves.   Gloves can insulate the hand from contact with extremely hot or cold objects. Asbestos gloves are no longer used in the United States; however, gloves made of materials such as Kevlar® and Zetex®, a new fabric made of silica, provide good heat resistance. Thick insulation aids in both heat and cold resistance. Gloves for use in cold environments should prevent water contact, which causes rapid heat loss.

Cloth and leather gloves.   Simple leather gloves provide excellent protection against abrasion and physical exertion. Leather is useful in keeping hands warm in cold environments. Cotton and polyester cloth gloves can protect against abrasion, exertion, and other minor hazards.

Rescue gloves.   Rescue and rope-handling gloves are thin and tight-fitting with reinforced padding across the palms. They protect the wearer from rope burns and abrasions while providing the high dexterity needed for rescue operations.

Use of gloves

Select gloves based on a task and hazard analysis. The glove types selected must strike the right balance between durability and dexterity. Thin gloves that fit tightly to the hand generally provide excellent dexterity and sense of touch. Thick, loosely fitting gloves are needed for more demanding physical tasks and greater chemical resistance.

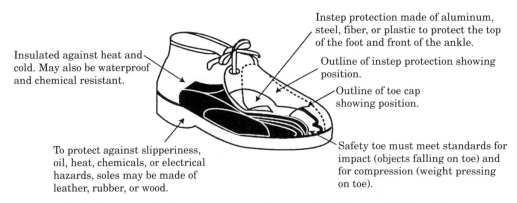

Instep protection made of aluminum, steel, fiber, or plastic to protect the top of the foot and front of the ankle.

Outline of instep protection showing position.

Insulated against heat and cold. May also be waterproof and chemical resistant.

Outline of toe cap showing position.

To protect against slipperiness, oil, heat, chemicals, or electrical hazards, soles may be made of leather, rubber, or wood.

Safety toe must meet standards for impact (objects falling on toe) and for compression (weight pressing on toe).

Figure 9.7 Safety footwear is designed to protect against a variety of hazards. (OSHA 2000)

Gloves should fit properly. Keep them clean and in good repair. Inspect them each time they are used, and replace them as needed following chemical permeation or degradation or physical damage.

Foot Protection

The feet can be injured by a number of hazards that may be found in confined spaces, such as falling objects, compression by heavy objects, and punctures. Properly selected safety shoes can prevent or reduce many of these injuries (Fig. 9.7).

Standards

Once again, the two primary standards regarding foot protection in industry come from OSHA and ANSI.

OSHA *Foot Protection Standard.* OSHA's *Foot Protection Standard* (29 CFR 1910.136) simply requires foot protection be worn by employees where foot injuries may result from rolling or falling objects, objects piercing the sole, or electrical hazards. It requires that all footwear comply with ANSI Z41-1991. The current revision of the ANSI standard was issued in 1999.

ANSI. ANSI Z41-1999, *American National Standard for Personal Protection— Protective Footwear,* provides the testing and performance requirements for industrial foot protection. The standard has requirements for classifying all footwear according to the minimum compression and impact protection at the toes. These classifications are based on the minimum force that the shoes will withstand without causing damage to the foot and are intended for impact protection:

- I/75: 75 ft·lb
- I/50: 50 ft·lb
- I/30: 30 ft·lb

For compression, the classifications and values are

- C/75: 2500 lb
- C/50: 1750 lb
- C/30: 1000 lb

The other sections of the standard address the following as they apply to shoes being tested:

- Metatarsal protection
- Conductive and static dissipative footwear to prevent static electricity buildup and discharge
- Electrical hazard protection
- Puncture resistance

ANSI-compliant footwear is labeled according to a specific format to communicate the relevant ratings. An example of the label might be:

ANSI Z41 PT 99

M I/75 C/75

Mt/75 EH

PR

This label indicates that this male (M) shoe was tested according to the protective (PT) section of ANSI Z41-1999. It received the highest ratings for impact (I/75), compression (C/75), and metatarsal (M/75) support. It also passed tests for electrical hazards (EH) and puncture resistance (PR) as dictated in the standard. This label is required to be stamped or sewn on the inside of the shoe and provides useful information to the wearer.

Features of protective footwear

Steel-toe shoes protect the toes of the wearer from injury by impact or compression by heavy objects. Some shoes also provide this protection for the metatarsal area of the foot as well. A steel cup in the front of the shoe provides this support. Puncture-resistant footwear includes a steel plate in the insole that prevents sharp objects from penetrating the sole and injuring the foot.

Electrical protective footwear is designed to prevent the flow of electricity through the shoe in the event of contact with an electrical conductor. The sole and heel of the shoe must be made of nonconductive material. The electrical hazard protection of the shoes deteriorates in wet environments and when there is excessive wear on the sole and heel.

Conductive and static dissipative shoes are designed to prevent the buildup of static electricity in the body. The soles of the shoes conduct any static electricity to the ground, where it dissipates. This footwear may be needed in flammable atmospheres where static discharge could ignite a fire and in work around electronics and other equipment that is sensitive to a static electricity discharge.

Rubber boots keep feet dry in wet and muddy environments. Some footwear includes design features to increase traction or reduce slipping on slippery floors. The shoes may be marked as oil-resistant if they do not absorb oils and become slippery. If rubber boots are selected for protection against chemical hazards, they must be properly selected according to the principles discussed in Chap. 8.

Summary

The hazards in confined spaces can cause injury literally anywhere from the head to the toe of an entrant. Employers must try to control these hazards by engineering controls and safe work procedures whenever feasible. Personal safety equipment provides extra protection that can prevent injury and, perhaps, save the life of the entrant. That's why confined space entrants should always select and use the proper equipment to protect against all hazards in the space.

Safe Use of Personal Protective Equipment

Entrants into a confined space must use personal protective equipment (PPE) only if there are no feasible engineering controls that can adequately eliminate or control the hazards present. Controls like those discussed in Chap. 6 should be the employer's first choice in protecting the entrant. The successful use of engineering controls such as ventilation removes chemical hazards before they become exposures to the worker. Prevention of exposure is always preferable to simply trying to protect workers that are being exposed.

In many instances, engineering controls alone cannot provide adequate protection. PPE, when used properly, reduces the level of exposure to the entrant and can make the difference for safety during the confined space operation. While they should never be the only form of protection used, PPE such as respirators, chemical protective clothing (CPC), and personal safety equipment are important components of the total protective measures.

The safe and proper use of PPE requires planning and an effective program. Among the personal protective equipment regulations found in 29 CFR 1910 Subpart I, only the respiratory protection standard requires a formally written program. Nonetheless, any employer that requires its employees to use PPE should have a program to organize and document its effective use.

This chapter discusses the general requirements for PPE use that are found in the OSHA standards. It looks at what a comprehensive PPE program should include. It also describes combining different protective equipment into levels of protection that provide adequate chemical protection without placing an undue burden on the wearer.

OSHA's PPE Standards

The preceding chapters each look at the relevant OSHA standards for specific types of PPE; however, OSHA precedes all of these standards in Subpart I with the *General Requirements Standard* (29 CFR 1910.132).

General requirements standard

In this standard, OSHA provides general guidelines that apply to employers that require employees to wear any forms of PPE. The requirements are brief and address topics that may be covered in more detail in the standards for specific equipment.

Hazard assessment and equipment selection. The employer is required to assess the workplace to identify any hazards that may require the use of PPE. For entry into a confined space, the entry supervisor performs a similar assessment when preparing the entry permit. The assessment identifies the hazards and estimates the risk and potential severity of harm they may pose to the entrants.

The OSHA standard requires the employer to produce a written certification that the hazard assessment has taken place. The certification must identify who performed the assessment and the dates on which it was performed. Perform and certify such a hazard assessment for every confined space entry, even if it indicates that no PPE is required. The entry permit documents the assessment for the conditions of the space at the time of the entry and does not address other areas and jobs of the workplace.

The employer must select the types of PPE that provide protection from those hazards identified in the assessment. In some cases, the selected PPE provides protection from more than one hazard, such as when a full-facepiece respirator provides respiratory protection as well as face and eye protection. The protection must be adequate for each hazard. Nonmandatory Appendix B to 29 CFR 1910 Subpart I (discussed below) gives more guidance on the hazard assessment and equipment selection process.

The employer communicates the selection decision to each affected employee and ensures that these employees use the required equipment at all times. The standard does not specifically require inspections, but these obviously will be necessary to ensure the proper use of the selected equipment.

Where the employer provides the PPE, it is designed and constructed to provide the protection safely. The employer ensures that employee-owned equipment provides adequate protection as well. The standard imposes the commonsense requirement that damaged or defective equipment shall not be used. The previous chapters on respirators, CPC, and personal safety equipment discuss the repair or disposal of damaged PPE in detail.

Training.   The employer must provide training to each employee who is required to use PPE. No reference is made in the standard as to how long the training must last. The specific content is not stated, except that the training must address at least the following:

- When PPE is necessary
- What PPE is necessary
- How to properly don, doff, adjust, and wear the PPE
- Limitations of the PPE
- The proper care, maintenance, useful life, and disposal of the PPE

The employee must show that she understands the training and can use the PPE properly before she is allowed to use it in the workplace. The training can take whatever format the employer chooses, but should include hands-on practice with the equipment and the opportunity to ask questions of the instructor.

The standard requires the employer to retrain employees whenever it is apparent that they do not understand how to use the PPE properly. The standard also requires retraining when

- Changes in the workplace render previous training obsolete.
- Changes in the types PPE used render previous training obsolete.
- Inadequacies in the employee's knowledge and use of the PPE indicate the need for retraining.

The employer must certify the training through a written certificate that shows the name of the employee, the date, and the subject of the training. The respiratory protection standard has more detailed requirements for the selection of the equipment and the training of users that supersede those sections of 29 CFR 1910.132.

Appendix B to Subpart I

OSHA provides a nonmandatory appendix to guide employers in meeting the hazard assessment and selection requirements of the PPE regulations. Employers can refer to the guidelines in the appendix when establishing their own PPE programs. The guidance in the appendix is not intended to add to or take away from the requirements of individual regulations, but provides suggestions and examples that may help employers better understand the requirements.

The appendix begins by pointing out that PPE should not be the only means relied on to protect workers from the hazards. It should be used in conjunction with engineering controls, guards, and sound manufacturing procedures to prevent exposure to the hazards in the first place. It cannot be overstated how important using these other hazard controls is to creating a safe workplace. PPE provides additional protection when engineering controls are not enough.

Assessment guidelines. Hazard assessment is the necessary first step to determining if PPE is needed. The appendix assigns the responsibility for hazard assessment to the "safety officer." OSHA does not elsewhere require that someone be named as a safety officer, so companies may not have formally designated such a person. In applying this appendix to confined space operations, the entry supervisor would be the person designated to ensure that the assessment is performed.

The assessment should include a walkthrough survey of the work area to look for sources of hazards to employees. Entry supervisors may not be able to "walk through" the confined space before an entry. They will need to gather and study any drawings, plans, or other information that will help them look for such hazard sources prior to entry.

The survey should look for hazards within these basic categories:

- Impact

- Penetration

- Compression (rollover)

- Chemical

- Heat

- Harmful dust

- Eye-damaging radiation

The appendix suggests a number of general sources of hazards from these categories. Confined spaces obviously may include sources of any of these types of hazards. Entry supervisors should identify these sources during the permit process.

The final steps of the assessment phase are to organize and analyze the data. If all entrants into a space will do the same work, the hazard assessment focuses on that work. If multiple jobs will be performed, a job hazard analysis is performed on each. The job hazard analysis procedure described in Chap. 3 can be used for this analysis.

Selection guidelines. The appendix further directs the safety officer to become familiar with the potential hazards and the types of PPE available. Reference books are available to describe the important characteristics, such as measurable properties, of the hazards. Understanding the hazards is vital to determining the best equipment to protect the entrant.

Safety equipment vendors and manufacturers are excellent sources of information about the types of protective equipment available. Printed catalogs and literature, Internet sites, and visits by sales representatives are useful ways to learn about the important features of various types of protective equipment. Obviously the more different sources that are consulted, the better. It is the job of each company to sell its products, and each will try to convince the safety officer that theirs is the best. By comparing the information from several sources, you can identify the products that best meet the needs of the entrant.

PPE that protects against most of the hazards faced by the entrants is not enough. Compare the hazards of the space to the capabilities of the available equipment. Select PPE that provides greater than the minimum degree of protection for all the identified hazards. The appendix describes some basic selection guidelines for each type of protective equipment that the standards cover. The guidelines are general in nature and include suggestions of some occupations that regularly require each type of PPE.

Fitting and adjusting equipment.   The appendix further guides the employer to fit the equipment properly to the user. Most types of PPE are available in different sizes. Poorly fitting PPE usually provides poor protection and can even prove to be dangerous. Entrants are more likely to continue to wear equipment that fits them comfortably.

Some equipment can be adjusted to fit the unique dimensions of each wearer. The manufacturer's instructions will describe the procedures for making the adjustments to produce the most comfort and protection. Follow the instructions, since improper fitting and positioning of the equipment may reduce its ability to protect the wearer. Train users to properly don and fit the equipment each time it is used.

Comprehensive Written PPE Program

OSHA's regulations do not require the employer to develop a comprehensive written PPE program. Among the PPE regulations, only the respiratory protection standard requires a written program be developed. The use of each type of equipment entails procedures for selection, fitting, use, storage, maintenance, and cleaning, and a comprehensive program can serve to ensure that all types of PPE are properly selected, used, and maintained. A complete written program ensures that all hazards in a workplace are addressed.

While not directly covering confined space operations, the HAZWOPER regulation can provide some guidance in what such a program should include. The section on personal protective equipment [29 CFR 1910.120(g)(5)] of the standard requires the employer to develop a written PPE program addressing at least these topics:

- Selection based on site hazards
- Use and limitations of the PPE
- Work mission duration
- Maintenance and storage
- Decontamination and disposal
- Training and proper fitting
- Donning and doffing procedures
- Equipment inspection procedures
- Program evaluation
- Temperature extremes and other medical considerations

These topics cover the important aspects of using PPE. Subsections under each heading address the specific procedures unique to each type of PPE (i.e., face and eye protection, head protection, hand protection, foot protection). In some cases, general procedures can be applied to all types. A single comprehensive program would allow an employer to consider the potential problems when more than one type of PPE is required, such as using hardhats and respirators with head harnesses.

All the topics listed above potentially apply to using PPE in confined spaces as much as they do to work on hazardous-waste sites. Confined space entries may involve exposure to a relatively few hazards that require few items of PPE. Nonetheless, the comprehensive written program can ensure that all the important issues related to the proper selection and use of even these few items are addressed.

Levels of Protection

Chemical hazards may enter or harm the body via contact with the skin, inhalation, or ingestion. The risk of exposure by each of these routes depends on the chemical and physical properties of the chemical, the type of work being done, and the conditions inside the space. Consequently, what constitutes adequate protection for each of these routes can vary.

Protection always comes at a cost to the wearer. The possible burdens or stresses placed on the wearer by chemical protective clothing and respiratory protection include

- Insulation that aggravates heat stress
- Decreased mobility
- More difficult communication
- Added weight and bulk
- Loss of dexterity and sense of touch
- Limited work time

Besides these stresses, equipment that is more protective is typically more expensive. The required maintenance and care procedures are more complex. While adequate protection is required, overprotection is not helpful and should be avoided.

The concept of levels of protection has been used for years to select adequate protection without overburdening the wearer. CPC and respiratory protection are combined into protective ensembles that are appropriate for different degrees and types of hazards. The two most frequently used systems of levels used in the hazardous-materials field are those developed by the EPA and the NFPA. While neither system was developed specifically for work in confined spaces, each is applied easily to potential chemical exposure in these work areas.

EPA levels of protection

The EPA uses a four-level system of PPE ensembles to assign protection for its employees as they work on hazardous-waste sites. The levels are assigned the letters A, B, C, and D in descending level of protection. Level A provides the highest degree of protection, while level D provides the least protection. The following discussion builds from the lowest level to the highest as hazards are added or increased.

Level D.   The least protective ensemble, level D, is a basic work uniform (Fig. 10.1). There is no respiratory protection and no chemical protective clothing. The only protective equipment used, if needed, is personal safety equipment such as safety shoes, head protection, safety glasses, and hearing protection. Level D is not appropriate for a confined space that contains a hazardous atmosphere or a skin hazard of any kind.

Level C.   The key features of level C are the air-purifying respirator (APR) and limited splash protection (Fig. 10.2). Level C can be used where there is a real or potential hazardous atmosphere and the conditions are acceptable for using APR:

- At least 19.5% oxygen
- Contaminant that has been identified
- Chemical concentration in air less than the IDLH
- Cartridge replacement schedule calculated

Skin protection with this level is limited to basic splash protection. It should be selected according to the principles discussed in Chap. 8.

The use of level C in confined space operations requires special attention to air monitoring to ensure that the conditions do not change from those described above. Given the enclosed nature of many spaces, the atmosphere may change more rapidly than an open, well-ventilated work area. Skin contact with chemical residues also may be more substantial because of the tight spaces.

Figure 10.1 Level D consists of a normal work uniform with safety equipment but no chemical protection.

Level B.   Level B is the minimum level that can be used in a space that contains a hazardous atmosphere, but that lacks any of the conditions necessary for the safe use of APRs. This level of protection uses the highest available respiratory protection, atmosphere-supplying respirators such as airline respirators and SCBAs, and splash protection for the skin (Fig. 10.3). It can be used where there are conditions such as oxygen deficiency and IDLH concentrations of chemicals that are not significant skin hazards. It is appropriate for spaces with unidentified contaminants, provided there is no significant liquid splashing that may lead to skin exposure.

Level A.   The highest level of protection, level A, completely seals the entrant inside a safe environment. In addition to the highest level of respiratory protection, this level uses the totally encapsulating chemical protective (TECP) suit that gives total skin protection (Fig. 10.4). The TECP suit builds a positive internal pressure as air is exhaled inside the suit, and this pressure forces internal air out through any leaks in the suit. Level A protection can be used in IDLH atmospheres as well as where unknowns are present that are found not to be flammable. This level is appropriate for situations with a high degree of splashing of a chemical known to harm the skin, provided the proper suit material is selected. The bulkiness and loss of mobility make the use of level A for confined space entry very difficult.

NFPA standards

The NFPA has published standards outlining three levels of protective ensembles used in hazardous-materials emergencies. These standards describe in detail the required test methods and performance requirements manufacturers must use to test the suitability of their products for hazardous-materials response. Equipment that meets the specifications can be marked as compliant with the relevant standard, giving the wearer assurance of the protection that he can expect.

The standards state that they do not apply to protective clothing for any firefighting applications. They do not specify the level of respiratory protection needed for each

Figure 10.2 The air-purifying respirator is the primary feature of level C protection.

ensemble. They do address the protection provided by the complete CPC ensemble in hazardous-materials emergency response applications.

NFPA 1991. *NFPA* 1991, *Standard on Vapor–Protective Ensembles for Hazardous Materials Emergencies* (NFPA 2000), addresses the requirements for all vapor-protective ensembles that include TECP suits. It requires that all components of the garment be tested for permeation by 21 representative chemicals and found to not break through in less than one hour. The material must pass tests for flame impingement, bursting strength, puncture propagation, and cold-weather performance. All components such as the suit material, the visor, boots, and gloves (if attached) must meet all the requirements. Optional certification testing can be conducted for protection from chemical flash fire, liquefied gas, both, or chemical and biological terrorism. The standard requires that the manufacturer put a label in the garment that identifies the suit, glove, boot, and visor materials. Additional instructions are to be included in a technical data package that accompanies the suit. This information can be useful in making a proper selection of the PPE.

Figure 10.3 Level B protection includes the highest level of respiratory protection, but allows some unprotected skin.

NFPA 1992. *NFPA 1992, Standard on Liquid Splash-Protective Ensembles and Clothing for Hazardous Materials Emergencies* (NFPA 2000), addresses the protective equipment used where the chemical is not harmful to the skin as a gas or a vapor. The ensembles provide basic splash protection, but can allow some degree of skin exposure. Like NFPA 1991, this standard outlines the testing and performance requirements for the suit material as well as the other components of the ensemble. These garments are tested for permeation by only seven chemicals and may be either encapsulating or nonencapsulating. For this standard, optional certification is available for flash fire protection only.

NFPA 1993. *NFPA 1993, Standard on Support Function Clothing for Hazardous Chemical Operations* (NFPA 1994), addresses the requirements for clothing worn by personnel supporting the response from outside the hot zone. It includes testing requirements for garments as well as visors, gloves, and foot protection. Clothing certified as NFPA 1993–compliant should not be used in chemical exposure situations where chemical vapor, liquid splash contact, or flash fire potential are of concern.

Figure 10.4 The totally encapsulating chemical protective suit is the defining feature of level A protection.

Summary

The proper use of PPE requires planning and organization. This chapter looked at OSHA's *General Requirements Standard* (29 CFR 1910.132) for its hazard assessment and training requirements for all PPE. OSHA's HAZWOPER standard provides a good example of a written PPE program that employers can use to manage their PPE usage, although the written program is not required by other OSHA regulations. To avoid unnecessary burden to the entrant, entry supervisors should use a level of protection approach to selecting PPE that provides adequate protection without overprotection.

Part

Equipment and Entry

Chapter

11

Ropes, Webbing, Harnesses, and Hardware

Two workers were hired by an industrial painting contractor to sandblast and paint the interior of a 48-ft-high 30-ft-diameter steel water tank. One worker climbed a ladder and welded steel brackets to the sidewalls at the top of the tank for scaffolding suspension. Using the brackets as anchors, the workers rigged nylon ropes through a block-and-tackle system at each end of the platform. The platform could then be positioned with a haul line and tied off to the platform handrails. The scaffolding was being raised for a second time and was approximately 40 ft above the floor when one of the nylon ropes supporting the scaffolding broke and the end of the scaffolding fell. One worker fell, landed on a horizontal water pipe at the bottom of the tank, and died from blunt-force trauma to the head and trunk. It was later learned that the nylon ropes had been lying in the bottom of the tank in an area where sparks fell while the brackets were being welded.

Various types of equipment are required for safe confined space operations. Getting into and out of spaces for permitted entries may require the use of equipment such as rope-and-pulley systems, portable anchor systems, and fall-arrest systems. Rescuers may need to use the same systems and others during confined space rescues. These systems typically consist of components such as ropes, webbing, harnesses, and various hardware items such as carabiners, pulleys, and rope-grabbing devices. In this chapter we will learn the basics of selection, use, and care for these components. In subsequent chapters we will learn how to assemble the components into systems for use in confined space operations.

Standards Related to Equipment

The types of equipment we covered in this chapter differ from other types of equipment in that their failure is likely to result in serious injury or death to the people using them. For this reason we must be assured that any piece of equipment will perform adequately the job it is designed to do. Minimum performance standards have been developed to provide this assurance. Relevant standards include those written by NFPA, OSHA, and ANSI. Overviews of some of the applicable standards were provided in Chap. 2, and details from them will be provided as needed in this chapter.

Overview of NFPA 1983

The first edition of NFPA 1983, *Standard on Fire Service Life Safety Rope and System Components,* was released in June 1985 (NFPA 2001). The original intent of the standard was to provide requirements for rope used by fire service personnel to perform rescues. During its development it was quickly realized that minimum performance requirements were needed also for harnesses and hardware used in rope rescue sys-

191

tems, since a system is only as strong as its weakest link. Since 1985 there have been three editions; the latest became effective on February 9, 2001 (Fig. 11.1). This standard establishes minimum levels of performance and a reasonable degree of safety for new life-safety rope and new system components used to support fire service personnel or other emergency services personnel and civilians during rescue, firefighting, and other emergency operations, or during training evolutions (Part 1.2.1).

NFPA 1983 describes design, construction, and performance requirements for life-safety rope, escape rope, life-safety harnesses, belts, auxiliary equipment, throwlines, and accessories that attach to any life-safety rope or system component. Auxiliary equipment includes, but is not limited to, any load-bearing system component that is to be used with life-safety rope and harnesses. Examples of these auxiliary system components are carabiners, snap links, webbing, ascending devices, descent control devices, pulley systems, and portable anchor systems. Examples of portable anchors include davits, tripods, quadpods, and cantilever devices.

Life-safety ropes are classified as general-use or light-use. Other components may have an escape-use classification in the 2001 edition of NFPA 1983. Previous editions of NFPA 1983 rated the same equipment as general-use, personal-use, or escape-use equipment. The term *light-use* replaced *personal-use* in the 2001 edition. General-use ropes are rated based on a two-person load at 300 lb per person with a 15:1 safety factor. This requires minimum breaking strengths (MBSs) of 9000 lb of force (lbf) for general-use-rated ropes, as illustrated by the following formula:

$$2 \times 300 \text{ lb} \times 15 = 9000 \text{ lbf MBS}$$

Personal-use ropes (now called *light-use life-safety rope*) are rated with respect to a one-person load with a 15:1 safety factor. This requires minimum breaking strengths of 4500 lbf for personal-use-rated ropes, as shown by the following formula.

$$1 \times 300 \text{ lb} \times 15 = 4500 \text{ lbf MBS}$$

Figure 11.1 NFPA 1983 provides performance requirements for rope, hardware, and harnesses used in rope rescue systems.

Escape-use rope and system components are used for self-rescue escape. Escape rope is a single-purpose, one-time-use rope that is not classified as a life-safety rope. It is rated with respect to a one-person load with a 10:1 safety factor. This requires minimum breaking strengths of 3000 lbf for escape ropes, as shown in the following formula:

$$1 \times 300 \text{ lb} \times 10 = 3000 \text{ lbf MBS}$$

The explanation of how the minimum breaking strengths were derived for rating ropes is not included in the 2001 edition of NFPA 1983. The newton (N) replaced the pound of force (lbf) as the primary unit used for specifying strength requirements in the 2001 edition; however, the pound of force (lbf) is still used as the secondary unit for specifying the requirements of the standard. For some components, the conversion from newtons (N) to pounds of force (lbf) caused minor changes in the numerical values stated in lbf, as compared to those seen in the previous version of the standard. This will be evident in the following discussions of individual components.

Be careful how you interpret the units in the 2001 edition of NFPA 1983. For example, general-use life-safety rope is required to have a minimum breaking strength of not less than 40 k/N (8992 lbf), where k stands for 1000. The slash is an error. The unit k/N normally would be interpreted as kilograms per newton, but what actually is meant is 40 kN (kilonewtons) or 40,000 N. There also are several typos, one of which can be found in NFPA 1983 Section 4.5.2.2. This section requires that light-use auxiliary equipment have a designed load of at least 133 k/N (300 lbf). This should read 1.33 kN or 1330 N, which equals approximately 300 lbf. Conversion factors are listed in Table 11.1.

A number of the types of equipment discussed in this chapter are classified as auxiliary equipment system components by NFPA. Examples include webbing, shock absorbers, and anchor plates. NFPA rates auxiliary equipment by *design load,* which is the load the piece of equipment is engineered to hold under normal static conditions. Section 4.5 of NFPA 1983 classifies auxiliary equipment with design loads of 1.33 kN (300 lbf) as light-use and auxiliary equipment with design loads of 2.67 kN (600 lbf) as general-use. The requirement for tensile strength or breaking strength is found in Section 5.5.6. This section requires that light-use auxiliary equipment have a minimum tensile strength of 22 kN (4946 lbf) and that general-use auxiliary equipment have a minimum tensile strength of 36 kN (8093 lbf). This differs from previous editions in that general-use auxiliary equipment was defined as equipment "intended for use where the system could be subjected to a two-person load" and personal-use auxiliary equipment as equipment "intended for use where the system could only be subjected to a one-person load." NFPA suggests using general-use auxiliary equipment for one-person load scenarios in which the load is a victim, and in cases where "unusual or extreme forces could be placed on the load" (Appendix A.4.5.2.2). Some auxiliary equipment must meet the additional requirements included in the discussions that follow.

Overview of OSHA requirements

OSHA requirements relevant to this chapter are related mainly to fall protection. At this time OSHA does not have a general industry standard that specifically addresses fall protection. Pieces of information dealing with fall protection are found in several general industry standards such as 29 CFR 1910 Subpart D, *Walking and Working Surfaces* and Subpart F, *Powered Platforms, Manlifts, and Vehicle-Mounted Work Platforms.* OSHA has published a comprehensive fall-protection standard for the construction industry (29 CFR 1926 Subpart M) that we will look to for guidance.

TABLE 11.1 Conversion Factors for Units of Force

SI units, N	English units, lb
1	0.225
4.45	1

Ropes

Ropes provide flexible connections in systems used for hauling or lowering loads and personal fall protection. Ropes may also be used as "tag lines," a term used to refer to the "lifelines" worn by confined space entrants, and also used to refer to the lines used to guide a suspended load. Ropes also are used for utility purposes, such as hauling or lowering tools and equipment. In any confined space operation it is important to understand the limitations of the equipment being used. Ropes are no exception.

Ropes should be selected according to the application in which they are to be used. A rope purchased to serve as a tag line in a horizontal entry may, or may not, be suitable for a vertical rescue operation where two-person loads could be expected.

Rope selection should be based on factors such as the following:

- Tensile strength or breaking strength

- Elasticity, elongation, or stretch

- Flexibility

- Weight per 100 ft

- Cost

NFPA 1983 requirements for rope

As explained earlier, rope is classified as either escape rope or life-safety rope by NFPA 1983 (2001). The classifications have different definitions and different uses.

Escape rope. *Escape rope* is a "single-purpose, one-time use, self-escape (self-rescue) rope" that is not classified as a life-safety rope (NFPA 1983 1.3.47.1). NFPA 1983 requires escape rope to be constructed of virgin, continuous filament fiber (block creel construction). Escape rope must have a minimum breaking strength of 13.5 kN (3034 lbf).

Life-safety rope. *Life-safety rope* is rope that is "dedicated solely for the purpose of supporting people during rescue, other emergency operations, or during training evolutions" (NFPA 1983 1.3.47.2). The standard requires life-safety rope to be constructed of virgin, continuous filament fiber (block creel construction) and to meet the requirements for class I, division I, hazardous locations of ANSI/UL 913, *Standard for Intrinsically Safe Apparatus and Associated Apparatus for Use in Class I, II and III, Division I, Hazardous (Classified) Locations*. Life-safety rope is rated either as light-use or general-use.

Light-use life-safety rope. *Light-use life-safety rope* should have a minimum breaking strength of 20 kN (4496 lbf).

General-use life-safety rope. *General-use life-safety rope* should have a minimum breaking strength of 40 kN (8992 lbf). General-use rope generally is recommended for confined space operations.

OSHA rope requirements

In any fall-arrest or fall-prevention system used in confined space operations, OSHA's *Fall Protection Standard* (29 CFR 1926 Subpart M) requires static ropes to have a minimum breaking strength of 5000 lbf (22.2 kN). Ropes rated as light-use life-safety ropes by NFPA (minimum breaking strength 4496 lbf) would not meet this requirement. Ropes rated as general-use life-safety ropes (minimum breaking strength 8992 lbf) would more than meet OSHA's requirements. General-use rated life-safety rope is recommended for all confined space operations where live loads are supported.

Construction Materials

Rope materials range from natural fibers such as hemp, manila, and cotton to such synthetic materials as nylon, polyester, and polypropylene. The newer synthetics have replaced natural fibers in most confined space applications.

Natural-fiber ropes.   Rope choices were limited to natural fibers until the advent of polymer technology. The most commonly used natural-fiber ropes were hemp, manila, cotton, and sisal. Hemp and manila were used for strength, cotton for manageability, and sisal because of its low cost. These have been replaced almost completely by synthetics (Bigon, 1982, p. 9). Several fire service organizations, including the NFPA and the International Association of Fire Fighters (IAFF), now prohibit the use of natural-fiber ropes in rescue or for any other life-essential operation.

Problems with natural-fiber ropes are many. Some relate to the fact that the fiber strands are not continuous from one end of the rope to the other, allowing for weak spots in the rope's construction that are undetectable until the rope fails under the stress of a load. Natural-fiber ropes are heavy and cumbersome compared to their synthetic counterparts, especially when they are wet. Some natural fibers can absorb as much as 100% of their weight in water. Once wet, an irreversible loss of strength occurs—up to 50% with some fibers. If not stored properly, natural-fiber ropes are susceptible to rotting and, even with proper storage, dry rot can occur, resulting in a reduction in strength.

Natural-fiber ropes such as hemp and manila cannot compete with the breaking strength of synthetic fiber ropes of the same diameter. In order to meet NFPA's strength requirement, a good-quality manila rope would need a very large diameter and would be so heavy, cumbersome, and difficult to tie that it would be impractical for use in confined space operations.

Synthetic-fiber ropes.   *Synthetic fibers* are, as the term suggests, human-made polymers derived from oil or coal. In their finished state they all tend to be shiny and nonabsorbent, and they do not tend to rot. Not all synthetic ropes are suitable for all types of operations, so it is important to understand the strengths and weaknesses of each fiber type. Rope construction affects the overall properties of the rope, as discussed in the next section of this chapter. Depending on how the fibers are put together to form a rope, it is possible to increase properties such as strength and reduce or increase elasticity, whichever is desired.

Common synthetic ropes are made of one or more of the following:

- Polyamide fibers (nylon)

- Polyester fibers

- Aramid fibers

- Polyolefin fibers [polypropylene fibers (olefin) and polyethylene]

Refer to Table 11.2 for a list of common names for these synthetic materials.

Chemical compatibility is a major concern when ropes are used in the industrial environment, since chemical degradation can severely weaken fibers. Table 11.3 contains very limited general information on chemical resistance of some of the fibers most commonly used in rope construction. Requests for specific chemical resistance information made to five major rope manufacturers or distributors in preparation for writing this text produced no results.

TABLE 11.2 Commercial Names of Synthetic Materials

Polyamide	Nylon, nylon 6, nylon 6.6, Perlon, Lilion, Enkalon
Polyester	Dacron, Seran, Terylene, Tergal, Terital, Trevira, Diolen
Aramid	Kevlar, Nomex, Arenka, Technora, Twaron

SOURCE: Adapted from *The Morrow Guide to Knots* (1982).

TABLE 11.3   Fiber Properties and Chemical Resistance

Characteristics	Polyamide nylon	Polyester	Polypropylene	Polyethylene	Aramid (Kevlar®)
Strength	2*	3*	4*	5*	1*
Shock-load ability	1*	3*	2*	4*	5*
Specific gravity	1.14	1.38	0.92	0.95	1.45
Elongation at break, %	18–25%	12–15%	15–25%	15–25%	2–4%
Abrasion resistance	2*	1*	5*	4*	3*
Resistance to sunlight	Good	Excellent	Poor	Fair	Good
Resistance to rot	Excellent	Excellent	Excellent	Excellent	Excellent
Resistance to					
Acids	Poor	Good	Good	Good	Poor
Alkalis	Good	Poor	Good	Good	Good
Oil and Gas	Good	Good	Good	Good	Good

Scale: best = 1; poorest = 5.
SOURCES: *Essentials of Fire Fighting*, 4th ed., Hall and Adams, 1998, Table 6.1, p. 149 and *CMC Rope Rescue Manual*, 3d ed. (Frank 1998), p. 15.

Polyamide fibers. *Polyamide fibers,* commonly known as *nylon,* combine tensile strength with elasticity, making these ropes most suitable for confined space operations. Two different nylon fibers are used in life-safety rope: nylon 6 (Perlon®) and nylon 6.6. Nylon 6.6 has a slightly higher melting point and better abrasion resistance than nylon 6. Melting points of nylon fibers range from 420 to 480°F. A number of rope manufacturers use a combination of nylon 6 and nylon 6.6 in their ropes to achieve desired properties. This combination also can be dependent on supply and, of course, cost.

Because of nylon's elongation characteristic, nylon ropes are able to absorb shock loads better than are other synthetic materials. Refer to Table 11.3 for a list of common fiber properties. Nylon ropes tend to resist abrasion, deploy easily, and hold knots well. They also offer good resistance to ultraviolet light, and excellent resistance to rot and mildew. Nylon ropes with different degrees of flexibility are available. A certain degree of flexibility is needed in order to knot and rig the rope, but a certain amount of strength or resistance to abrasion is given up to gain this flexibility. Brands such as Pigeon Mountain Industries' (PMI) EZ Bend™ use flexibility as a selling point; however, compared to PMI's Max-Wear™, some durability and abrasion resistance is lost to gain this flexibility. Both of these ropes are made up of 100% nylon fibers.

Nylon's specific gravity is greater than 1.0, so it does not float. It loses 10 to 15% of its strength when wet; however, after drying it regains the lost strength (Frank 1998, p. 15).

Polyester fibers. Polyester fibers are comparable to nylon in strength but not in elasticity. Polyester fibers will stretch only 12 to 15% before breaking. By comparison, nylon fibers will stretch 18 to 25%. This low stretch characteristic makes polyester less suitable than nylon for absorbing shock loads or for handling repeated loading. Polyester has a melting point between 490 and 500°F, good abrasion resistance, and excellent resistance to ultraviolet light, rot, and mildew. Polyester's specific gravity is greater than 1.0, so it does not float. When wet it retains approximately 100% of its strength (Frank 1998, p. 15). Sterling Rope's high-tenacity polyester (HTP™) static rope is an example of a rope made entirely of polyester.

Polyamide/polyester ropes. Many manufacturers produce a combination nylon/polyester rope. Nylon fibers make up the core of the rope, giving the rope its strength and ability to handle shock loads, and polyester fibers make up the outer sheath, providing protection from water, ultraviolet light, abrasion, and other environmental factors. PMI's IMPACT P6 and Blue Water® Rigger LE™ are examples of this type of rope.

Aramid fibers. Aramid fibers are used because of their very high melting points and decomposition temperatures. Aramid fibers tend to be very coarse and difficult to cut. Because of their coarseness, some aramid ropes are difficult to knot and can break easily at bends after repeated use. They are very costly, and some, such as the aramid rope

Kevlar®, will stretch only 2 to 4% before it breaks. This characteristic makes aramid ropes unsuitable for absorbing shock loads. They are not commonly used in confined space operations.

Polyolefin fibers. *Polypropylene* is a synthetic fiber material commonly used in throwlines for water rescue operations where floatation is needed. These fibers abrade easily and tend to break down rapidly with exposure to sunlight. They have very low strength and are a poor choice for confined space operations.

Some manufacturers produce a combination rope of nylon and polypropylene. Sterling Rope's WaterLine™ is made up of a nylon core for strength with a polypropylene outer sheath for flotation. This rope was developed for water rescue applications and, even with the added strength the nylon provides, should never be used in confined space operations. Polyethylene fiber ropes, such as the ones used in water skiing, should be avoided. They are weak and do not hold knots well.

Methods of rope construction

A variety of methods are used to assemble any of the various types of fibers into ropes. The construction method and the material of construction are important factors in determining the suitability of a rope for various uses.

Laid and twisted rope. Laid rope is constructed of fibers that have been twisted repeatedly. First the fibers are twisted together to form yarn. The twisting of the yarn is an important step and affects the "hand" of the rope or its handling characteristics. The yarn is then twisted together to form strands, and finally three or more of the strands are twisted to form the rope itself. This operation, known as "laying up," is one of the oldest methods of rope construction. A weakness of this type of rope for confined space operations is that the load-bearing fibers are exposed, and therefore are more susceptible to abrasion (Fig. 11.2).

Braided rope. The most common type of braided rope is the braid-on-braid, or double-braided, rope. Ropes are formed by braiding yarn together to form a core braid inside an outer braid called the *sheath*. The sheaths of these ropes seldom are woven as tightly as sheaths on kernmantle ropes, reducing their resistance to abrasion. This limits the application of braided rope in confined space operations, since up to 50% of the rope's strength may lie in the outer sheath (depending on the manufacturer). See Fig. 11.3 for an illustration of braided rope.

Figure 11.2 Laid or twisted rope is composed of fibers that are twisted together a number of times, each in opposite direction to the previous, to form the yarn, then the strands, and finally the rope itself.

Figure 11.3 Braid-on-braid rope, as the name implies, has a braided core and braided jacket.

Some rappel ropes are made using this type of construction. In rappelling, the rope is stationary and not subjected to the abrasion potential seen in moving systems such as lowering and hauling systems that may be used in confined space operations. Ropes should be used only in ways consistent with the manufacturer's approvals.

Kernmantle rope. The word *kernmantle* is of German origin, where *kern* means core and *mantle* means sheath (Fig. 11.4). The core construction can vary depending on the amount of elongation the rope will be required to achieve.

High-stretch or "dynamic" kernmantle rope's core fiber yarns may be significantly twisted or even braided to give up to 30% more stretch at failure than "static" or low-stretch ropes whose core fibers are only slightly twisted or even parallel. Dynamic kernmantle ropes are used in applications such as rock climbing where shock loading is expected. Shock loads occur in these activities because lead climbers are required to climb well above existing anchor points to set each new anchor. In such a situation, a climber could fall a significant distance before being caught by the existing anchor point. The elasticity of dynamic rope provides a built-in shock-absorbing effect that prevents shock loading of the system. Otherwise the resulting shock could injure the falling climber and cause failure of anchors, rope, or other system components.

Static kernmantle is the best choice for use in confined space operations and structural rescue operations. The core construction of the static kernmantle rope offers added control for lifting, lowering, and positioning loads because of the low-stretch characteristic. Dynamic rope is not required for confined space operations because as a general rule one does not work above anchor points. Rope systems used in confined space operations typically have very little or no slack in them during use, making significant shock loads very unlikely. Excessive slack may occasionally build up in safety belay systems, especially during initial training. For this reason the use of shock absorbers is recommended in safety belay lines, as discussed below and in Chap. 13.

The sheath construction of kernmantle rope can vary (Fig. 11.5). The percentage of the load on the core varies with sheath construction from 70 to 90%. Stated differently, the percentage of the load on the sheath varies from 10 to 30%, depending on the sheath thickness. Kernmantle ropes have a 16-strand sheath, a 32-strand sheath, or a 48-strand sheath. The 16-strand sheath has twice the thickness of a 32-strand sheath and 3 times the thickness of a 48-strand sheath. One manufacturer advertises that the 16-strand sheath gives ropes greater durability, abrasion, and resistance to cuts. Another manufacturer states that the tighter 48-strand sheath allows less yarn to be exposed to abrasion and that the smaller thickness of the sheath allows room to put more yarn in the core for strength. You will have to decide which is best for your application.

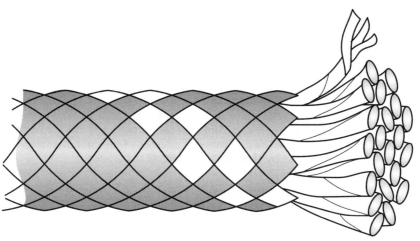

Figure 11.4 Static kernmantle rope has a low-stretch core that offers added control for lifting or lowering and a sheath that protects core fibers from abrasion.

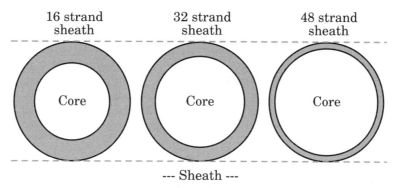

Figure 11.5 Kernmantle ropes can be purchased with a 16-, 32-, or 48-strand sheath.

Block creel construction. Block creel construction is required by NFPA 1983 for all escape and life-safety ropes. In block creel construction the synthetic fibers run continuously from one end of the rope to the other. This method of construction gives the rope added strength and eliminates weak spots due to fiber splicing where fibers end and begin in the middle of the rope.

Utility rope. Utility rope can be purchased as twisted/laid, braided, or kernmantle rope. We suggest that utility ropes be of the twisted or laid construction so that they can be distinguished easily from life-safety kernmantle ropes. Never use utility ropes to support a human load, and never use life-safety rope to haul and lower equipment.

Selection and use considerations for rope

Rope diameter. Most manufacturers sell static kernmantle ropes in the following diameters: $^3/_8$ in (10 mm), $^7/_{16}$ in (11 mm), $^1/_2$ in (12.5 mm), and $^5/_8$ in (16 mm). As diameter increases, so do strength, weight, and bulkiness. Most manufacturers' $^1/_2$-in static kernmantle rope is strong enough to meet both the NFPA 1983 general-use life-safety rope and the OSHA requirements when used in a single-line configuration. Static kernmantle ropes with smaller diameters meet the requirement for NFPA's general-use rope only when used in a multiple-line configuration such as mechanical-advantage systems. Half-inch static kernmantle rope weighs approximately 7 lb per 100 ft and is available in flexible grades that allow for easy handling and knotting.

Rope length.   Rope length should be determined when the spaces at your facility or in your jurisdiction are preplanned. Make sure that consideration is given to the types of hauling and lowering systems used. If you are using mechanical-advantage systems, multiply the length of the drop by 2, 3, or 4, depending on which system is being used, and add footage to compensate for the length of the haul line, anchoring, and knots.

Specific rope lengths can be purchased from most vendors; however, it may be more cost-effective to purchase the rope by the spool and cut the rope to the desired length. If this is the case, take care in the cutting process. A hot cutter or hot knife that will fuse the sheath to the core as the rope is being cut is best for synthetic kernmantle ropes (Fig. 11.6). There are a number of different types of hot cutters on the market, and these usually are sold by the same vendors that sell rope.

NFPA 1983 A.5.1 requires that any rope end terminations be made in accordance with the manufacturer's instructions. NFPA 1983 A.3.1.2.1 requires that, when life-safety ropes or escape ropes are purchased in longer lengths and cut, the user agency or employer must photocopy the product label and attach a copy to each life-safety rope when it is sent into service. The end users (the people making the entries) should keep the copy of the product label and have it readily available so that it can be viewed by all potential users.

Rope color.   Synthetic-fiber ropes can be purchased in many basic colors (Fig. 11.7). The sheath of an average static kernmantle rope comes with a base color and a tracer color. The tracer color is woven into the sheath at manufacturing. The dying process has the effect of preshrinking the nylon fibers, so a rope with a white nylon sheath, in which the fibers that make up the sheath were not dyed, may be subject to excessive shrinkage of the sheath. This can result in stiffening of the rope or cause the sheath to pull away from the core at the ends of the rope. It is also reported that when white is used as the tracer color in a rope, the shrinkage of the undyed white filaments woven into the colored sheath can cause a bumpy surface to develop on the sheath of the rope.

Figure 11.6   Hot knives fuse the sheath of the rope to the core as the rope is being cut.

The manufacturer should be consulted to make sure that these fibers have been preshrunk in the manufacturing process. Some experts suggest avoiding ropes having a white sheath with a white core because this color combination may make sheath damage hard to detect.

Color can be used to indicate a number of different things; most often it is used for entry management. Keeping up with which rope is the main line, safety line, or tag line is easier if the three ropes are different colors. In some operations you may have more than one entrant, each on a different line. If these lines are different colors, management of the entry will be a lot easier for the attendant. It is easier to convey information to rescue personnel on the functions of the different ropes in the event of an emergency.

Inspecting ropes.   A qualified person should inspect every inch of a rope by sight and touch before and after each use. The inspector looks for damage such as significant sheath abrasion or fraying, exposed core fibers, discoloration, rope hardening, or slick spots that could indicate exposure to heat or chemicals, and soft spots or lumps in the rope that could indicate core damage. High stresses and impact loading can damage ropes. Inspection procedures, provided by the manufacturer or found in ASTM F1740, *Guide for Inspection of Nylon, Polyester and / or Nylon / Polyester Blend Kernmantle Rope* (ATM 1996), should be followed. It is the ultimate responsibility of the user to inspect the rope before putting it into service, to keep a history of the rope, and to make decisions about when to retire the rope.

Rope retirement.   If you ever suspect that there may be a problem with a rope, it should be retired and destroyed. NFPA 1983 Appendix A suggests that if a rope fails to pass the inspection, or if it has been impact-loaded or overstressed, it should be destroyed immediately. The destruction of a life-safety rope means altering it in such a way that it cannot be mistaken for life-safety rope. This could mean disposal, or cutting the rope

Figure 11.7   Many base and tracer color combinations are available for static kernmantle ropes.

into short lengths for utility use. To destroy an escape rope, cut it into pieces less than one foot long, or burn it, or damage it in some other way to prevent its reuse.

Care of rope.   There are practices you can use to protect and increase the service life of your ropes. Most of these practices also apply when using other software items such as accessory cord, webbing, and harnesses.

Do not allow anyone to step on your ropes. This can drive abrasive dirt and grit into the fibers or pinch the rope between a boot or shoe and abrasive flooring such as diamond grating or nonslip grit coatings that commonly are used in industry. Always pad sharp or jagged edges to prevent rope abrasion. Avoid deploying ropes by dropping bags from heights. The resulting impact can damage the fibers of the rope. It is a good idea to untie knots in ropes after use, as they may hold acute bends in the rope that could fatigue the rope fibers over time.

Do not allow anyone to smoke around ropes or any other type of software, including webbing and harnesses. Embers from a cigarette can burn through the sheath of a rope or through webbing quickly, causing irreversible damage. Be careful not to expose ropes or any other type of software to hot-work operations (welding, soldering, brazing, burning, or cutting). Be mindful of chimneys and hot piping that might contact ropes or other software. If a pipe or other equipment is too hot for you to hold your hand on, then it is too hot for your rope and other software. Do not rig systems in such a way that rope is exposed to a significant amount of heat-causing friction. For example, avoid situations where loaded ropes run in contact with each other.

Rope identification.   Mark each rope with an identifying code that allows you to keep track of the rope's history. Typically the rope is marked at the end near the product label. Vendors sell a number of different items that can be used for this application. Rope-marking pens (made of inks that do not harm nylon, polyester, or polypropylene), self-laminating labels, or vinyl marking tape are some of the alternatives available. Some vendors sell a liquid vinyl compound that can be used to coat the marked information for protection, or purchase a clear heat-shrink tubing to fit the size of rope you are marking (Fig. 11.8).

Rope logs.   Rope logs are used to keep track of the rope's history. Each rope in service should have its own log. At the top of the log, list the rope manufacturer, rope identification marking, color, length, date into service, and other pertinent information. Record information each time the rope is inspected and used. Record specifics such as date, preuse inspection results, where the rope was used, the type of application the rope was used in, whether the rope was wet, the occurrence of any high stress loading or impact loading, chemical exposures, postuse inspection results, and the name of the person who inspected the rope in the log (Fig. 11.9). Inspection prior to use is important, especially if the rope has not been used for a long time, to ensure that no damage or deterioration has occurred while the rope was in storage.

Washing and drying practices.   Keeping ropes clean and free from dirt and grit will help increase rope life by decreasing abrasion. The manufacturer should provide specific washing and drying instructions. Most suggest washing ropes only occasionally with cold water and a small amount of a mild detergent that does not contain any whiteners or brighteners. Agitate the ropes with your hand. Be careful to use only a small amount of detergent! Rinse the rope in several baths of clean cold water, again agitating by hand, to remove the soap residue. A large container such as a clean garbage can dedicated solely to this purpose works well.

Rope-washing devices are available from vendors of ropes and related equipment. These are small plastic tubular items that can be connected to a garden hose (Fig. 11.10). The flow of water from the hose flushes away contaminants as the rope is pulled through the device. Some rope washers contain internal brushes that can be adjusted to fit different rope diameters and provide a brushing action in addition to the flushing action of the water for removing contaminants.

Figure 11.8 Ropes should be marked for easy identification.

ROPE LOG

Owner: _____ Identifying Markings: _____ Date of Purchase: _____

Manufacturer: _____ Type: _____ Material of Construction: Core_____ Mantle _____

Color: _____ Diameter: _____ Length: _____ Date Out of Service:_____

DATE	LOCATION	APPLICATION USED	COMMENTS	Initial if OK

Figure 11.9 Rope logs are used to track the history of a rope. Each rope should have its own log.

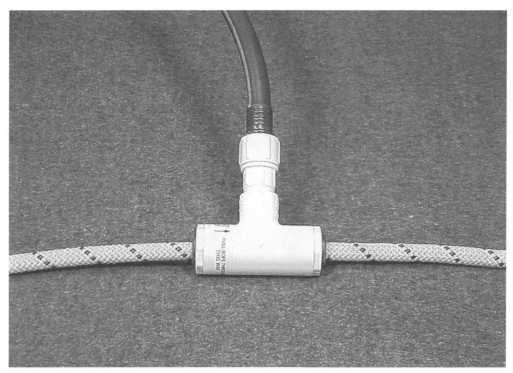

Figure 11.10   Rope-washing devices, such as the one seen here, are used to flush away contaminants or detergents as the rope is pulled through the device.

After washing the ropes, air-dry them by loosely coiling and hanging them in the shade away from sunlight or other heat sources. Never place rope directly onto concrete to dry, as the alkaline agents in the concrete can wick into the rope and cause damage to the fibers. Never dry ropes in a dryer. This can glaze ropes and cause irreversible damage when the sheath hardens and cracks.

Rope storage.   The most common, easy, and practical way of storing rope is to use a rope bag (Fig. 11.11). Nylon rope bags are inexpensive and can be purchased in a number of different sizes, capacities, and colors to fit your specific needs. Rope can be stuffed quickly into the bags hand over hand. Attaching the rope bag to your harness' front waist D ring with a carabiner and clipping the rope through the carabiner makes this process faster and easier. Bagging the rope offers a good opportunity for visual and tactile inspection (by touch with an ungloved hand) as the rope is being bagged. Bagged rope is easy to deploy without tangling.

Carrying straps are presewn to most bags for easy mobility, and cord locks on drawstrings at the top of the bags keep them securely closed. Most rope bags have grommets in the bottom of the bag to allow for water drainage and allow for a way to secure the rope to the bag. Stopper knots placed just inside and outside the grommet achieve this and prevent the rope from feeding out the bottom. After stuffing the bag, place a stopper knot on the top end of the rope to make finding the end of the rope easier when it is needed.

Rope bag color can signify length, type of rope, or anything else you might want to show, and rope-marking pens can be used to write identifying markings or rope lengths directly onto the bag for easy recognition.

Bagged rope should be stored away from direct sunlight in an atmosphere that is cool, dry, and free from chemicals (solids, liquids, gases, vapors, and mists). If your ropes are stored on a truck, make sure that engine exhaust does not leak into the storage compartment. Ropes and other types of software should never share compartments with acid batteries, any gasoline- or oil-powered equipment, or equipment that is greased. To prevent the possibility of acute bends, never store heavy equipment on top of your rope.

Figure 11.11 Rope can be stuffed quickly hand over hand into a rope bag for storage and easily deployed from the bag without tangling.

Accessory Cord

Accessory cord is small-diameter kernmantle rope intended for accessory uses. The care and use considerations for accessory cord are essentially like those for rope. Accessory cord can be purchased in sizes ranging from 3 to 11 mm in diameter and lengths up to 300 m (328 ft). The 6- to 10-mm accessory cords are the sizes most commonly used for confined space operations. Accessory cord can be used to form Prusik loops and hitches in soft rope grab or belaying applications (see Chap. 13) or for ascending a fixed rope (see Chap. 17). In addition, accessory cord can be used for everything from tag lines to shoelaces during a rescue operation.

Webbing

Webbing is used in many applications and in many accessory products that are useful in confined space operations (Fig. 11.12). Prefabricated harnesses are constructed using webbing, and improvised harnesses can be tied using webbing. Webbing commonly is used to lash patients to backboards and basket litters.

Because of its flat construction, webbing can handle acute bends that might fatigue a rope. For this reason it is excellent for use around anchor points. Webbing is used in anchor slings that are either an endless loop or have loops or "eyes" sewn into each end (Fig. 11.13), anchor straps that have D rings sewn into each end, or utility belts that have D rings sewn in each end and are adjustable (Fig. 11.14).

NFPA 1983 requirements for webbing

Webbing is classified as "auxiliary equipment" by NFPA 1983. Minimum performance requirements for auxiliary equipment components were discussed in the first section of this chapter.

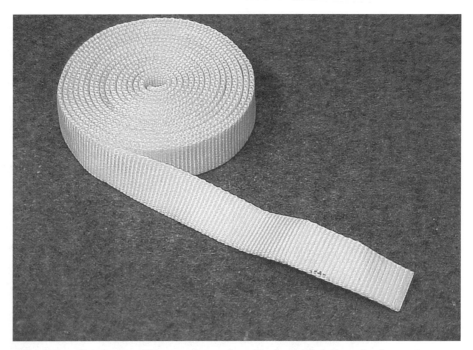

Figure 11.12 Flat-weave webbing can be purchased in desired lengths or by the spool.

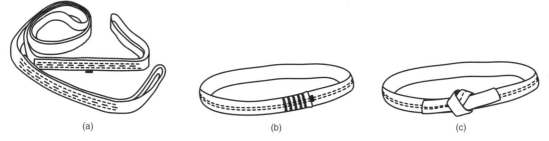

(a)   (b)   (c)

Figure 11.13 (*a*) Prefabricated "eye" slings have sewn loops at each end; (*b*) prefabricated "endless" loop slings are available in a number of different webbing widths and loop diameters; (*c*) loop slings can be tied using webbing and a water knot.

NFPA 1983 requires that webbing be constructed with virgin, synthetic, continuous filament fiber and that the ends be secured by heat sealing to prevent unraveling. In presewn webbing slings, or other software equipment made with webbing, the sewing thread should be compatible with the webbing and allow for easy inspection by a person with normal vision.

OSHA requirements for webbing

OSHA's *Fall Protection Standard* (29 CFR 1926.502) states that the static webbing used in lanyards, lifelines, and strength components of body belts and body harnesses must be made from synthetic fibers and have a minimum breaking strength of 5000 lbf.

Materials and methods of webbing construction

There are several types of webbing on the market. Construction materials include nylon and polyester. The same advantages and disadvantages of these fibers noted in discussing ropes also apply to webbing.

Webbing is tubular or flat-weave. Looking at its construction makes the distinction between the two clear; *tubular webbing,* as the name implies, forms a tube with a hollow center, while *flat-weave webbing* is one solid piece with no hollow center (Fig. 11.15).

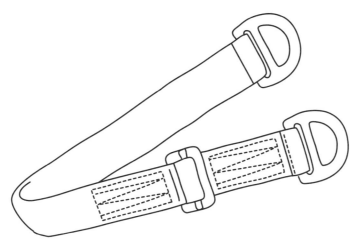

Figure 11.14 Utility belts are anchor straps that are length-adjustable.

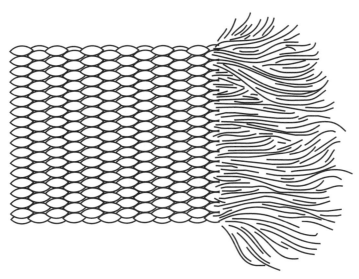

Figure 11.15 Flat-weave webbing is a solid piece construction with no hollow center.

Tubular webbing is made with either shuttle loom or needle loom construction (Fig. 11.16). Shuttle loom webbing is stitched in a continuous spiral, allowing seamless construction of the webbing. Needle loom webbing is made with a flat piece of webbing that is folded double and chain-stitched along the seam to form the tubular construction. If you open up the needle loom webbing, you will see the seam. Most needle loom types of tubular webbing do not meet the necessary strength requirements, and are not used in confined space applications.

For a long time, most people preferred one-inch tubular shuttle loom webbing for confined space operations. Traditionally it was softer and easier to tie than the stiffer flat-weave webbing; however, with advances in webbing manufacturing, flat-weave webbing has become softer and more manageable and now is used widely. Flat-weave webbing has the advantage of greater strength: One-inch flat-weave webbing provides 6000 lbf MBS, compared to 4000 lbf MBS for one-inch shuttle loom tubular webbing.

Webbing can be purchased from most manufacturers or distributors by the yard or in spool lengths of up to 300 ft. Length requirements depend, as with ropes, on the application in which the webbing is used. To tie most improvised harnesses, you will need 12- to 24-ft lengths of webbing. Lashing patients to backboards or into stokes baskets

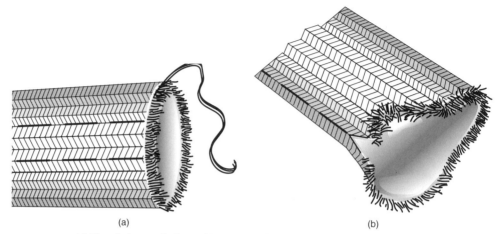

Figure 11.16 (*a*) Shuttle loom tubular webbing is stitched in a continuous spiral, so it is seamless; (*b*) needle loom tubular webbing is made with a flat piece of webbing that is folded and seamed.

requires a 30-ft length, while lengths as short as 10 to 12 ft may be appropriate for anchoring. Decisions on application will help you choose the correct lengths.

Webbing width ranges from $^1/_2$ to 3 in. In most confined space applications, one-inch webbing is preferred as long as it is used in a configuration that allows you to meet NFPA and OSHA requirements. Most harnesses are constructed with 1.75-in flat-weave webbing, and presewn anchor straps with 1.75- or 2-in flat-weave webbing. These are averages, and variations exist from manufacturer to manufacturer. Be sure that the webbing you use meets the appropriate regulatory requirements.

Care and maintenance of webbing

Inspection and care of webbing are similar to those for rope. A qualified person should inspect the webbing by sight and touch before and after each use. Any webbing that is excessively frayed, contaminated with chemicals, hardened, or discolored, or has been shock-loaded, overstressed, or exposed to any other condition that is questionable should be retired. Follow manufacturers' guidelines for washing and drying. Keep your webbing clean and dry and store it properly to increase its service life.

Harnesses and belts

Most prefabricated harnesses are constructed of nylon flat-weave webbing that has a minimum breaking strength of 6000 lb. Buckles and D rings are incorporated for adjusting the harness and attaching it to rope systems or other system components. Some harnesses may have padding added for comfort. Some rescue situations necessitate the need to improvise a harness on a patient or a rescuer. Methods for tying improvised harnesses are discussed in Chap. 18.

There are two classification systems for belts and harnesses: ANSI and NFPA. It is important to understand both systems.

ANSI classification of harnesses and belts

The American National Standards Institute standard ANSI Z10.14-1991 classifies body belts and harnesses into four major categories according to use:

Class I: body belts used to retain a worker in a hazardous work position and reduce probability of falls

Class II: chest harnesses used in horizontal applications where there are only limited fall hazards and for retrieval purposes

Class III: full-body harnesses used for fall arrest in situations where significant free falls may occur

Class IV: suspension belts used to suspend or support the worker

It is apparent in looking at the ANSI classification that only the class III full-body harnesses would be appropriate for any vertical application.

NFPA 1983 classification of harnesses and belts

NFPA classifies belts and harnesses differently than ANSI. NFPA distinguishes between belts and harnesses, and further distinguishes between two types of belts and three classes of harnesses.

Belts. Belts are separated into two categories in the 2001 edition of NFPA 1983: They are either ladder belts or escape belts. Ladder belts fasten only around the waist (Fig. 11.17) and are intended for use only as a positioning device to prevent a person on a ladder from falling. Escape belts also fasten only around the waist (Fig. 11.18) but are intended to be used for self-rescue, such as when the wearer needs to "bail out" of a dangerous situation by doing an emergency rappel down an escape line. Escape belts are not intended for use as positioning devices.

Life-safety harnesses. Life-safety harnesses are divided into three classes by NFPA 1983.

Class I: seat harnesses designed for emergency escape purposes only. The harness should fasten around the waist and thighs or under the buttocks of the wearer and should have a minimum design load of 1330 N or 300 lbf.

Class II: seat harnesses designed for rescue. The harness should fasten around the waist and thighs or under the buttocks of the wearer and should have a minimum design load of 2670 N or 600 lbf (Fig. 11.19).

Class III: full-body harnesses designed for rescue. The harness should fasten around the waist and thighs or under the buttocks, and over the shoulders of the wearer and should have a minimum design load of 2670 N or 600 lbf (Fig. 11.20).

Class III harnesses may consist of one or more parts. The CMC/ROCO harness is an example of a two-part harness, a class II seat harness, and a chest-only harness that can be connected and worn together as a class III harness (Fig. 11.21).

In previous editions of NFPA 1983, class I harnesses were described as being used with one person loads, class II with two person loads, and class III with two person loads or any time inverting may occur. These definitions were replaced in the 2001 edition to the standard with the classifications described above.

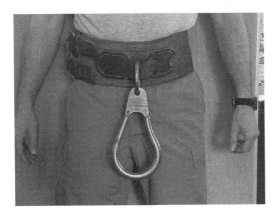

Figure 11.17 A ladder belt is used as a positioning device for a person on a ladder.

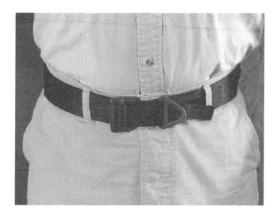

Figure 11.18 An escape belt is used by the wearer as an emergency self-rescue device.

Figure 11.19 The CMC®/Roco™ seat harness is classified as a NFPA clas II harness.

Figure 11.20 The Riggers full-body harness by Yates is classified as a NFPA class III harness and is designed with one-part construction.

Figure 11.21 The CMC®/Roco™ seat harness used with the chest harness is an example of a harness with two-part construction that can be used together as a NFPA class III harness.

Section 5.3 of NFPA 1983 lists a number of performance requirements that harnesses must meet in order to be certified. Several static and dynamic drop tests must be performed and passed before the harness is given NFPA 1983 certification. The testing requirements for harnesses are listed in Section 6.3 of the standard.

OSHA classification of harnesses and belts

OSHA's fall-protection standard (29 CFR 1926.500 Subpart M) defines a body harness as "straps which may be secured about the employee in a manner that will distribute the fall arrest forces over at least the thighs, pelvis, waist, chest, and shoulders with means for attaching it to other components of a personal fall arrest system." Only the ANSI class III or the NFPA class III full-body harness meets this definition.

The same standard defines a body belt or safety belt as "a strap with means both for securing it about the waist and for attaching it to a lanyard, lifeline, or deceleration device." Revisions to OSHA's fall-protection regulations allow the use of body belts only for fixed-position systems. They cannot be used in fall-arrest systems. OSHA specifies that the belts be at least $1^5/_8$ in wide.

Harness use in confined space operations

In confined space operations a full-body harness should be used because of the possible need to extract someone vertically through a small opening. If the entrant wears only a seat harness, the attachment point would be near waist level, putting the entrant in a more horizontal position if he is unconscious and needs to be lifted by the retrieval line. This could make rescue from outside the space very difficult. For this reason the PRCS standard requires that retrieval lines be attached to the harness in order to provide the most favorable orientation for nonentry rescue should the entrant become incapacitated. For vertical situations this requires attaching the retrieval line to the highest point of attachment available on the harness.

Some harnesses have D rings built into the shoulder straps and positioned at the tops of the wearer's shoulders. These provide points of attachment for a spreader bar above the wearer's head. The retrieval line is attached at the midpoint of the spreader bar. This provides a completely vertical orientation when the wearer is suspended from the retrieval system.

Only full-body harnesses should be used with personal fall-arrest systems or safety belay systems, as described in Chap. 13. This allows the shock of the arresting force to be distributed over a significant part of the wearer's body. Belts and seat harnesses should never be used with fall-arrest or belay systems, because any shock produced in arresting the fall would be focused at waist level. The resulting shock could produce a "head meets toe" effect on the person, resulting in serious injuries.

Inspection and care of harnesses

Inspection, care, and maintenance procedures for software components of harnesses are like those previously described for ropes and webbing. Inspection and maintenance procedures for hard components of harnesses such as buckles and D rings are like those described below for hardware. As with any other type of equipment, a record of the harness' history should be kept. Placing identification markings on the harnesses in a way that is acceptable to the manufacturer allows the use of each harness to be tracked. Keep harnesses clean and dry in storage. Follow all manufacturer guidelines for specific maintenance, inspection, and retirement procedures.

Shock Absorbers and Lanyards

In confined space operations the use of lanyards and/or shock absorbers may be required with fall-protection or safety belay systems, as described in Chap. 13. Shock absorbers (Fig. 11.22) are prefabricated load limiters designed to prevent the shock effect that otherwise might occur when a fall is arrested. The use of shock absorbers is intended to prevent shock-loading the system, which could result in severe personal injury and damage or failure of anchors, ropes, and other system components.

Lanyards are short lengths of rope or webbing with snap hooks prefabricated at each end that connect a worker's safety belt or harness to an anchor point (Fig. 11.23). They are commonly available in 2-, 4-, or 6-ft lengths. Lanyards with built-in shock absorbers are also available (Fig. 11.24). Retractable lanyards can be used where extra mobility is needed (Fig. 11.25). These may have built-in inertial brakes or shock absorbers.

NFPA 1983 requirements for lanyards and shock absorbers

NFPA 1983 does not specifically address requirements for lanyards and shock absorbers. These items fall under the heading of "auxiliary equipment" in NFPA 1983 and must meet the same design requirements as webbing and other auxiliary equipment components. Those requirements were discussed in the first section of this chapter.

As an example of how these requirements might apply, the Zorber™ shock absorber by Elk River Safety Belt Company is reported to have a minimum tensile strength of 6000 lbf. It is not NFPA 1983–approved, but its tensile strength would place it in the light-use category, indicating that it is suitable for use between a system and a one-person load. This shock absorber and others with tensile strengths less than 8093 lbf should never be placed between a system and a two-person load.

OSHA requirements for lanyards and shock absorbers

OSHA's fall protection standard (29 CFR 1926.502) requires that lanyards limit the free fall to a maximum of 6 ft and have a minimum breaking strength of 22.2 kN (5000 lbf). Self-retracting lanyards designed to limit free-fall distances to 2 ft or less must be able to sustain a minimum tensile load of 13.3 kN (3000 lbf). Self-retracting lanyards that do not limit free-fall distance to 2 ft or less, such as longer lanyards with shock absorbers prefabricated into the webbing, should be capable of sustaining a minimum tensile load of 22.2 kN (5000 lbf).

Shock absorber designs and ratings

Several designs of shock absorbers are available commercially. The most common design is constructed with webbing that has been looped and sewn with successive rows of stitching or bar-tacked. This shock absorber dissipates the energy of a fall arrest by the progressive tearing of the stitching in the webbing. These shock absorbers are some-

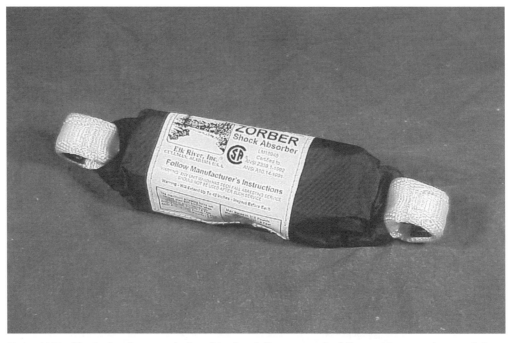

Figure 11.22   Shock absorbers are designed to absorb the energy of a fall arrest to prevent severe injury.

Figure 11.23   This 4-ft webbing lanyard is used to connect a worker's harness to an anchor.

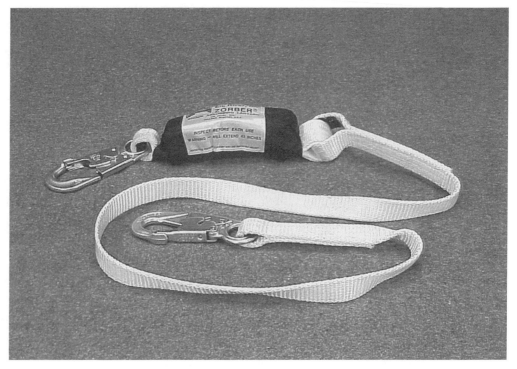

Figure 11.24 A lanyard with an integral shock absorber is used to connect a worker's harness to an anchor and has the added shock absorber to absorb the energy of a fall arrest if the worker falls.

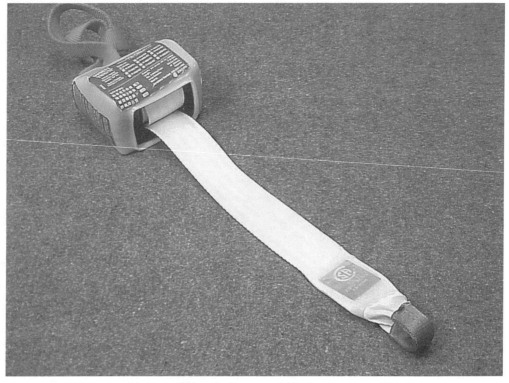

Figure 11.25 This retractable lanyard has a built-in inertial brake to arrest a fall.

times called "screamers" because of the loud noise produced when the shock absorber deploys. They may have a webbing eye at each end for connection to the system, or come with D rings or snap links prefabricated to the ends.

Most shock absorbers activate, or initiate tearing, at around 400 to 500 lbf. Shock absorbers also are rated with a maximum arresting force, tensile strength, and maximum extension distance. Shock absorbers meeting ANSI, OSHA, and the Canadian Standards Authority (CSA) requirements are found in several vendor catalogs.

Use considerations for shock absorbers

Shock absorbers should always be used between static ropes and live loads whenever the possibility of shock-loading exists, especially during training. A perfect example of this is safety belay lines, discussed in Chap. 13. The purpose of a safety line is to catch the load if the primary system fails. Significant impact forces could be placed on the person on the end of the line and everything else in the belay system without the use of the shock absorber. Most manufacturers recommend placing the shock absorber between the rope and the load, at the point where impact forces are transferred from the system to the person on the end of the line.

Individuals should be trained in the proper use of shock-absorbing devices in fall protection systems and safety belay systems before they are required to use them in the field. Any shock absorber or shock-absorbing lanyard that has been subjected to a fall arrest must be removed from service and replaced.

Inspection and care of lanyards and shock absorbers

Shock absorbers and lanyards should be inspected by a qualified person at regular intervals as suggested by the manufacturer, and records of inspections should be maintained. Prior to each use the user should visually inspect the device for cuts, hardness or stiffness, alterations in length, abrasions, knots, chemical damage, heat damage, or other sign of excessive wear. Some shock absorbers have a label or red mark on the webbing that shows if the absorber has been activated. If the shock absorber or lanyard fails the inspection, or if it has previously been activated, it must be removed from service immediately.

As with other types of software, shock absorbers and lanyards must be kept free of grease, dirt, and oil and stored in a clean, dry place away from the possibility of contamination.

Hardware and Accessories

Hardware and accessory components are vital parts of systems used in confined space operations. Examples of these components include carabiners, quick links, pulleys, descent control devices, rope-grabbing devices, and portable anchor systems. Hardware components used to handle live loads in confined space operations should be life-safety-rated, that is, designed for use with life-safety rope and harnesses. They should meet all specifications and strength requirements of applicable NFPA, OSHA, and/or ANSI standards. This ensures that only the highest-quality equipment is being used to support live loads.

In some cases, both life-safety-rated hardware components and hardware components rated for utility uses, such as hauling equipment, may be available for use in confined space operations. In such cases, the different types of equipment should be marked and stored in such a way as to prevent life-safety-rated equipment from being used for utility purposes and to prevent utility hardware from being used to support live loads. Everyone who uses the equipment must know and understand the differences.

General procedures for inspection, care, and maintenance of hardware

All hardware components have similar recommendations regarding inspection, care, and maintenance procedures. The common recommendations that apply to all hardware items will be covered here. Recommendations specific to individual hardware items will be incorporated into the discussion of each item.

Inspect hardware items before they are put into service and after each use. Be sure that each item is clean, dry, and free of scratches, gouges, sharp edges, corrosion, distortion, and excessive wear. Check that all warning labels and user information sheets are present. These inspections should follow manufacturers' guidelines and should be logged on a history card for each individual component. The history cards should track the specifics of what, when, where, and how the equipment has been used.

Store hardware components correctly according to the manufacturers' recommendations and keep them clean, dry, and away from chemicals or their vapors. If they get wet, dry them thoroughly before storing. Some items may require periodic lubrication. In such cases the lubricant must be applied precisely. Excessive lubrication can result in contamination of ropes or software components that could cause degradation of components.

Always consult the manufacturer for maintenance and upkeep procedures. Never make alterations or repairs that do not follow manufacturer guidelines or recommendations. Any unauthorized alterations could void equipment certification and place a certain amount of liability on the person doing the alteration. Before marking equipment by stamping or any other means, consult with the manufacturer as this is an alteration.

General considerations for using hardware

Hardware and accessory components must be used in such a way that the design load or working load of the equipment is never exceeded. This will be discussed further in Chap. 13. Never use a hardware component for any purpose other than the one for which it was designed without first gaining approval from the manufacturer.

Any hardware component that falls any significant distance onto a hard surface or has been severely shock-loaded should be evaluated for hairline cracks and other types of damage that would impair operation. Some types of damage are internal or invisible to the naked eye. X rays and other tests can rule out damage, but may be more costly than equipment replacement. Always consult the manufacturer if there is any question of the equipment's integrity, and follow all recommendations for testing, retirement, and replacement.

Carabiners

Carabiners and snap links function as mechanical knots. They offer quick, strong connections to D rings on harnesses, knots in ropes, anchor points, litter rails, pulleys, and many other components of a confined space rope system (Fig. 11.26).

NFPA 1983 requirements for carabiners. NFPA 1983 requires that all carabiners and snap links have self-closing gates with some type of locking mechanism. They are considered auxiliary equipment system components and are defined as "load-bearing connectors." Load-bearing hardware should be constructed of forged, machined, stamped, extruded, or cast metal.

Carabiners are classified as light-use or general-use based on minimum breaking strengths along the major axis with the gate closed (Fig. 11.27), along the major axis with gate open, and along the minor axis (Fig. 11.28), using testing specifications laid out in Section 6.5.1 of NFPA 1983.

Light-use carabiners. Light-use carabiners should have the following minimum breaking strengths:

Major axis with gate closed	27 kN (6069 lbf)
Major axis with gate open	7 kN (1574 lbf)
Minor axis	7 kN (1574 lbf)

General-use carabiners. General-use carabiners should have the following minimum breaking strengths:

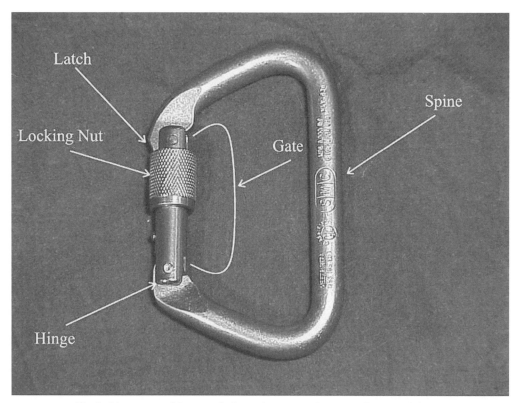

Figure 11.26 The locking nut on this screw-lock carabiner must be manually rotated down over the latch to lock the carabiner. This type of carabiner should be used in a system so that the locking nut fastens downward when the carabiner is loaded.

Major axis with gate closed	40 kN (8992 lbf)
Major axis with gate open	11 kN (2473 lbf)
Minor axis	11 kN (2473 lbf)

If your team has both general-use and light-use carabiners, take steps to ensure that team members can tell the difference between them.

OSHA requirements for carabiners. Paragraph (d) of OSHA's fall-protection standard (29 CFR 1926.502) requires that all connectors be made of drop-forged, pressed, or formed steel or equivalent materials and have a corrosion-resistant finish; and that all surfaces and edges shall be smooth to prevent damage to interfacing parts of the system. Snap hooks should have a minimum tensile strength of 22.2 kN (5000 lbf) and a minimum tensile load of 16 kN (3600 lbf). A snap hook is defined as

> A connector comprised of a hook-shaped member with a normally closed keeper, or similar arrangement, which may be opened to permit the hook to receive an object and, when released, automatically closes to retain the object.

Snap hooks are generally one of two types: the locking type (Fig. 11.29) with a self-closing, self-locking keeper which remains closed and locked until unlocked and pressed open for connection or disconnection; or the nonlocking type with a self-closing keeper which remains closed until pressed open for connection or disconnection. As of January 1, 1998, the use of a nonlocking snap hook as part of personal fall arrest systems and positioning device systems is prohibited.

Shape and size. Carabiners can be purchased in a number of different shapes and sizes. Shape is an important aspect of carabiner design that dictates where the strength of the carabiner lies and how it should be loaded. Carabiners are pear-shaped, oval, D-

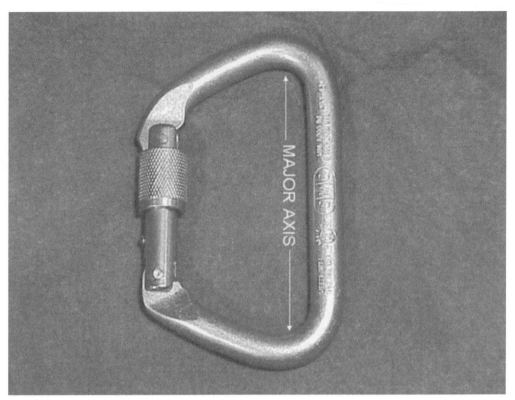

Figure 11.27 A steel screw-lock carabiner is pictured with its major axis labeled.

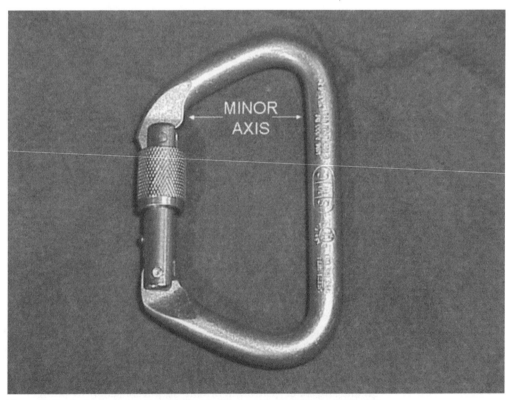

Figure 11.28 A steel screw-lock carabiner is shown with its minor axis labeled.

Figure 11.29   The snap hook or "lemon squeezer" has a self-closing, self-locking keeper.

shaped, or what is called the "offset D" shape (Fig. 11.30). The D and the offset D are the carabiners generally used in confined space operations. The strength of these two designs lies primarily in the spine of the carabiner (its major axis), and the shape allows for loading along this high-strength area (Fig. 11.31). These carabiners are designed to handle two-way pulls along the spine. If for any reason the load forces are placed anywhere else except along the major axis (Fig. 11.32), or if the carabiner is triloaded (Fig. 11.33), the carabiner's working strength is significantly reduced.

Carabiners of two sizes are used in confined space operations: the large and extra large carabiners. Large carabiners are the most common size seen in industrial confined space applications. They are big enough to accommodate knots tied with $\frac{1}{2}$-in kernmantle rope and will fit over the rails of most basket litters. They are accepted by most other general-use-rated hardware components, and can be purchased to meet NFPA 1983 general-use strength recommendations. One example is SMC's large steel carabiner, which has a gate opening of approximately 1.2 in and an inside dimension along the major axis of approximately 4 in.

Extralarge carabiners are preferred for certain applications, including operating a Munter hitch for a belay system (as we will see in Chap. 13) and applications that require the larger gate opening. The gate opening of SMC's extralarge steel carabiner is approximately 1.45 in and the inside diameter along the major axis is approximately 4.5 in (Fig. 11.34).

Construction materials.   Carabiners commonly are constructed of steel or aluminum. Some manufacturers also use steel alloys and stainless steel. The gates of SMC's steel carabiners are constructed of stainless steel to allow for better resistance to corrosion and higher strength.

Steel carabiners are stronger and more rugged than aluminum carabiners, and typically have a general-use rating; therefore, steel carabiners are recommended for general confined space applications. In climbing applications or applications where

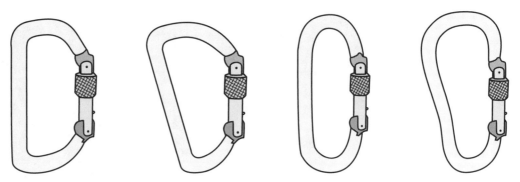

Figure 11.30 Common carabiner shapes are D, offset D, oval, and pear.

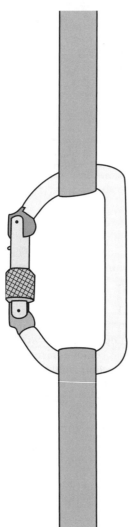

Figure 11.31 A carabiner should be loaded along its major axis.

equipment must be carried for long distances, aluminum carabiners may be used because of their light weight. Aluminum carabiners are typically light-use-rated and are therefore not favored for confined space operations.

Latches and gates.   Two latch designs commonly are seen on large steel carabiners: the pin-and-notch latch (Fig. 11.35) and the claw latch. The pin-and-notch latch is preferred

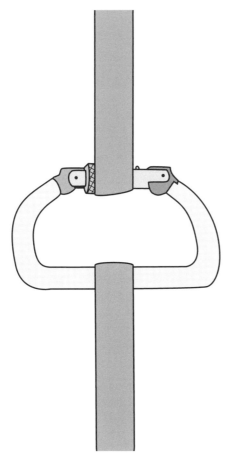

Figure 11.32 Care should be taken to prevent carabiners from becoming side-loaded, along its minor axis, in a rope system.

because when a carabiner with this type of latch is loaded, the notch slides over the pin, locking it into place and holding the gate closed (Fig. 11.36). The claw latch (Fig. 11.37*b*) is made of two machined notches, one on the gate and the other on the end of the carabiner body. Reportedly some types of claw latches are machined at such an angle that when they are loaded their gates are forced open unless the locking nut is engaged to hold the gate closed (Frank 1998, p. 33).

Carabiners with both locking and nonlocking gates are available. Nonlocking carabiners are not recommended for use in live-load applications because of the possibility of "rollout" (Fig. 11.37). Rollout can occur when a nonlocking carabiner becomes side-loaded or the software attached to it gets twisted around the carabiner, thereby forcing the gate open and allowing the load to be released. Nonlocking carabiners should be used only for utility purposes and other situations where live loads are not involved.

Both OSHA and NFPA require the use of locking gates. Three basic types of locking mechanisms are available on large steel carabiners: the screw lock, the autolock, and the safety lock.

Screw-lock carabiners. Figure 11.26 shows a screw-lock carabiner. The locking nut on the gate of this type of carabiner has to be manually rotated down over the latch to lock the carabiner.

Locking nuts should never be overtightened, or they may be very difficult to remove after the operation is completed. Once a carabiner is loaded, it will elongate slightly, making it possible to tighten the locking nut further. This should be avoided because after the carabiner is unloaded, it will return to its previous length, leaving the locking nut hopelessly jammed. In some cases movement due to vibration, gravity, or other causes may have the effect of tightening the nut while the carabiner is loaded. Jammed locking nuts can be loosened by reloading the carabiner or by using accessory

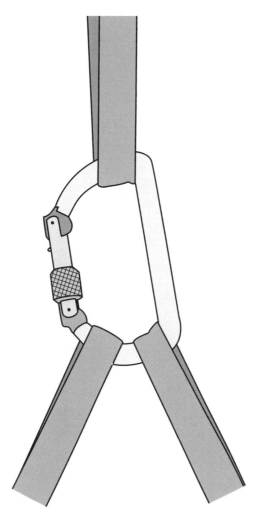

Figure 11.33 Avoid three-way pulls using carabiners.

cord or webbing to grip the nut. In extreme cases pliers may be required to free a stuck locking nut.

Establish a habit of checking to make sure that the gates are closed and locked when using this type of carabiner. If feasible, the carabiner should be positioned in the system in such a way that the locking nut fastens downward when loaded. If the locking nut screws upward, vibration and gravitational forces can work to unlock the nut. Always position a carabiner so that the gate faces the load to reduce the chance that movement in contact with edges or ropes will unlock the locking nut or push the gate open.

Autolocking carabiners. The locking nut on the autolocking carabiner requires a twisting motion to open the nut and a pushing motion to open the gate (Fig. 11.38). Once the gate is released, it closes and locks by itself under spring pressure.

Autolocking carabiners are quick and convenient to use. As long as the carabiner is in good working condition, if the gate is closed, then the carabiner is locked. Some rescuers have expressed concerns that contact of the locking nut with an edge or a rope in motion could unlock the gate easily since only a fraction of a turn of the nut is required to do so.

Safety-lock carabiners. The locking nut on the safety lock carabiner requires two separate motions to unlock the nut and a pushing motion to open the gate (Fig. 11.39). Unlocking the gate requires that the lock nut first be moved along the length of the gate, then rotated to complete the unlocking process. Just as with the autolocking carabiner, once the gate is released, it closes and locks under spring pressure. The safety-lock carabiner was designed to provide all the advantages of the autolocking type while elimi-

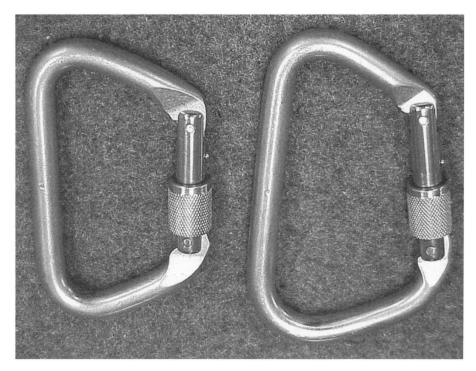

Figure 11.34 SMC's large and extralarge steel carabiners commonly used in confined space operations.

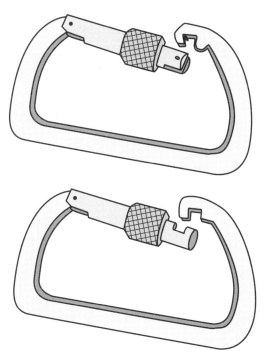

Figure 11.35 Common latch designs seen on large steel carabiners are the pin-and-notch latch and the claw latch.

nating any possibility of the gate accidentally being unlocked while in use, but some users find the double-locking mechanism difficult to manipulate.

Use, care, and maintenance of carabiners. Try to avoid the following situations when using carabiners:

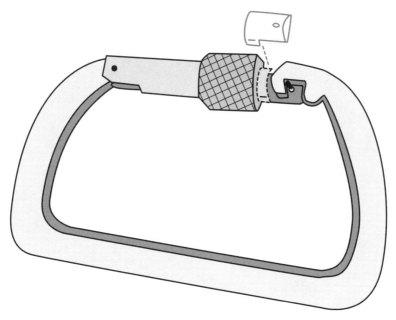

Figure 11.36 When the pin-and-notch-latch-type carabiners are loaded, the notch slides over the pin, preventing the gate from opening.

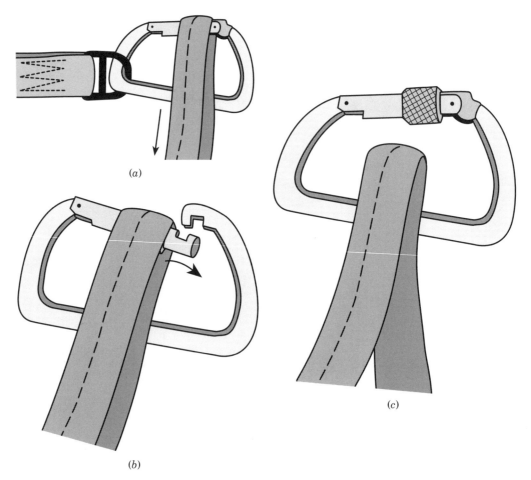

Figure 11.37 (*a*) Rollout can occur with nonlocking or unlocked carabiners when software becomes twisted. (*b*) This can force the gate open (*c*) Releasing the load.

Figure 11.38 The locking nut of the autolocking carabiner requires two motions to open the gate and, once the gate is released, closes and locks itself under spring pressure.

Figure 11.39 The locking nut of the safety-lock carabiner requires three motions to open the gate. Once the gate is released, it closes and locks itself under spring pressure.

- Dropping carabiners onto hard surfaces

- Overtightening screw locks

- Improper loading (loading any other way besides along the major axis)

- Triloading or multidirectional pulls

Some manufacturers' guidelines suggest lubricating the gate and nut periodically to protect from corrosion. Be careful not to overlubricate. Wipe any excess lubricant off before storing carabiners to prevent contaminating ropes or other software components.

Quick links

Quick links are load-bearing connectors with screw-type sleeves instead of gates (Fig. 11.40). They commonly are used in situations where carabiners are not recommended, such as when the possibility of multiloading exists. Quick links can be purchased in the following shapes:

- Triangular (delta)

- Oval

- Half-moon (semicircle)

The shape most useful in confined space operations is the triangular or delta shape, which is designed to be triaxially loaded. The "trilinks" are a good alternative to carabiners for applications such as anchor systems where the carabiners may be subjected to heavy, compounded, three-way loading (see Chap. 13).

Quick links are classified as auxiliary equipment in NFPA 1983 (Section 5.5.6) and are rated as light-use or general-use. Materials of construction include aluminum, galvanized steel, and stainless steel. Steel quick links commonly are available in $7/16$- and $1/2$-in diameters, both of which exceed the NFPA 1983 general-use strength requirements.

NFPA 1983 also specifies that auxiliary equipment with metal parts such as quick links must be tested for corrosion resistance. All testing procedures are listed in Chap. 6 of the standard.

When using quick links, be sure that the gate is fully closed. Quick links are slower to open and close than are carabiners, but are more secure when closed properly and cost less than carabiners with equivalent strength capacities. Avoid overtightening the screw sleeves—they should be finger-tightened only.

Pulleys

Pulleys are used in confined space operations to build mechanical-advantage systems and to allow for a change of direction in a moving rope system (Fig. 11.41). A more specific discussion of the pulley's role in confined space rope systems is found in Chap. 13.

NFPA 1983 requirements for pulleys. NFPA 1983 lists requirements for pulley system components under auxiliary system components in Section 5.5.5, and specifies that they be tested under requirements listed in Section 6.5.5. Like other system components, pulleys are rated for light use or general use.

Light-use-rated pulleys. Pulleys that are rated for light use should have a minimum tensile strength of at least 5 kN (1124 lbf) without permanent damage to the pulley or without damage to the rope, and should have a minimum tensile strength of 22 kN (4946 lbf) without failure. The becket, a central plate that extends below the pulley to form a connector on a light-use pulley, should have a minimum tensile strength of at least 12 kN (3709 lbf) without failure.

General-use-rated pulleys. Pulleys that are rated as general-use should have a minimum tensile strength of at least 22 kN (4946 lbf) without permanent damage to the pul-

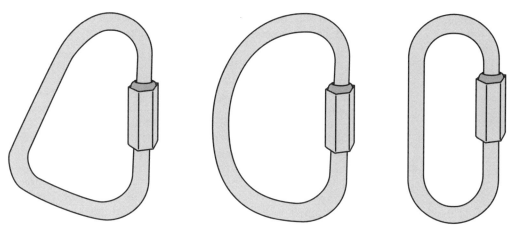

Figure 11.40 The triangular (delta), half-moon (semicircle), and oval are common quick-link shapes.

ley or the rope, and should have a minimum tensile strength of 36 kN (8093 lbf) without failure. Beckets on general-use pulleys should have a minimum tensile strength of at least 19.5 kN (6070 lbf).

Construction materials. Most pulleys that meet NFPA 1983 general-use rating have stainless-steel sideplates, axle, and nut, and some type of aluminum alloy sheave. Avoid pulleys with plastic sheaves, as the sheaves can distort under a load and cause added friction in the system. Some manufacturers offer double-locked axle nuts marked with witness lines that make loosening detectable at a glance.

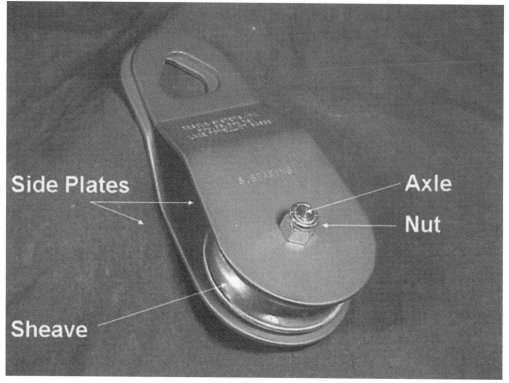

Figure 11.41 A single-sheave pulley is pictured with the basic parts labeled.

Pulleys can be purchased with either sealed ball bearings or Oilite™ bronze bushings. Table 11.4 lists the advantage and disadvantages of each type.

Use considerations for pulleys. Pulleys can be purchased with either single or double sheaves and with or without beckets (Fig. 11.42). The application that the pulley will be used for will dictate how many sheaves are needed and whether a becket is needed. Most confined space entry and rescue teams have both types. When selecting a pulley for purchase, several other factors should be considered.

To begin with, make sure that the width of the sheave is large enough to accept the diameter of the rope you will be using. In most instances the pulleys used need to be able to accept $\frac{1}{2}$-in kernmantle rope. The tread diameter of the sheave should be at least 4 times the diameter of the rope in order to avoid acute bends in the rope (Fig. 11.43) and to provide for more efficient operation of the system. This is called the *four-to-one rule* and applies to any anchor or piece of equipment that a rope will pass around or through.

Keep in mind that the diameter of the tread (the slot machined into the sheave to accept the rope) is smaller than the outside diameter of the sheave. A 2-in pulley has a tread diameter of only $1\frac{1}{2}$ in and would not meet the four-to-one rule for use with $\frac{1}{2}$-in kernmantle rope. By comparison, a 3-in pulley has a tread diameter of $2\frac{3}{8}$ in, which would be well over 4 times the diameter of the $\frac{1}{2}$-in kernmantle rope. As a general rule, as the diameter of the sheave increases, so does the efficiency of the pulley. This is because the larger sheave diameter provides more leverage to the rope moving around the outside of the sheave for overcoming friction at the axle in order to turn the sheave. This is why 4-in pulleys are preferred by some people working in confined space operations. Obvious disadvantages are the added weight, bulk, and expense of larger pulleys.

Keep in mind that when mechanical-advantage systems are used, ropes with diameters smaller than that of $\frac{1}{2}$-in kernmantle rope can be used while still meeting general-use strength requirements because the load is divided evenly over multiple lines. If smaller-diameter ropes are used, then smaller-diameter pulleys can be used while still meeting the four-to-one rule. This is one of many things that make small, prebuilt mechanical-advantage systems attractive for confined space operations. We will discuss this further in Chap. 13.

Specialized pulleys. Several types of pulleys with special design features are available. Examples include Prusik minding pulleys, pulleys with built-in rope grabs, and knot-passing pulleys. These design features give the pulleys special operating capabilities.

Prusik minding pulleys. Prusik minding pulleys were originally designed for use with systems in which the Prusik hitches were used as soft rope grabs for safety belay systems, as we will see in Chap. 13. These pulleys are designed with extended sideplate cor-

TABLE 11.4 Sealed Ball Bearings versus Oilite Bronze Bushings

	Pros	Cons
Sealed ball bearing	Lower friction, more efficient	Cost significantly more
	Bearings permanently sealed against contamination with dirt, water, etc.	If bearings do become contaminated, they can't be readily cleaned and lubricated
		Not as strong as bushings
Oilite bronze bushings	Cheaper	More friction, not as efficient
	Stronger	Can readily pick up dirt, water, etc.
	Can be easily disassembled, cleaned, and lubricated	

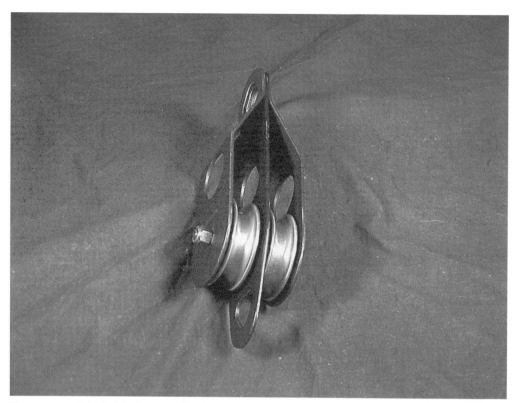

Figure 11.42   This double-sheave pulley has a becket.

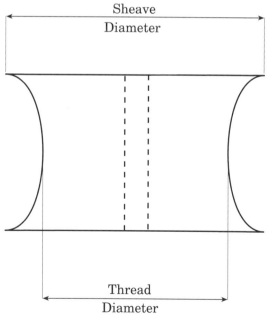

Figure 11.43   The tread diameter of the sheave should be at least 4 times the diameter of the rope.

ners intended to prevent a Prusik hitch from feeding into the pulley and jamming during hauling operations (Figs. 11.44 and 11.45). Pulleys of this design also work well with mechanical rope-grab devices, such as the industrial rope grab or the Gibbs ascender. This feature eliminates the need of having to have someone assigned to "mind" the rope-grabbing device when hauling.

Pulleys with built-in rope grabs. Pulleys such as the Haul Safe™ pulley by RSI are double-sheave pulleys that include a cam prefabricated to an extended middle plate (Fig. 11.46). These pulleys are primarily used in four-to-one mechanical-advantage systems for vertical access to confined spaces. They are marketed as self-tending pulleys, and the cam usually can be locked in the open position if needed. Several manufacturers, including RSI, CMC, and SFT, offer variations of this option that are rated for general use.

Knot-passing pulleys. Knot-passing pulleys offer up to 2.5 in between sideplates, which allows enough space for a knot in $\frac{1}{2}$-in kernmantle rope to pass through. These pulleys are designed with flat bottoms that give them the ability to double as edge rollers. They are supplied with two pins that allow the sheave to be locked into place so that the pulley can be used as a tie-off point.

Care and maintenance of pulleys. When inspecting pulleys, check the sideplates for damage, including cracks or bends, elongation of the carabiner hole or becket, or any other type of distortion. The sideplates should rotate freely. Check nuts to be sure that they are properly positioned and tightened. Sealed ball bearings should not allow dirt and grit to enter unless damage has occurred. Oilite bushings can be disassembled, cleaned, and lubricated if needed. Keep records tracking the dates and results of each inspection.

Descent control devices

Descent control devices are used to control the friction on ropes used in lowering operations, belays, and rappels. Controlling friction allows the rate of descent to be controlled. The two most commonly used descent control devices in confined space operations are the figure 8 descender with ears and the brake bar rack. Descent control devices typically are not used in routine confined space entry operations but frequently are used in rescue operations. We will introduce these devices here and elaborate on their use in rappels and lowering systems in Chaps. 17 and 18.

NFPA 1983 requirements for descent control devices. NFPA 1983 requires that all descent control devices be tested in accordance with Section 6.5.3. Descent control devices are rated as escape, light-use, or general-use.

Escape-rated descent control devices. Descent control devices rated only for escape should withstand a minimum test load of 5 kN (1124 lbf) without permanent damage or deformation to the device or damage to the rope, and a minimum test load of 13.5 kN (3034 lbf) without failure.

Light-use-rated descent control devices. Descent control devices rated as light-use should withstand a minimum test load of 5 kN (1124 lbf) without permanent damage or deformation to the device or damage to the rope, and a minimum test load of 13.5 kN (3034 lbf) without failure.

General-use-rated descent control devices. Descent control devices rated as general-use should withstand a minimum test load of 5 kN (1124 lbf) without permanent damage or deformation to the device or damage to the rope, and a minimum test load of 22 kN (4946 lbf) without failure.

Figure 8 descender with ears. The figure 8 devices are available in several variations and sizes, both with and without ears. They are used for descent control in rappelling

Figure 11.44 The Prusik minding pulley has squared-off sideplates to prevent the Prusik hitch from jamming the pulley.

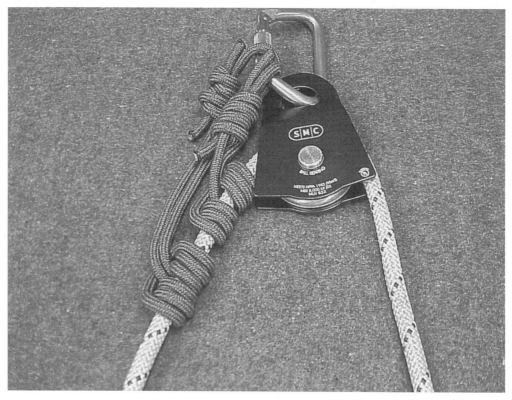

Figure 11.45 A Prusik minding pulley can be used with tandem triple-wrapped Prusik hitches.

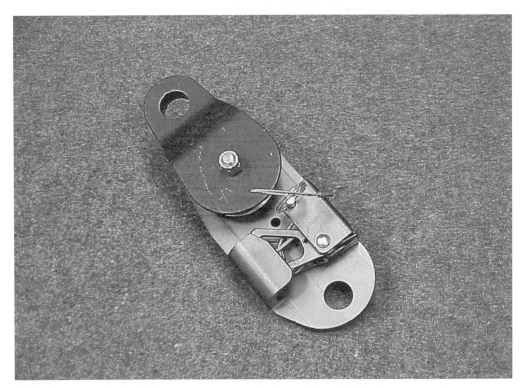

Figure 11.46 The Haul Safe Pulley by RSI is a self-tending pulley with a built-in rope grab.

and for lowering and belay systems. Ears were added to the basic figure 8 descender (Fig. 11.47) to prevent the rope from sliding up and creating a girth hitch at the top of the device. The ears also make the device easier to tie off, as we will see in Chap. 17. Figure 8 devices are small, light, relatively inexpensive, and easy to use. These are all reasons why the figure 8 descender with ears is commonly used.

Construction materials. Most figure 8s are made from drop-forged or machined anodized aluminum that is both strong and lightweight. Some manufacturers produce nonanodized aluminum versions, and even steel versions, depending on the wear characteristic desired. Anodizing the aluminum adds a protective oxide coating to the aluminum and increases the hardness at the surface to prevent wear while decreasing the amount of friction between the device and the rope. This decreased friction requires a stronger grip on the rope to control the descent. Steel figure 8 devices are even harder and have even less friction, thus requiring an even stronger grip on the rope for control of the descent.

Use considerations. Most recommendations suggest that the figure 8 descender be used to perform only short one-person rappels or lowers. Use is limited because the friction is controlled only by the tension that the rappeller or operator has on the control rope. If the tension on the control rope is released, the figure 8 allows the rappeller or the load to drop rapidly. Another limitation of the figure 8 is that it causes the rope to twist or "pigtail" on long rappels, causing rope management problems. Longer rappels or fast descents can cause the points where the rope touches the device to heat up so severely that damage to the rope occurs.

The user should understand specific rope-in and tie-off procedures before attempting a rappel or lower. Chapter 17 addresses these procedures. See Figs. 11.48 and 11.49 for illustrations of the figure 8 descent control device.

Brake bar rack. The brake bar rack consists of a steel frame or rack that contains six metal bars (Fig. 11.50). It can be converted to rappel or to lower either one- or two-per-

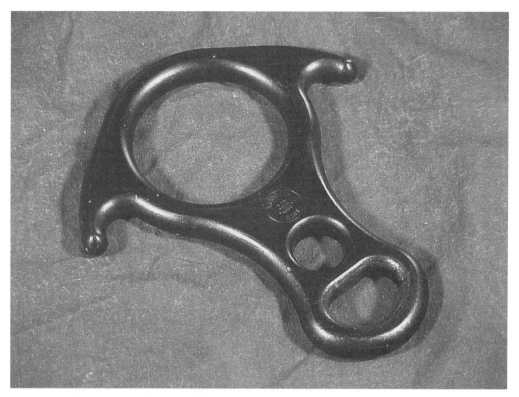

Figure 11.47 The figure 8 with ears is a descent control device used in belays, as well as short one-person rappels and lowering operations.

son loads simply by changing the number of bars in the rack, which can be done easily under load. The brake bar rack is the recommended device for descent control because it provides a very wide range of friction control. It works well for long rappels and lowers because it does not tend to heat up as quickly as the figure 8 descender or to twist or pigtail the rope as does the figure 8 device. The brake bar rack can be used in single-line lowering systems with one- or two-person loads, or two racks can be used in a double-line system to lower one-, two-, or three-person loads.

Design and Construction Materials.   There are several variations of brake bar racks on the market. Those rated for general use typically have a stainless-steel frame with a welded eye that is rated by most manufacturers for over 44 kN (10,000 lbf). The bars are constructed of stainless steel or aluminum and are solid or U-shaped (hollow on the underside of the bar). The top bar is grooved for easy rope alignment, and on some versions the second bar is grooved also. One of the new features on new-style racks is the extended tie-off bar as seen in Fig. 11.50. Brake bar racks can be configured for right-handed or left-handed users simply by reversing the orientation of the bars.

Use considerations.   The user must understand specific procedures for roping in, operating, and tying off the rack before attempting a rappel or lower. These procedures and a description of the use of the brake bar rack in rescue are specifically addressed in Chap. 17. See Figs. 11.51 and 11.52, where brake bar racks are pictured.

Rope-grabbing devices

In rigging rope systems it is sometimes advantageous to be able to create a point of attachment on a rope that can be moved along the rope and reset as needed. Rope-grabbing devices serve this function. Some rope-grabbing devices also can be used to grip a belay line in order to stop a falling load.

Figure 11.48 The figure 8–type devices are easily and quickly roped in.

Rope-grabbing devices can be of either the soft type, such as a loop of accessory cord Prusik-hitched to a rope, or the mechanical type. The discussion here will be limited to the mechanical rope-grabbing devices. The Prusik hitch is shown in Chap. 12. Its applications in belay systems are discussed in Chap. 13, and its use in ascending a fixed line is described in Chap. 17.

Mechanical rope-grabbing devices (Fig. 11.53) are commonly used as a ratchet cam in a hauling system, as the haul cam or point of attachment of a mechanical-advantage system to a main line, in fall-protection systems, and for ascending a rope. There are several types of mechanical devices on the market, including the Gibbs ascender, the Petzl Rescucender™, MIO Industrial rope grab, and several handle ascenders such as the CMI Expedition™ and the Petzl ascender. Each has advantages and disadvantages, and not all can be used for every purpose stated above. Handle ascenders are not used in confined space operations and will not be discussed here. We will limit our discussion to the other examples of mechanical ascenders mentioned above.

NFPA 1983 requirements for mechanical rope grabs.   NFPA requires ascending and rope-grab devices to be tested in accordance with Section 6.5.2 of 1983. Ascending devices should withstand a minimum test load of 5 kN (1124 lbf) without permanent damage to the device or damage to the rope. Rope-grab devices should withstand a minimum test load of 11 kN (2473 lbf) without permanent damage to the device or damage to the rope to be rated for general use.

Rope grab design and construction materials.   Most ascender components are made of forged or machined aluminum, anodized aluminum, or stainless steel (Fig. 11.54). Ascenders with stainless-steel shells offer more protection and resistance to hostile environments, but do not necessarily increase the strength of the device.

Figure 11.49   Figure 8s can be tied off to prevent slippage of the system during a rappel or lower.

As the term implies, ascenders were originally intended for use in ascending fixed lines. Items such as the Gibbs ascender were converted later for use as cams in hauling systems. Problems have occurred in the past when these systems were shock-loaded or overloaded and the rope was damaged, or in extreme cases severed, by the pinching action of the ascender. Design modifications have been made to reduce the likelihood of this type of problem, but it continues to be a major cause for concern.

The larger the contact area the cam has with the rope, the less chance of rope damage occurring. The newer-model Gibbs ascenders (Fig. 11.55) have two dents along the back of the shell that are intended to form a curve in the rope that will match roughly the curvature of the cam as the cam locks down on the rope. The curve allows for a larger surface area of the rope to be pressed against the shell by the cam. The increased surface area provides a more effective braking action and decreases the chance that the rope will be damaged by the cam. A more effective means of achieving this effect is evident in the Petzl Rescucender™, which is designed with an aluminum shell machined to match the curvature of the cam. Cams in the Gibbs ascender and Petzl Rescucender™ are designed with smooth, rounded teeth to reduce the potential for damaging the rope.

Avoid ascenders having cams with sharp teeth. If the system is shock-loaded or overloaded, the sharp teeth can damage or sever the rope. Some manufacturers apply a coating such as polyurethane to the cam to increase grip and prevent the rope from sliding through the cam when loaded. Others do not coat their cams so they will slip on the rope instead of failing the rope in the event of a shock load.

Some cams are spring-loaded and will support their own weight when attached to a rope, which prevents them from sliding out of reach of the user when rigged onto a vertical line. Cams that are not spring-loaded are referred to as "free-running" and may have advantages for applications such as haul cams in rescue systems (see Chap. 17).

Ascenders are available in different sizes to accept specific diameters of rope. The Gibbs ascender is available in a $\frac{1}{2}$-in size that will accept rope diameters of $\frac{3}{8}$ to $\frac{1}{2}$-in

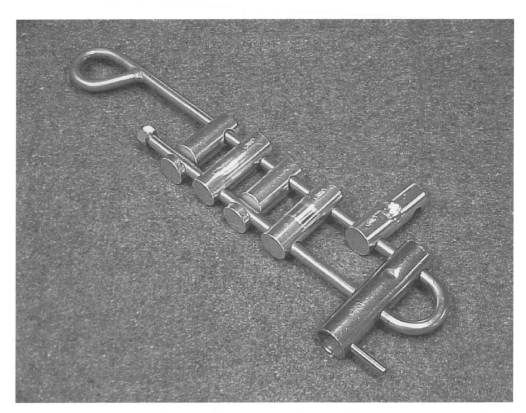

Figure 11.50   The brake bar rack is another descent control device used for one- or two-person rappels and lowers.

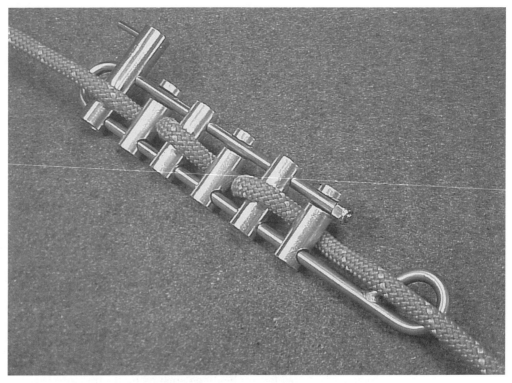

Figure 11.51   The brake bar rack must be roped in correctly.

Figure 11.52 A brake bar rack can be tied off to prevent slippage of the system during a rappel or lower or any time that the operator needs free hands.

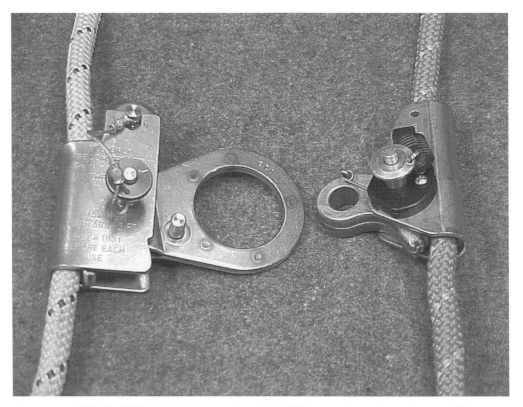

Figure 11.53 Both the MIO industrial rope grab and the Gibbs ascender can be used as a ratchet cam or a haul cam.

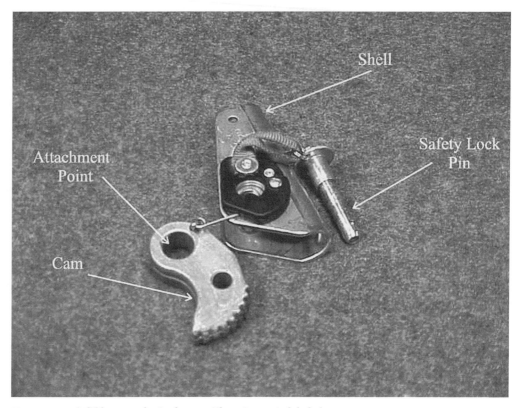

Figure 11.54 A Gibbs ascender is shown with major parts labeled.

and a $^3/_4$-in size that will accept a $^5/_8$-in-diameter rope. The Petzl Rescucender is designed to accept either $^7/_{16}$- or $^1/_2$-in-diameter rope. The MIO industrial rope grab is available in a $^1/_2$-in size that will accept $^1/_2$-in-diameter rope and a $^5/_8$-in size that will accept a $^5/_8$-in-diameter rope.

The MIO industrial rope grab differs from items such as the Gibbs ascender in that it was designed as a rope grab for use in fall-protection systems, not for ascending a rope. For this reason it is designed to accommodate a shock load without damaging the rope. Toward this end it uses a flat jaw rather than a curved cam to grab the rope. The MIO industrial rope grab is designed to slip and gradually lock onto the rope if a shock load occurs.

Some rope manufacturers have tested and approved the use of specific industrial rope-grab devices with their ropes. PMI, for example, has specifically approved the use of the MIO industrial rope grab with the PMI $^1/_2$-in EZ Bend and Max-Wear rope.

Use considerations for mechanical rope grabs. Mechanical rope-grab devices function by pressing the rope between the curve in the shell and the cam. Making sure that the rope grab is installed properly and pointing in the correct direction is crucial to its operation. The pin of the rope grab must be snapped completely into place after the device has been placed on the rope. If both grooves of the pin are not in place, the pin can back its way out, allowing the release of the cam. Some pins have either a button- or spring-type safety lock.

The etched arrow on the shell of the rope grab should be installed pointing toward the load in most applications. Always test the rope grab to make sure that it grabs the rope to ensure that it has been installed properly.

Mechanical rope grabs may damage the rope to which they are attached if a shock load occurs, and mechanical grabs used as haul cams in hauling systems may damage the rope to which they are attached if subjected to excessive force. These potential problems and precautions to help prevent them are discussed further in Chap. 13. Because of concerns about

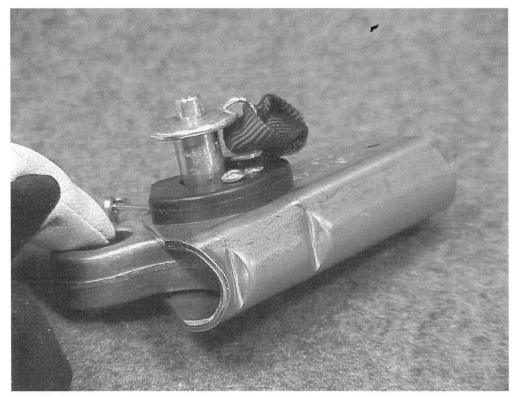

Figure 11.55 The two dents shown here on the newer-model Gibbs ascender are intended to form a curve in the rope that will roughly match the curvature of the cam to decrease the chance of the rope being damaged by the cam.

the capability of mechanical rope grabs to damage ropes, some rescue organizations do not use mechanical grabs at all. As an alternative, they use only soft rope-grabbing devices.

Care and maintenance of mechanical rope grabs.   Most manufacturers suggest the following minimum inspection for rope-grabbing devices. Check the cam assembly to make sure that it is clean, smooth, and moving freely. Release the safety-locking pin and pull out the pin, which should pull out smoothly and quickly. Swing the cam up and away so that the rope channel is visible. The rope channel's inner surface should be smooth and clear of any distortion or scratches. Swing the cam back into place and reinsert the pin, which should go back into place without sticking or jamming. All manufacturer's labels and warnings should be legible.

If a rope-grabbing device has been shock-loaded, most manufacturers suggest immediately removing it and the rope from service. After every use clean and dry all parts of the rope-grab assembly to remove dirt and moisture. Store in a clean, dry place away from the possibility of chemical contamination.

Miscellaneous Equipment

In addition to the types of equipment already described, a variety of other items are handy for rigging. Examples include anchor plates, swivels, edge pads and rollers, and rigging rings.

Anchor plates

Anchor plates, also known as *rigging plates* and "bear paws" (Fig. 11.56), are used to organize different system components at an anchor location. This allows for easier safety checks because the different components are easier to see and touch. It also allows for a faster changeover from one system to another, as we will see in Chap. 17.

Anchor plates are manufactured from aluminum or stainless steel. As one example, CMC's aluminum anchor plate has a tensile strength or breaking strength of 39 kN (8900 lbf), while the strength of the stainless-steel version is 40 kN (9150 lbf).

Swivels

Life-safety-rated swivels, such as the Petzl swivel (Fig. 11.57), are useful in applications where lines can twist, such as in stairwell lowers (see Chap. 18). The Petzl swivel uses a sealed ball bearing that rotates freely when loaded. The top of the swivel is triangular in shape and accepts up to three carabiners. The Petzl swivel is rated for a working load of 5 kN (1125 lbf) and has a tensile strength of 40 kN (9000 lbf).

Edge pads, rollers, and sliders

Edge pads protect rope, webbing, anchor straps, and other types of software from sharp edges and rough surfaces that could cut or abrade. Edge pads also are used to keep ropes and other types of software out of dirt to prolong the service life of the equipment. Edge pads protect the edge or anchor point itself from being cut into by the rope or webbing. This reduces the possibility of loosening materials that can fall down on people below.

Edge pads can be prefabricated or homemade, but need to be constructed of a heavy, stiff material such as canvas. It is always a good idea to pad any questionable edges. Edge pads can be wrapped around anchor points and held in place with duct tape to protect the anchor system. Figure 11.58 shows edge pads in use.

Edge rollers and edge-sliding devices are used to reduce the friction created when moving ropes in a system move over an edge. They are available in many designs. The typical roller design holds the rope 1 or 2 in off the surface and should be used in conjunction with a second roller that is linked with the first roller (Fig. 11.59). Edge rollers should be anchored to the edge to prevent them from falling off the structure or flipping if the rope moves off center. If you do not have an edge roller, some flat-bottomed knot-passing pulleys can double as edge rollers, or a directional pulley can be used to pull the rope from the edge. Edge pads should be placed under edge rollers for added protection in case the rope comes off the roller.

Edge-sliding devices are made of a slippery plastic surface that reduces friction and provides edge protection. Edge-sliding devices are not commonly used in industrial applications, but work particularly well in climbing and caving applications.

Never use a single edge-protection device for both a stationary rope and a moving rope. Using a separate edge-protection device for each rope will ensure that the ropes do not come into contact with each other. This prevents rope-on-rope wear. Remember that a moving loaded rope can cut through a fixed, stationary rope very quickly.

Rigging rings

Forged steel and steel alloy rigging rings can be used anywhere in a system in which multidirectional pulls may occur or anywhere you need one strong connection point (Fig. 11.60). For example, a rigging ring might be attached at the end of a lowering line to serve as the point of attachment for multiple carabiners connecting an improvised litter bridle to the lowering line. There are endless possibilities for using rigging rings in confined space operations. They are inexpensive and strong. Half-inch rigging rings have tensile strengths up to 25,000 lbf, while $9/_{16}$-inch rigging rings have tensile strengths up to 40,000 lbf.

Portable Anchor Systems

Portable anchor systems serve as temporary anchors when engineered anchor points are not available. Tripod types, quadpod types, and boom types are all examples of portable

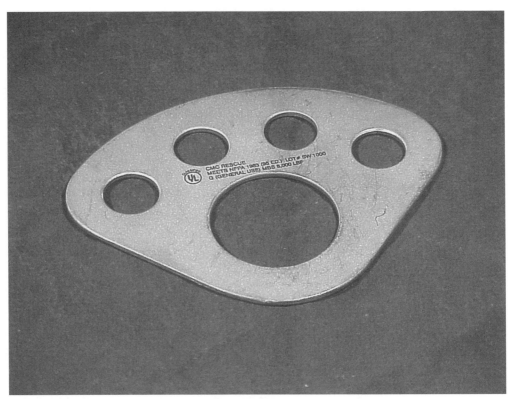

Figure 11.56 Anchor plates are used to provide multiple connection points at the anchors.

Figure 11.57 The Petzl swivel is useful in applications where lines can twist.

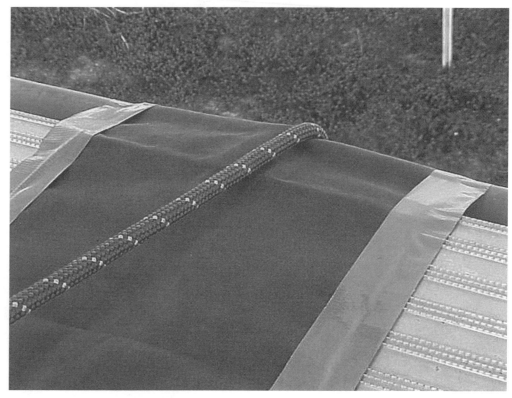

Figure 11.58 Edge pads are used to protect any software component of a system from sharp or rough surfaces that could cut or abrade.

anchor systems that can be positioned over an opening to allow for vertical lifts (Figs. 11.61 and 11.62). The tripod and quadpod devices usually have telescoping legs that can be adjusted to various lengths to accommodate different entry positions. The working load of these devices usually decreases with height, while the distance the legs must extend outward to maintain stability increases. In rescue, the device needs to be high enough to pull packaged patients clear of the entrance to the space. Tripods that have a maximum extension of 7 ft or less are not sufficient for some rescue applications.

Not all portable anchor systems are rated for human loads. Make sure that if you plan to use your tripod to support human loads, it is rated for such loads. Never exceed the working load of the device.

Boom-type devices can be freestanding or mounted on existing supports, tank openings, or truck trailer hitches. There are preassembled boom-type devices available that are configured either for horizontal or vertical pulls.

Most devices can be used with mechanical-advantage systems rigged with ropes (see Chap. 13) or can be purchased with a winch that typically comes with $1/4$-in galvanized- or stainless-steel wire rope. Only winches rated for human loads should be used to transfer people during confined space operations. Both powered and manually operated winches are available; however, powered winches should never be used in confined space operations because of the possibility of personal injury resulting if a body part becomes hung up while someone is being transferred with it.

Always follow manufacturers' guidelines for the assembly, use, care, maintenance, and retirement of portable anchor devices. Portable anchor systems and winch operations are covered more fully in Chap. 13.

NFPA requirements for portable anchor systems

NFPA 1983 rates portable anchor systems for either light use or general use when tested in accordance with Section 6.5.4 of the standard.

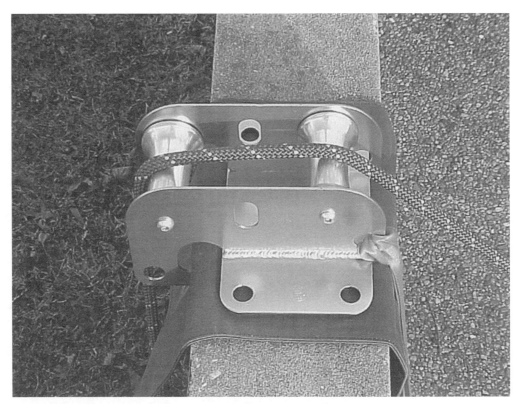

Figure 11.59   Edge rollers reduce friction created by ropes moving over an edge.

Figure 11.60   This $\frac{1}{2}$-in rigging ring has a tensile strength of 25,000 lbf and can be used anywhere in the system where a strong connection point is needed.

Figure 11.61 A tripod-type portable anchor system with telescoping legs can be adjusted to various lengths to accommodate different entry positions.

Light-use portable anchor systems. Portable anchor systems rated for light use should withstand a minimum load of 5 kN (1124 lbf) without permanent damage or visible deformation to the general shape of the device and should withstand a minimum load of 22 kN (4946 lbf) without failure.

General-use portable anchor systems. Portable anchor systems rated for general use should withstand a minimum load of 13 kN (2923 lbf) without permanent damage or visible deformation to the general shape of the device and should withstand a minimum load of 36 kN (8093 lbf) without failure.

Figure 11.62 A boom-type portable anchor system is positioned over an opening for a vertical lift. (*Photograph courtesy of Unique Concepts Ltd.*)

Summary

In this chapter we introduced various types of equipment that may be used during confined space operations. Most of these items may be used for confined space entry, and all may be used for confined space rescue. They are considered the basic building blocks for the systems we will learn how to assemble in later chapters. It is important that each individual component be properly selected, cared for, and used, since a system is only as strong as its weakest component.

Tying Knots

A group of firefighters was conducting rappel training on the side of a tall hotel. No harnesses were available, so the firefighters were using the *Swiss seat,* an improvised harness tied with webbing. Rigging the Swiss seat should be completed by using a square knot to connect the webbing ends, then adding two safety knots on each side of the square knot. During one of the rappels, a firefighter's Swiss seat suddenly failed, and he fell most of the way down the side of the building onto a lower roof. Although very seriously injured, the firefighter miraculously survived. An investigation revealed that a granny knot instead of a square knot was used to complete the Swiss seat and that no safety knots were used.

There are hundreds of different types of knots used in thousands of different applications. Assembled in this chapter is a collection of knots commonly used in confined space operations. The intent of this chapter is instructional. A brief description is included of each knot's application in confined space entry and rescue, but the focus will be on how to tie the knot. Later chapters will describe and show applications in more depth.

A knot may be known by several different names. The names used here are the ones most commonly used in other confined space rescue references. Confined space entry and rescue teams should decide on common terminology for knots and standardize their applications, thereby eliminating confusion during entry or rescue operations.

Only a few knots are necessary for most confined space operations. It is more critical for team members to be proficient with these knots than to know partially a large number of knots they will not remember how to tie in time-critical situations. Standardizing the knots to be used, and the applications where they will be used, will reduce the number of knots team members must learn and will make safety checks easier.

Knot-tying is an individual skill that must be practiced. Repetition is the mother of skill. As you spend more time practicing and training with these knots, you will become more confident until tying and using them becomes second-nature. This will eliminate mistakes and confusion when you are facing an emergency situation.

Strength Reduction Due to Knots

Whenever a kernmantle rope is knotted, an overall reduction of strength occurs. The sharper the bends needed to form a particular knot, the more the overall strength of the system is reduced. Remember the four-to-one rule—anytime a rope bends around another object (another rope, a carabiner, etc.), that object should be at least 4 times larger than the diameter of the rope, or there will be a reduction in strength of the system. This reduction occurs with bends less than 4 times the diameter of the rope because the core fibers on the outside of the bend are stretched and forced to hold most of the load, while the core fibers on the inside of the bend are puckered and loose. When a rope is knotted, less than 100% of the core fibers are involved in supporting the load. The percentage of strength loss due to the configuration of the knot in the rope is proportional to the percentage of fibers not supporting the load. Knots in the figure 8 family have easier bends that allow a larger percentage of the core fibers to be involved in holding or supporting the load, making these knots preferable to other types of knots when supporting human loads. See Table 12.1 for a list of knots and the percentages of rope system strength reduction they cause.

Common Terminology

There are certain words and phrases you need to know and understand before using the instructions in this chapter:

Working end—the part of the rope used to form the knot.

Running end—the part of the rope used for hauling or belaying.

Standing part—the part of the rope between the running end and the working end.

Bight—formed by simply bending the rope back on itself 180° to form a U shape in the end (Fig. 12.1).

Loop—formed by crossing the running end of the bight over the standing part to form a 360° turn or loop (Fig. 12.2).

Round turn—formed by continuing the running end of the loop around so that both ends are emerging from the same side (Fig. 12.3).

TABLE 12.1   Percentages of Strength Lost for Various Knots

Knot	Percentage of strength lost, %
Figure 8 on a bight	20
Figure 8 bend	19
Figure 8 followthrough	19
Double-loop figure 8	18
Bowline	33
Butterfly knot	25
Water knot in webbing	36
Double fisherman's knot	21

SOURCE: *CMC Rope Rescue Manual*, 3d ed. (Frank 1998), p. 56.

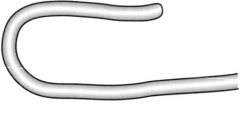

Figure 12.1 A bight of rope is formed by bending the rope back on itself to form a U shape.

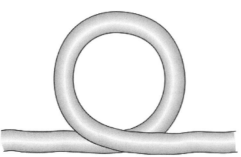

Figure 12.2 A loop in a rope is a 360° turn.

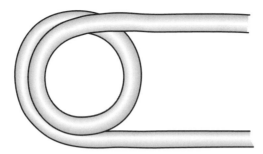

Figure 12.3 To form a round turn, continue the running end of the loop around so that both ends are emerging from the same side.

Tying Procedure

An experienced rescue trainer likes to say that repetition is the mother of skill. Get in the habit of tying your knots following the same procedure each time. We suggest that you follow these steps each time you tie a knot:

1. *Tie.* Properly tie the knot. Keep loops as small as possible. This will keep the rope system as compact as possible. This is important in confined space operations when working with fixed-height anchors.

2. *Dress.* Dress the knot. Remove any slack, puckers, or crossovers. Make the knot easily recognizable at a glance. This will help with safety checks on systems.

3. *Set.* Set or load the knot by pulling the rope ends and loops the same way that the rope will be pulled when loaded. This removes the remainder of the slack and makes the knot more compact.

4. *Safety.* A safety knot secures the emerging tail or tails around the standing part of the rope to prevent them from slipping back through the knot when the system is loaded. A knot that requires a safety is not completely tied until the safety knots have been added.

The four steps in this procedure should be performed every time you tie a knot. It is important that you do this when practicing so that the motions become instinctive.

The figure 8 family of knots

Knots of the figure 8 family are the knots most commonly used in kernmantle ropes for confined space entry and rescue applications. They hold well, are easy to tie and untie, and do not cause an excessive reduction in the strength of the rope.

Stopper knot (simple 8).   The simple figure 8, also known as the "stopper knot," is the basis of all the knots in the figure 8 family (Fig. 12.4). Stopper knots are used in the bottom of rope bags to prevent the rope from slipping out of the bag. In rappelling applications where the rope does not reach the ground, a stopper knot should be placed in the end of the rope to prevent anyone from rappelling off the end of the rope. The stopper knot is used in lowering applications for the same reason, to prevent the end of the rope from slipping through a brake bar rack used for descent control. Since this knot is used in applications where it does not see direct stress, a safety is not required in the stopper knot.

Figure 8 on a bight. Primary use: anchor knot.   This is the most common knot used for tying a loop at the end of a kernmantle rope (Fig. 12.5). The knot is easy and quick to tie, easy to spot when it has been tied incorrectly, and is easy to untie after being loaded. The figure 8 on a bight requires a safety knot.

Figure 8 bend. Primary use: to connect two ropes of equal diameter.   The figure 8 bend is used to connect two ropes of equal diameter together end to end (Fig. 12.6). This is a very secure knot that will self-tighten when loaded. Two safety knots should be placed as close as possible to each side of the knot to secure the loose tails to prevent slippage.

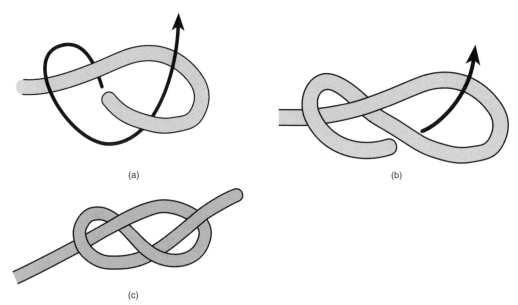

Figure 12.4   Stopper knot. (*a*) Start with a loop in the rope and bring the working end around the standing part of the rope one full turn. (*b*) Bring the working end back through the loop and pull. (*c*) This forms the distinctive figure 8 shape in the knot as shown here.

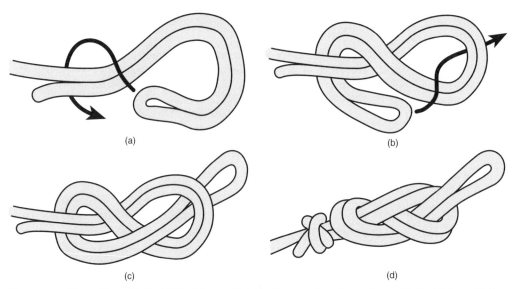

Figure 12.5   Figure 8 on a bight. (*a*) First form a bight in the rope; then form a loop with the bight and bring the end of the bight around the standing part of the rope one full turn. (*b*) Put the end of the bight through the loop. (*c*) Dress the knot and set by pulling the loop of the knot and the rope ends simultaneously in opposite directions as seen here. (*d*) To safety or secure the tail of the knot, place an overhand knot or preferably a barrel knot, as shown here, as close as possible to the base of the figure 8 knot.

Figure 8 followthrough. Primary use: anchor knot around a fixed anchor.   This knot is used when you need a loop around a fixed anchor point or in some other type of closed system (Fig. 12.7). After the figure 8 followthrough is tied, it looks the same as a figure 8 on a bight; the difference is in how it is tied. A safety knot should be placed in the tail of the figure 8 followthrough knot to prevent slippage.

Double-loop figure 8. Primary use: anchor knot.   The double-loop figure 8 is preferred by some rescuers for attaching to anchors or loads. Having two loops reduces the wear and the strength loss due to acute bends around a carabiner because the load is split between the two loops (Fig. 12.8). This knot is a little more difficult to tie than the other figure 8

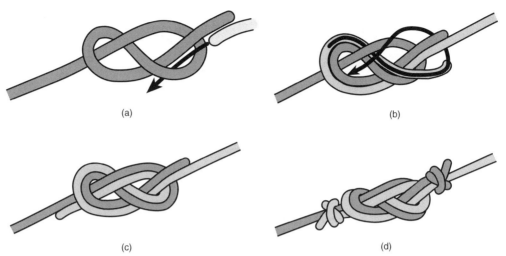

Figure 12.6 Figure 8 bend. (*a*) Place a loose stopper knot in rope one leaving enough tail at the working end to tie a safety knot. Rope 2 traces the stopper knot starting at the working end of rope 1. (*b*) Trace the stopper knot in rope 1 with rope 2, as do the arrows in the diagram. This forms the distinctive figure 8 shape. Pull ample slack through so that there will be enough tail to tie a safety knot. (*c*) The working end of rope 2 should end up pointing in the direction of the running part of rope 1 and vice versa. Dress and set the knot. (*d*) Place safety knots, either the overhand or barrel knot, as close as possible to both sides of the primary knot to secure the tails.

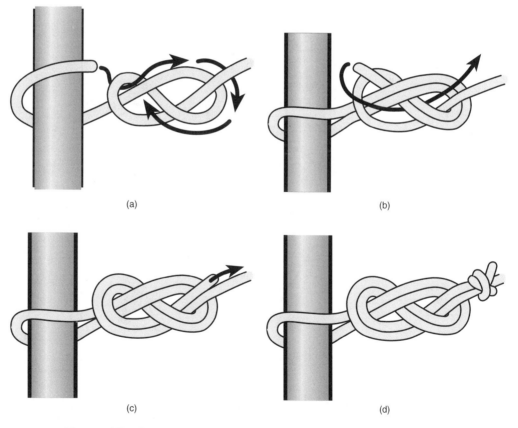

Figure 12.7 Figure 8 followthrough. (*a*) Place a loose stopper knot in the rope, pulling enough length on the working end so that it will go around the anchor and complete the knot and safety knot. (*b*) After you bring the working end around the anchor, trace the stopper knot starting at the end closest to the anchor. (*c*) Dress and set the knot, making sure that the distinctive figure 8 shape is seen in the knot. (*d*) Place a safety knot, an overhand or barrel knot, as close as possible to the base of the primary knot to secure the tail.

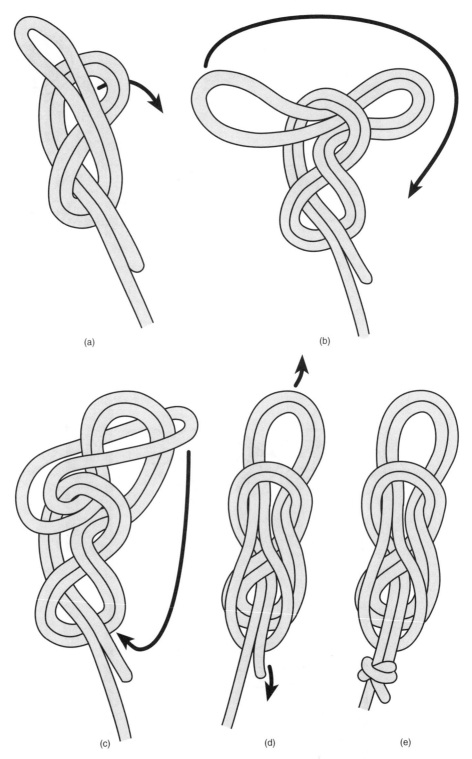

(a)

(b)

(c)

(d)

(e)

Figure 12.8   Double-loop figure 8. (*a*) Start as if you are tying a figure 8 on a bight, but don't put the end of the bight through the loop. Extend the bight well above the top of the loop of the figure 8 as shown. (*b*) Reach through the top loop of the figure 8, grab the midsection of the bight, and pull it partially through the loop. Now pull the top of the bight over the entire knot while still holding the midsection of the bight. (*c*) The top of the bight should now rest at the base of the knot, and the midsection of the bight should form the double loops. (*d*) Dress the knot and set it by pulling the double loops and the ropes extending from the base of the knot simultaneously in opposite directions. (*e*) Place a safety knot, an overhand or barrel knot, as close as possible to the base of the primary knot to secure the tail.

family knots, so some teams opt to use the figure 8 on a bight instead. A safety knot is required to prevent slippage of the double-loop figure 8.

Safety knots

Safety knots are used to provide an extra margin of safety. It is critical that they stay tied. The two most frequently used safety knots are the overhand and the barrel knot. Using a half-hitch or even two half-hitches is not recommended because they can work themselves loose.

Overhand knot. Primary use: safety knot.   The *overhand knot* has two major applications in confined space operations. It can be used as a safety knot to secure the loose tail of a primary knot around the standing part of the rope or webbing to prevent slippage when loaded (Fig. 12.9). When used as a safety knot, the overhand knot should be tied so that it rests directly against the primary knot. The overhand also is used as a basis for the water knot, the knot most commonly tied in webbing.

Barrel knot (half of a double fisherman's or double overhand). Primary use: safety knot.   The *barrel knot* is the preferred safety knot. Like the overhand, it is used to secure the loose tail of the primary knot around the standing part of the rope to prevent slippage (Fig. 12.10). The barrel knot is a stronger, more secure knot than the overhand; however, it requires more rope. The barrel knot should be tied so that it rests directly against the primary knot.

Other knots

Bowline knot. Primary use: anchor knot and for forming a loop that will not constrict.   The *bowline knot* was the preferred anchor knot until the advent of synthetic fiber ropes.

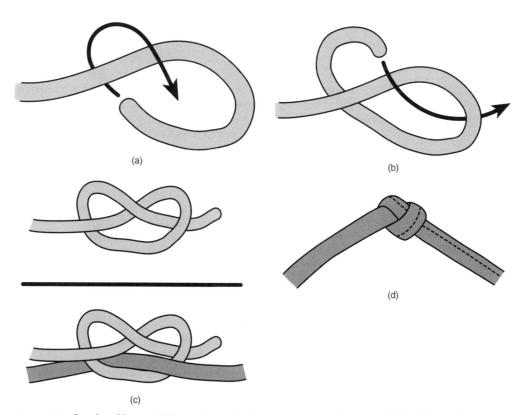

(a)

(b)

(c)

(d)

Figure 12.9 Overhand knot. (*a*) Form a loop with the working end of the rope. (*b*) Pull the working end of the rope around to the opposite side of the loop and place it through the loop. (*c*) This is the basic overhand knot. This knot can also be tied around the standing part of the rope as a safety knot. (*d*) This is the basic overhand knot tied in webbing. This is the foundation of the water knot.

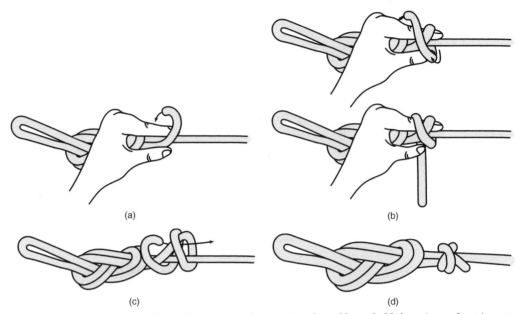

(a)

(b)

(c)

(d)

Figure 12.10 Barrel knot. (*a*) To safety a primary knot using a barrel knot, hold the primary knot in your hand with the index finger pointed toward the running part of the rope. (*b*) Wrap both the index finger and the rope at least twice with the tail of the knot. Wrap toward the wrist. (*c*) Remove the index finger and slide the working-end tail through the two wraps from the side of the loops closest to the primary knot. (*d*) Hold the wraps and pull the tail to set.

When a bowline knot is used on synthetic fiber ropes such as nylon kernmantle ropes, a safety knot must be properly placed to prevent the tail of the rope from slipping back through the knot (Fig. 12.11). Because of the acute bends in a bowline, the rope system loses up to 33% of its original strength. This is the major disadvantage of the bowline. The advantages of the bowline knot are that it can be untied easily and that it is a good knot for forming a single loop that will not constrict.

Square knot. Primary use: connecting two ropes of equal diameter. Not used in load-bearing applications.   The *square knot* is used to connect two rope ends of the same diameter and should not be used in load-bearing applications. This knot has two tails that should both emerge from the same half of the knot, either the top half or the bottom half (Fig. 12.12). The tails must be secured with an overhand or barrel safety knot.

It is very important that everyone on your team know how to tie the square knot properly and understands the importance of the safety knots. The square knot is very similar to the granny knot, which will slip and release when loaded. If the safety knots are not added and a team member mistakenly ties a granny knot, then failure of the system will occur. We will use the square knot in two applications discussed in this book—when tying a Swiss seat (hasty harness), a type of improvised harness using webbing, and when securing the bridle on a SKED or Stokes litter rigged for vertical lifts or lowers.

Butterfly knot. Primary use: midline knot.   The *butterfly knot* is a midline knot that can be used to tie a loop at any point along a rope (Figs. 12.13 and 12.14). The butterfly loop can accommodate pulls in three directions. When using the butterfly knot to establish a multipoint anchor (between two anchors), compound forces must be considered. A discussion of this follows in Chap. 13. The butterfly knot requires no safety since it is a midline knot.

Double-loop butterfly knot.   Primary use: midline knot.   The *double-loop butterfly* is a midline knot with two loops (Fig. 12.15). It is used in the same way as the single-loop butter-

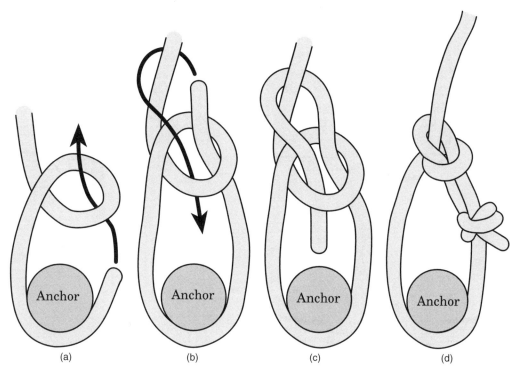

Figure 12.11 Bowline. (*a*) Make a loop in the working end of the rope, leaving enough length to extend around the anchor and complete the knot and safety knot. (*b*) Carry the working end around the anchor and back through the opposite side of the loop. (*c*) Now take the working end around the standing part of the rope and back through the loop. Dress and set the knot by pulling the working and standing parts of the rope in opposite directions. (*d*) Place a safety knot in the working end of the rope around the closest side of the loop.

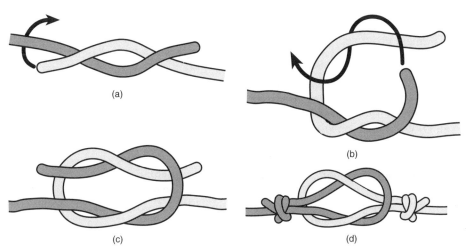

Figure 12.12 Square knot. (*a*) Hold one end of the rope in the right hand and the other in the left. Take the rope in the right hand over the left, down and around one complete turn. (*b*) Now take the rope on the left side and place it over the rope on the right, down, around, and through the loop. (*c*) Pull the ends to set. Notice that the tails should both be coming out the top half of the knot as in the diagram. Once set, if the tails are coming out opposing, one from one side and the other from the other side, you have tied a granny knot and not a square knot. (*d*) Place safety knots as close as possible to both sides of the square knot. The safety knot is essential.

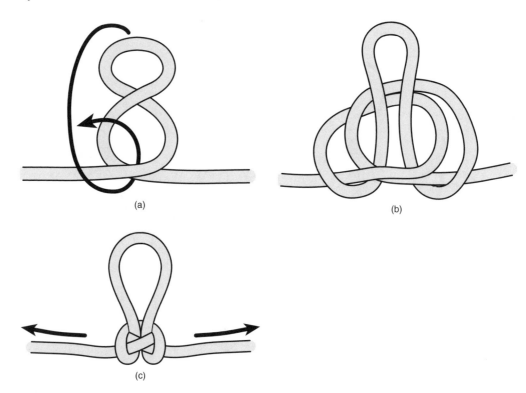

Figure 12.13 Butterfly knot (method 1). (*a*) Find the position in the rope where you would like the midline loop to rest. Twist the rope to form two stacked loops as shown here. Bring the top loop forward. (*b*) Push the top loop under, behind, and through the lower loop. Dress the knot by pulling the loop and the two ends of the rope in opposite directions. (*c*) Set by pulling the standing ropes in opposite directions.

fly. As with the double-loop figure 8, having two loops reduces the wear and the strength loss from acute bends around a carabiner because the load is split between the two loops.

Water knot. Primary use: to connect webbing together. The *water knot* can be used to tie a single strand of webbing into a loop or to tie two lengths of webbing together. The tails of the webbing should emerge from the knot running in different directions. Some schools teach that if the tails of the water knot are at least 3 in in length when the knot is set, safety knots are not needed to prevent the tails from slipping back through the knot as it is loaded. Others teach placing overhand knots against each side of the water knot for safety. We show both variations in Fig. 12.16. This is the best knot to tie in webbing to support human loads.

Double fisherman's knot. Primary use: to join two ropes of equal diameter. The *double fisherman's knot* is used to make Prusik loops, which will be discussed in Chaps. 13 and 17. It is a small, compact knot that is self-locking (Fig. 12.17); therefore, it does not require a safety if properly tied and set.

Hitches

Hitches are used for attaching ropes to other objects. Hitches differ from knots in that hitches are always formed around something, and they depend on what they are formed around to maintain their configuration.

Half-hitch and two half-hitches. Primary use: utility knots. The *half-hitch and two half-hitches* are very quick and easy to tie (Fig. 12.18). As the name suggests, making one half-hitch and then adding another forms two half-hitches. The use of two half-hitches as an anchor or a safety is not recommended for life-safety applications. Other safety

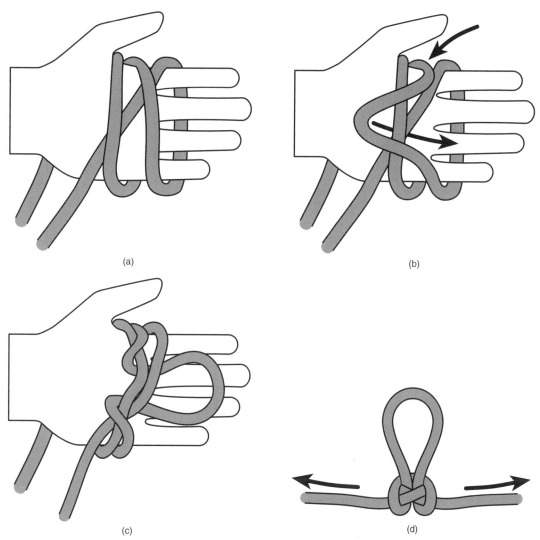

(a)

(b)

(c)

(d)

Figure 12.14 Butterfly knot (method 2). (*a*) Turn your left hand with your palm facing you. Wrap the rope once over the face of your palm from left to right. Then wrap the rope right to left forming an X in your palm. Wrap a second time right to left toward your wrist. (*b*) Pull the top right wrap over toward your wrist and under the two other wraps. (*c*) Remove your hand and pull the loop and two rope ends in opposite directions. (*d*) Set the knot by pulling the standing ropes in opposite directions as shown here.

knots, including the overhand and barrel knots, are more secure and will not loosen up like half-hitches tend to do.

Munter hitch. Primary use: safety belay.   The *Munter hitch* (Fig. 12.19) is commonly used in safety belay applications. It is not recommended for lowering a load because the "nylon on nylon" effect can damage the rope due to heat buildup. We will describe how to use the Munter hitch on a safety belay line in Chap. 13.

Clove hitch. Primary use: anchor knot.   The *clove* commonly is tied in rope or webbing. The clove hitch tightens down on itself, gripping whatever it is tied around (Figs. 12.20 to 12.22). It works well as an anchor knot, for securing patient lashing to litters or backboard, and for placing around certain types of equipment for lifting and lowering. It requires properly placed safety knots.

Girth hitch. Primary use: anchor knot.   The *girth hitch* can be tied in rope or webbing and is an easy way to attach to an anchor (Fig. 12.23); however, the use of the girth hitch

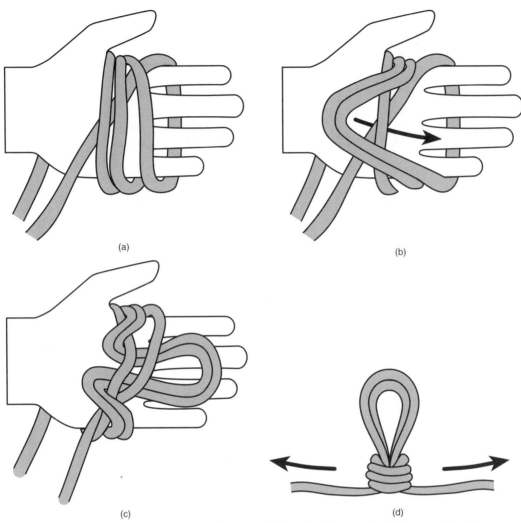

(a)

(b)

(c)

(d)

Figure 12.15   Double-loop butterfly knot. (*a*) Turn your left hand with your palm facing you. Wrap the rope once over the face of your palm from left to right. Next wrap the rope right to left forming an X in the middle of your palm. Then wrap a second and third time right to left toward your wrist. (*b*) Take the top two wraps closest to the fingers and pull them up over the X and toward the wrist. (*c*) Push the wraps under the X and back toward the fingers, grabbing them with your free hand. Remove your hand and pull the two loops and the rope ends in opposite directions. Dress. (*d*) Set the knot by pulling the standing ropes in opposite directions.

results in a 20% loss of strength in the rope or webbing. Its use is recommended only in configurations that will not place the anchor system below strength requirements.

Prusik hitch. Primary use: soft rope-grabbing devices for uses such as safety belays, rope grabs in hauling systems, and ascending fixed ropes.   The Prusik hitch is tied with a loop of accessory cord around a larger-diameter rope (Fig. 12.24). Both double-wrapped and triple-wrapped Prusik hitches are used, depending on the application. When unloaded, the Prusik hitch allows the rope it is tied around to slide through unhindered. When loaded, the Prusik hitch grips the rope tightly. The use of Prusik hitches in safety belay systems will be discussed in Chap. 13. Their use in ascending fixed ropes will be discussed in Chap 17.

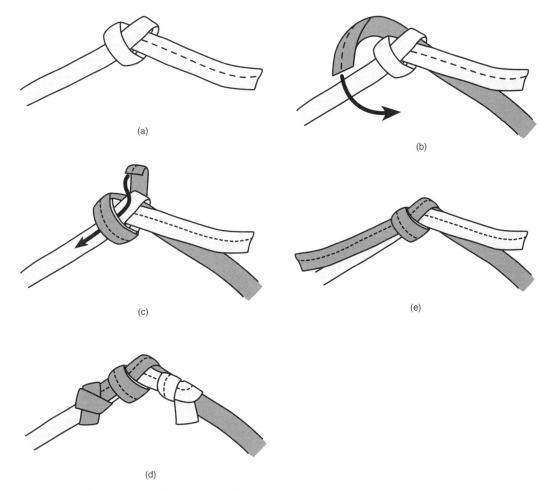

(a)

(b)

(c)

(e)

(d)

Figure 12.16   Water knot. (*a*) Tie an overhand knot in one end of the webbing leaving at least 3 in of tail. (*b*) Take the opposite end of the webbing and trace the overhand knot, starting at the tail end. (*c*) Continue tracing the overhand knot as seen here. (*d*) If saftey knots are desired, on both sides, safety knots are not required. (*e*) If the knot has less than 3 in of tail, then safety overhand knots should be placed on both sides of the primary knot to prevent slippage.

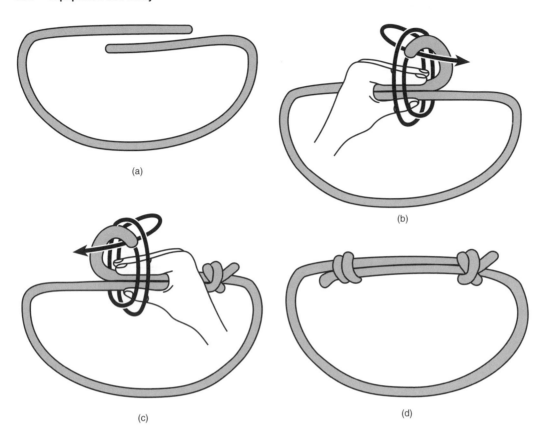

(a)

(b)

(c)

(d)

Figure 12.17   Figure 12.17 (*a*) Double fisherman's knot (two barrel knots). Form an inverted loop with the ends crossing at the top. (*b*) Holding the top of the loop with your left hand, point your left index finger to the right and wrap both your finger and the right standing part of the rope 2 times toward your wrist. Remove your left index finger and place the working-end tail through the two wraps from left to right. Set as you would a barrel knot. (*c*) Holding the top of the loop with your right hand, point your right index finger to the left and wrap both your finger and the left standing part of the rope 2 times toward your wrist. Remove your index finger and place the working-end tail through the two loops from right to left. Set as you would a barrel knot. (*d*) Pull the loop in opposite directions to close the knots together.

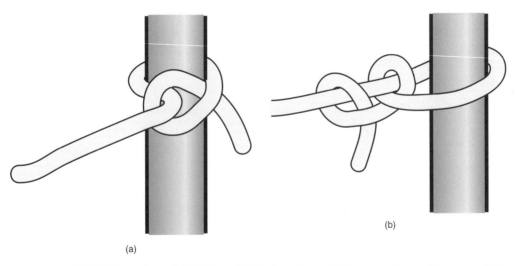

(a)

(b)

Figure 12.18 Half-hitch and two half-hitches. (*a*) To form the half-hitch, run the working end of the rope around an object, under the standing part of the rope, and back through the top of the loop. (*b*) Add one more half-hitch to form two half-hitches by bringing the working end of the rope under and around the running part of the rope and back through the loop just formed.

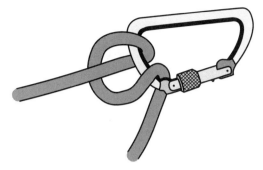

Figure 12.19 Munter hitch. Drape the rope over your left hand. Reach around the left side of the front rope with your right hand and grab the rear rope. Pull the rear rope forward and wrap it underneath and over your fingers on the right side of the front rope. Lock the carabiner around the two ropes at the top of your hand.

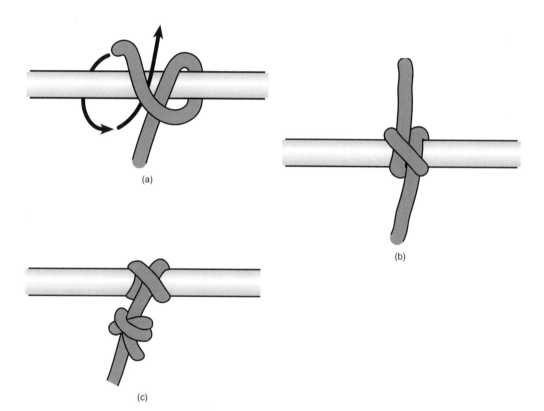

(a)

(b)

(c)

Figure 12.20 Clove hitch (method 1: tied around an anchor). (*a*) Make a turn around an anchor, crossing the working end over the standing part of the rope. Continue the working end around the anchor again and tuck under the second turn. (*b*) Set the hitch by pulling both ends taut. (*c*) Safety the clove hitch by continuing the working end around the anchor and tying an overhand or barrel knot around the standing part of the rope.

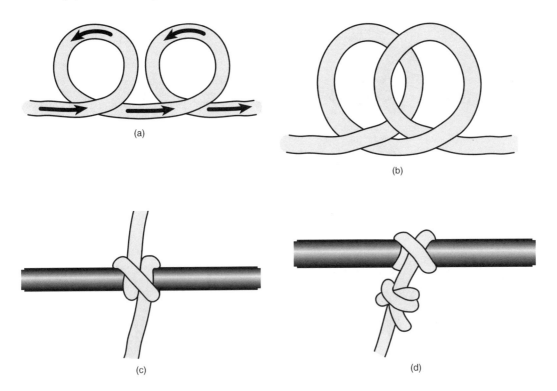

Figure 12.21 Clove hitch (method 2: used to slip over an open anchor). (a) Make a large loop in the standing part of the rope. Hold the loop securely with your left hand while making a second loop. (*Note:* Both loops in the diagram are counterclockwise.) (b) Place the loop on the right on top of the loop on the left without inverting the loop. (c) Slip the two loops over the anchor and set by pulling the ends taut. (d) Safety the clove hitch by continuing the working end around the anchor and placing an overhand or barrel knot around the standing part of the rope. (c) Slip the two loops over the anchor and set by pulling the ends taut. (d) Safety the clove hitch by continuing the working end around the anchor and placing an overhand or barrel knot around the standing part of the rope.

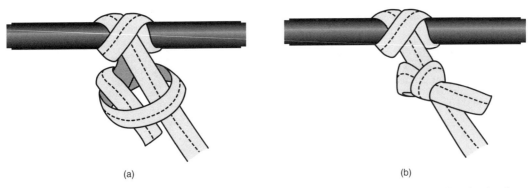

Figure 12.22 Clove hitch tied with webbing. (a) Tie the clove hitch with webbing. Tie the overhand safety knot with the tail end around the standing end. (b) The completed clove hitch and overhand safety knot should look like this.

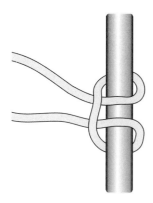

Figure 12.23 Girth hitch. Form a bight using either rope or webbing. Wrap the bight around an anchor, pull the ends of the rope or webbing through the bight, and pull.

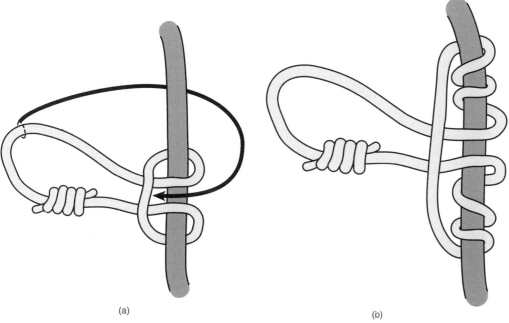

(a)

(b)

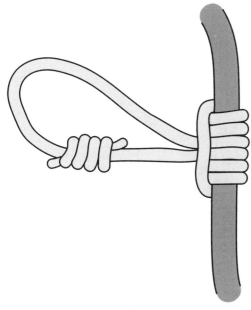

(c)

Figure 12.24 Prusik hitch. (*a*) Wrap a Prusik loop around a rope in a girth hitch type of motion but do not cinch taut. (*Note:* The double fisherman's knot used to form the Prusik loop should end up somewhere between the end of the loop and the hitch on the side of the loop.) (*b*) Continue wrapping the loop end through the girth 1 or 2 more times depending on the application in which it will be used. Set by pulling the loop taut. (*c*) Dress the hitch so that no ropes in the wraps are crossing. If the Prusik hitch is not dressed properly, it will not grab the rope properly when needed.

Summary

Knots are critical building blocks in rope systems used in confined space operations. They can be used in a wide variety of applications, including both life-safety applications and utility applications. Since knots are a critical link in any system in which they are used, it is important that the proper knot be utilized for each application and that the knot be tied correctly. In Chap. 13 we will explore combining the basic equipment discussed in Chap. 11 by using the knots described in this chapter in order to rig basic systems for use on confined space operations.

Rigging and Using Basic Systems for Confined Space Operations

The term "rigging" as used in this text refers to the building of load-handling systems using components such as ropes, hardware, and other equipment. This chapter will focus on simple systems that can be used for both routine confined space entries and confined space rescues.

For routine entries into confined spaces, basic systems such as those described in this chapter allow us to lower workers into spaces, haul them out of spaces, protect workers from falls, and perform nonentry rescues. The simple systems described here are considered sufficient for permitted entries because we know in advance that such entries will be required. This allows us to make advance preparations for the use of basic systems in carrying out routine entries.

For simple confined space rescues, basic systems such as those described here may be adequate to complete the rescue. More challenging rescues may require us to rig and use more versatile and complicated rescue systems, such as those described in Chap. 17. Keep in mind that basic rigging will serve as the foundation on which to build the more advanced systems.

Many different systems will work for any given confined space operation. The basic systems discussed in this chapter are by no means the only ones that should be used. The authors believe that everyone working in confined space entry or rescue operations should be able to rig and use these basic systems competently. Learning what systems would be best for each confined space you deal with comes with practice and experience and should be addressed while preplanning your spaces. Remember the KISS method, and keep it simple and safe, no matter the operation or systems used.

Overview of Basic Rigging

For many generations human beings have been rigging systems that allow heavy loads to be safely handled. These systems are attached between the load and a fixed anchor, and must be designed for the operation to be performed.

In this chapter, we will use the anchor as the point of beginning for rigging basic systems because the anchor is a common component of all systems. We will then explore techniques for rigging simple systems that we can use to lower a load and to lift or haul a load. We will cover the rigging of belay systems used to prevent the load from falling while being moved. We also will consider tripod operations, since tripods are used very commonly in confined space entry and rescue. All the systems described here can be rigged using the basic components covered in Chap. 11.

In this chapter, we will assume that the load to be handled is a confined space entrant or rescuer. Keep in mind that the techniques described here also can be used to move equipment and supplies if the systems are rigged using utility rated ropes, hardware, and other components. Never use life-safety-rated equipment for utility purposes, and never use utility-rated equipment for human loads.

Establishing Anchors

An *anchor* is the point of attachment for the load-handling system. Anchors consist of individual anchor points and the items of software and hardware used for connecting the load-handling system to them. The connecting items are rigged to the anchor points to form an anchor system. The anchor system is a critical component because it forms the foundation of the load-handling system as a whole.

Types of anchor points

Various types of anchor points may be suitable for confined space operations. We can categorize anchor points in several ways.

Stationary versus portable anchor points.   Some anchor points are permanent and stationary. Major support members of a structure are an example of this type as shown in Fig. 13.1. Other anchor points are temporary and portable, and are established before an operation and removed afterward. Tripods are commonly used to provide temporary anchor points for confined space operations (Fig. 13.2). Other examples of temporary anchors include clamps designed to attach to steel I-beams to provide temporary anchor points and portable hoisting frames.

Engineered versus nonengineered anchor points.   Anchor points also can be classified as engineered or nonengineered. An engineered anchor point is created specifically to serve as an anchor point and is designed to meet a specified strength requirement. An example of an engineered anchor point is a large eyebolt installed in a major structural

Figure 13.1 A major support member of a structure is an example of a nonengineered anchor point.

Figure 13.2   Tripods are temporary portable anchors engineered to hold a specific working load.

component under the direction of a registered professional engineer or other qualified professional (Fig. 13.3).

Structural components and other items that were not specifically designed to serve as anchor points and have no specified strength rating can be classified as nonengineered anchor points (see Fig. 13.1). No matter what type of anchor is used, it must be strong enough to handle any potential load that may be placed on it by the load-handling system.

"Bombproof" anchor points.   When using nonengineered anchor points, select "bombproof" anchor points if they are available. The term "bombproof" can be defined in several ways:

- According to some authorities, a bombproof anchor point is one in which we have complete confidence in its ability to withstand the entire potential load (Frank 1998, p. 67). The classic example of a bombproof anchor point in the nonindustrial setting according to this definition is a BFR (initials used to indicate very large rocks or other items that are equally immovable).

- Some structural rescue authorities classify an anchor point as bombproof if its failure would result in a major failure of the structure of which it is a part (Roop et al. 1998, p. 90). One example of a bombproof anchor point according to this definition would be a major vertical support member of a structure supporting a vessel being entered.

- A third definition holds that a bombproof anchor point is one that is strong enough to survive a bomb blast; hence the term "bombproof." This definition can be considered a variant of the first definition.

Figure 13.3 Engineered anchor points are established under the supervision of a registered professional engineer or other qualified person to safely handle a specified working load.

When establishing anchor systems, always select bombproof anchor points if they are available. Any nonengineered anchor point that is not bombproof must be considered a questionable anchor point. Anytime you are not 100% confident in an anchor point, or if any team member doubts the integrity of the anchor point, back it up. Questionable anchor points should be backed up to another anchor point that is either engineered or obviously bombproof. Methods used to back up anchors are discussed later in this chapter. Consider this an ironclad rule, since failure of an anchor point can cause failure of the entire system with disastrous results.

Selection and designation of anchor points

Selecting good anchor points can be one of the most challenging aspects of basic rigging. This is especially true when systems are being rigged in preparation for confined space rescue operations. Preplanning the spaces at your facility or in your jurisdiction will give you the opportunity to locate or establish acceptable anchor points or to bring in an appropriate temporary anchor device that will work well for the space to be entered.

Selecting permanent anchor points.   Entries for routine work in confined spaces should afford enough lead time to allow bombproof anchor points to be identified or permanent engineered anchor points to be installed before entry. The same advantage is available to rescuers, but only if preemergency planning and preparation are carried out.

An excellent practice is to load-test any nonengineered anchor point that is not obviously bombproof unless the anchor point will be backed up. Perform testing well in advance of routine entries or during the preplanning phase in preparation for rescue operations. Load testing should be conducted by a registered professional engineer or other qualified person in order to determine that the anchor points in question are strong enough for the intended purpose. It makes no sense to build systems in which the anchor points are weaker than the components attached to them.

The best practice is to select and identify acceptable anchor points well in advance of the time when they are needed. If feasible, clearly identify the anchor points by some method such as permanently labeling them with paint. A qualified person must inspect the designated anchor points carefully for damage periodically and before each use. Once anchor points are designated, do not use other points of attachment for rigging.

Objects that may be suitable anchor points.   Strong sturdy anchor points are often readily available in confined space operations. Points that may serve as anchor points include the following:

- Steel structural components that are well supported
- Uninsulated large, heavy steel tubing
- Structural components of reinforced concrete
- Large masonry components

- Large machine supports

- Large rocks

- Large, healthy, well-rooted trees

Caution is always required when selecting anchor points. For example, steel tubing that appears strong from the outside may be heavily corroded from the inside out. A thick coat of paint may hide significant damage to structural components.

Items such as handrails are frequently available when anchor points are needed, but should be viewed with suspicion. While some handrails may provide adequate anchor points for some operations, many will not. As a general rule, always back up handrails and handrail supports used as anchor points.

When searching for anchor points, look for the "mother source" of strength. For example, never attach to a minor structural member when you can attach to the major structural member it is bolted to. As a general rule, the corners of most structures are the strongest, best-supported parts. As a simple example, corner supports of handrail systems are usually stronger than the handrail sections on either side of the corner.

Objects to avoid as anchor points. During confined space operations you may encounter items that appear be adequate anchor points but actually are not. Objects to avoid using as anchor points include the following:

- Flimsy handrails

- Extremely hot or cold items that can damage rope or webbing

- Fire hydrants, since hydrant barrels are designed to shear away from the base when a lateral force is applied to them

- Cast-iron pipes or fittings, since they tend to be brittle

- Small, light, or thin tubing or frame members that may bend or collapse if loaded

- Insulated tubing, since the insulation indicates hot or cold contents and some insulation may break into shards that can damage software components

- Small masonry or wooden structures that have no major structural support members or can be pulled off a foundation

- Anything that is corroded or damaged

When evaluating anchor points, consider how loading configurations affect strength. For example, the load limit of eyebolts is rated for a pulling force that is gradually applied parallel to the shank. Eyebolt strength is significantly less under shock loads and when angular forces are applied.

Using temporary anchor points. As previously discussed, temporary anchors are ones that are established to carry out a confined space operation and are then removed. Several approaches are used to provide temporary anchors.

Portable anchor systems. Examples of temporary anchors include portable anchor systems such as tripods (Fig. 13.2) and hoisting frames (Fig. 13.4). Portable anchor systems are rated according to a specified working load (the maximum amount of weight that the device was designed to carry without permanent damage to the device) and breaking strength (the maximum amount of weight that the device was designed to carry without failure of the device). Never exceed the working load rating of the device. Some tripods are rated strictly for utility use. If the portable anchor system will be used to support human loads, make sure that it is rated for that purpose, because some are not. The use of tripods and other portable anchor systems will be discussed more fully later in this chapter.

Using mobile equipment and vehicles as anchors. Large vehicles, boom trucks, cranes, and other types of mobile equipment have been successfully used to provide anchor points

Figure 13.4 The hoisting frame is another example of a temporary portable anchor system. (*Photograph courtesy of Unique Concepts Ltd.*)

for confined space operations; however, their use is controversial, to say the least. To begin with, powered equipment should never be used for moving people into or out of confined spaces, since a minor operator error could severely injure or kill the entrant. The same effect may result if mobile equipment being used as an anchor is accidentally activated or moved during an entry.

For routine confined space entries, the lead time available should allow safer types of anchor points to be identified or established. The best recommendation is to plan ahead to avoid the use of vehicles or mobile equipment as anchor points for these entries.

In some confined space rescues, rescuers may be forced to use available equipment such as fire trucks or cranes as anchors in order to save a victim's life. Safety precautions for using these types of anchors during rescue operations are discussed in Chap. 17.

Improvised anchor points. In some confined space rescue operations, rescuers are forced to use improvised elevated anchor points. Examples include A-frame and gin pole setups, both of which are described in Chap. 17. Like mobile equipment and vehicles, these improvised anchors are not ideal but may be the only option available to rescuers in order to save a life. Never use improvised anchors for routine entry into a confined space. As an alternative, work out something better before entry is made.

Anchor systems

Once individual anchor points have been identified, rig items of software and hardware to them to form anchor systems to which load-handling systems are connected. Both webbing and rope commonly are used for rigging anchor systems. Both types of software are subject to damage or failure if loaded in contact with sharp corners, rough edges, or abrasive surfaces on anchor points. To prevent this, "soften" the anchor points with padding before placing software items around them.

Rigging anchor systems with webbing. Webbing is very effective for rigging anchor systems. Because it has a flat construction, webbing is not as susceptible as kernmantle rope to significant strength loss when it forms acute bends around an anchor point. Webbing is economical, strong, and lightweight, and has high abrasion resistance in comparison to kernmantle rope. In many applications, webbing's wide surface area provides a good grip on the anchor.

Options for rigging with webbing. Webbing can be purchased or rigged in a versatile array of configurations. As discussed in Chap. 11, a length of webbing can be tied end to end with a water knot to form an improvised loop sling or purchased as a presewn endless loop sling. Presewn webbing slings can be purchased with a sewn eye at each end. Webbing anchor straps are available with D rings sewn into each end. Utility belts are length-adjustable anchor straps that provide quick and easy anchor connections.

Any of these connectors will work if used in a configuration that meets the requirements of NFPA 1983 for rescue systems or relevant OSHA standards for fall-protection systems. Most authorities recommend that webbing in anchor systems for rescue operations be used in a configuration providing a minimum breaking strength (MBS) greater than 9000 lbf for general use. Webbing in anchor systems should be configured to provide at least 4500 lbf MBS for personal use in rescue operations. Webbing in anchor systems used for worker fall-protection systems meeting OSHA requirements should provide at least 5000 lbf MBS. Otherwise, the anchor systems may be weaker than the components of the systems attached to them. The requirements for various system components under both NFPA and OSHA standards were described in Chap. 11.

Webbing strength considerations. As discussed in Chap. 11, one-inch flat-weave webbing used in confined space operations typically has a tensile strength of approximately 6000 lbf while one-inch tubular webbing typically has a tensile strength of approximately 4000 lbf. Tensile strength is the amount of force that would be required to break a length of webbing pulled from end to end.

When a length of the one-inch flat-weave webbing is connected end to end with a water knot, one might assume that the tensile strength of the resulting loop sling would double to approximately 12,000 lbf. This is not true. In Chap. 12 we found that the strength reduction due to a water knot being tied in the webbing is 36%. For the sake of safety and simplicity, let's assume that this means that 36% or 4320 lbf of the entire strength of the tied loop sling is lost. This means that if the loop sling tied with flat-weave webbing is used in a vertical configuration (Fig. 13.5), it will have a calculated MBS of 7680 lbf. Note that if one-inch tubular webbing had been used to form the tied loop sling, its calculated MBS in the vertical configuration would be 5120 lbf. Neither tied loop sling would meet the 9000-lb MBS requirement for general use in rescue operations when used in the vertical configuration.

Using the basket hitch to connect to anchor points. If a loop sling is basket-hitched around the anchor point (Fig. 13.5), its efficiency doubles. This means that the calculated MBS for the loop sling tied with one-inch flat-weave webbing would increase to just over 15,360 lbf. The calculated MBS for the sling tied with tubular webbing would increase to 10,240 lbf when used in the basket configuration. In addition to being an efficient anchor connection, the basket hitch is quick and easy to rig. Any of the anchor connectors described above can be used in this configuration.

If the tied loop sling is doubled before being basket-hitched around the anchor, the efficiency will be approximately doubled again. This is a good standard practice to follow when rigging anchor systems. The MBSs calculated above are theoretical calculations, and actual performance in the real world may vary significantly. Doubling the sling before the basket hitch is rigged is a quick and easy way to double the effective strength of the anchor connection for an extra margin of safety.

Another good practice is to make a round turn around the anchor point with the sling, strap, or utility belt when forming the basket hitch (Fig. 13.6). This places extra webbing around the side of the anchor point that sees most of the load. It also prevents the sling from moving along the anchor point during the operation.

When attaching to basket-hitched loops or slings placed around wide anchor points, avoid triloading carabiners as shown in Fig. 13.7. Using a trilink is an easy way to eliminate this problem, since trilinks are designed to handle a three-way pull (Fig. 13.8). If

Vertical
100% Efficiency

Basket Hitch
200% Efficiency

Girth Hitch
80% Efficiency

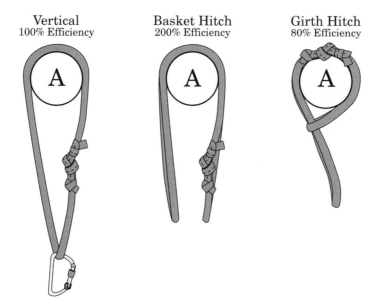

Figure 13.5 These are common sling configurations used to make attachments to the anchor. In this example webbing loops tied with a water knot are used in three different configurations: vertical, basket hitch, and girth hitch. The webbing loop used in the vertical configuration will have an efficiency of 100%. The webbing loop used in the basket-hitch configuration will have an efficiency of approximately 200% or twice that of the vertical configuration. The webbing loop used in the girth-hitch configuration will have an efficiency of approximately 80% of the vertical configuration.

Figure 13.6 Adjustable utility belts are quick and easy to basket-hitch around anchors. A quick round turn with the belt increases the friction in the anchor system, which increases the efficiency of the anchor attachment.

a trilink is not available, then use some other configuration such as the one shown in Fig. 13.9.

Using the girth hitch to connect to anchor points. When slings are girth-hitched around anchor points (Fig. 13.5), they constrict under load to grip the anchor point. For this reason, they are sometimes referred to as "chokers." This makes them handy for applications in which other configurations might allow the sling to slide along the anchor point

Figure 13.7 Avoid triloading carabiners when basket hitching around wider anchors.

Figure 13.8 Using a trilink eliminates the problem of triloading carabiners when attaching to anchor systems.

Figure 13.9 This is one method that can be used to avoid triloading carabiners when making anchor attachments.

during the operation; however, using a girth hitch will reduce the efficiency of the anchor attachment by about 20%. This means that the loop sling tied with one-inch flat webbing, which has a calculated MBS of 7680 lbf in the vertical configuration, would have only a 6144 lbf calculated MBS when girth-hitched around the anchor point. In the same application, the calculated MBS of the tubular webbing would be only 4096 lbf. For this reason it is critical that one-inch webbing slings be doubled at least once, if not twice, before being girth-hitched around an anchor point.

Using the "wrap 3, pull 2" method.   Another technique commonly used to attach webbing to anchor points is the "wrap 3, pull 2" method (Fig. 13.10). This method theoretically doubles the efficiency of the webbing. To tie a "wrap 3, pull 2," start with a length of webbing long enough to wrap completely around the anchor 3 times with enough webbing left over to tie a proper water knot with the ends, and proceed as shown in Fig. 13.11. This is a very efficient configuration because the water knot, the weakest point in the system, sees very little force when the system is loaded.

Three wraps are the minimum, not the maximum in using this technique. A "wrap 4, pull 3" method will be even stronger and better than the "wrap 3, pull 2" method. Use as many wraps as the length of webbing and size of the anchor point allow.

Rigging anchor systems with rope.   Kernmantle rope is commonly used to attach to anchor points. Rope can be tied in a number of configurations to form anchor bridles and other types of anchor attachments. Rope works well in attaching to rounded anchor points. Keep in mind the four-to-one (4:1) rule. The diameter of the anchor point ideally should be at least 4 times that of the rope to prevent acute bending that will result in a strength reduction of the rope. For example, when $\frac{1}{2}$-in rope is used, the anchor point should be at least 2 in in diameter to avoid acute bends in the rope. Avoid using rope around anchor points with angular corners unless they are well padded. As always, be sure that all rough edges and abrasive surfaces that the rope may contact are properly softened with padding.

Several knots and hitches that were discussed in Chap. 12 can be used for attaching ropes to anchor points. Examples include the bowline (Fig. 12.11) and the clove hitch (Figs. 12.20 and 12.21). Another method, which is used most commonly, is the tensionless tie-off (Fig. 13.12).

Bowline.   Although convenient to use in rigging anchor systems, the bowline is not a very efficient knot (Fig. 13.13). As noted in Chap. 12, a strength loss of 33% occurs when the bowline is tied in a rope. For this reason, before the bowline is used, the configuration of the anchor system must be evaluated to determine if this efficiency loss is acceptable given the intended load. The bowline does not constrict around the anchor point. In some situations this may be desirable, but in many situations the potential for slippage

Figure 13.10 "Wrap 3, pull 2" is another technique used to attach webbing to anchor points.

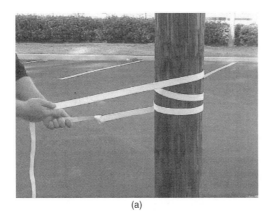

(a)

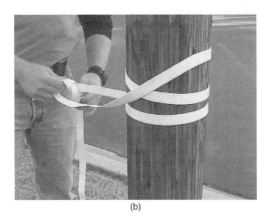

(b)

(c)

(d)

Figure 13.11 Tying a "wrap 3, pull 2." (*a*) Stand facing the anchor point on the load side of the anchor. Wrap the webbing around the anchor point two complete turns. (*b*) Tie a water knot with the ends of the webbing to close the loop. (*c*) While holding the knot in one hand, pull the two wraps of webbing that are against the anchor toward you. This will pull the water knot to the anchor. (*d*) Attach the carabiner to both the wraps of webbing to establish the connection point.

Figure 13.12 The tensionless tie-off is one method that can be used for attaching ropes to anchor points. This tensionless tie-off is wrapped 4 times around the anchor, and the end of the rope is tied back to the standing part of the rope with a barrel knot.

Figure 13.13 The bowline knot can be used to make the attachment to the anchor point. Remember that anytime the bowline is used, there is an approximately 33% loss of strength in the rope.

along the anchor point may be a cause for concern. The bowline typically is not the best knot for rigging anchor systems.

Clove hitch.   The clove hitch provides an efficient and convenient way of attaching rope systems to anchor points (Fig. 13.14). The use of the clove hitch does not result in a significant amount of strength reduction in the rope as long as the anchor point does not produce acute bends in the rope. Following the 4:1 rule and the guidelines for anchor point selection will prevent acute bends. It is easy to loosen the clove hitch and feed rope into or out of the standing line in the anchor system. This ease of adjustment makes the clove hitch convenient to use when rigging anchors.

Tensionless tie-off.   The tensionless tie-off is generally considered the most efficient method to use for connecting rope to an anchor point. This is contingent on good anchor point selection and following the 4:1 rule. To rig the tensionless tie-off, wrap the end of the rope at least 4 times around the anchor point. Finish it by tying the end back to the standing part of the rope just to the load side of the anchor point (Fig. 13.12) or by using a carabiner to make the attachment (Fig. 13.15). When properly rigged, the finishing knot or carabiner sees almost no force when the system is loaded, hence the name "tensionless tie-off."

Rigging single-point anchor systems.   Single-point anchors are simple anchor systems built around one anchor point. It is crucial that this anchor be either engineered or bombproof. Theoretically these anchor points should not have to be backed up. It is, however, a widely used practice to use two separate anchor points for rope systems that support human loads. When training, when working with team members with little experience, or when time is not critical, single-point anchors should always be backed up.

In some confined space operations, anchor points that are neither engineered nor bombproof may be very convenient to use because of their location and configuration. If anchor points such as these are used, they must be backed up as described below.

Figure 13.14 The clove hitch is used to attach to the anchor point in this example. The use of a safety knot with the clove hitch is critical.

Figure 13.15 The tensionless tie-off can also be completed by tying a figure 8 on a bight in the end of the rope. After the wraps are made around the anchor point, attach the end of the rope to the standing part of the rope with a carabiner.

To establish a backup system that includes a second independent anchor, several points should be considered:

- The secondary anchor must be of sufficient strength to provide a suitable backup. It makes no sense to use secondary anchors that are weaker than the primary anchor being backed up.

- Ideally the secondary anchor is positioned as much as possible directly behind and in line with the primary anchor (Fig. 13.16). This configuration will eliminate the possibility of a pendulum swing occurring in the main line if the primary anchor fails as shown in Fig. 13.17a. Pendulum swings can damage, cut, or abrade the rope. If the second anchor point is off to one side of the primary, then all edges that the rope could slide across if the primary fails should be padded. In a backed-up single-point anchor system, the primary anchor supports the entire load. In other words, the secondary anchor should become loaded only if the primary fails. It is there as a backup.

- Some slack will be present in the secondary attachment, but keep the slack to an absolute minimum. If the primary anchor fails, this reduces the potential for shock-loading the secondary anchor point, the anchor attachment, the rope system, and the person being moved with the system (see Fig. 13.18).

- The secondary attachment or the backup attachment should be made directly to the knot that supports the rope system, as shown in Fig. 13.19.

Keep in mind that backing up anchor points is not done in an effort to tie together weak anchor points to produce a system of adequate strength. That requires the connectors between the anchor points to be pretensioned to remove all slack so that the load will be applied to all points simultaneously. If this is not done, the anchor points can fail one after another like a row of dominos when the system is loaded. This can occur as a result of stretch in the system even though the connectors between the anchor points appear

Figure 13.16 The position of the secondary or backup anchor is important. Having the backup anchor directly behind and in line with the primary anchor prevents severe pendulum swings if the primary anchor fails.

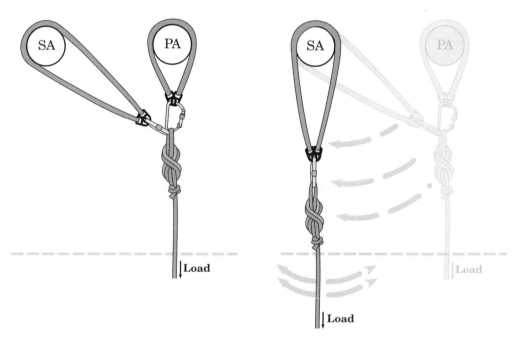

Figure 13.17a If the primary anchor fails, pendulum swings can occur as the load is shifting to the secondary anchor position.

tight. In the confined space setting, the best bet is to select adequate anchor points to begin with, rather than trying to tie weak points together.

It may be possible for two anchor points to back each other up (Fig. 13.20). This can have the effect of reducing the amount of time and equipment needed to rig the anchor systems. It is also possible to use a common backup for two anchor points in a three-point system (Fig. 13.21). Note that these still are classified as single-point systems.

When working with anchor points that have no chance of failure, the anchor attachment can still be backed up, as shown in Fig. 13.22. In this example, the tied webbing loop that is used for the primary attachment is backed up with a second webbing loop placed around the same anchor point. This secondary attachment is there to catch the

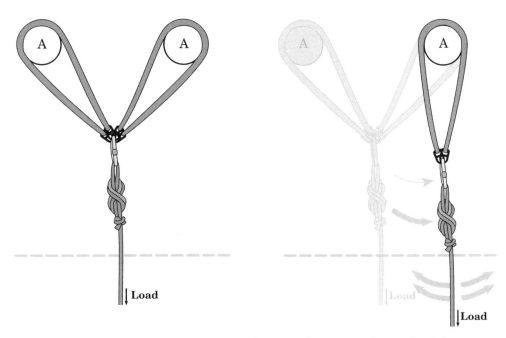

Figure 13.17b   Pendulum swings can occur in multipoint anchor systems if one anchor fails.

Figure 13.18 Excessive slack between the primary and secondary anchors, as seen here, will severely shock-load the system if the primary anchor fails.

system if the primary attachment fails. Basically this backs up the webbing and water knot in the primary attachment. As always, minimize slack in the backup attachment to reduce shock loading of the secondary attachment if the primary attachment fails.

Rigging multipoint anchor systems.   Multipoint anchor systems are built around more than one anchor point. The focus in this chapter will be on the establishment of multipoint anchor systems that are built with two primary anchor points. The practices shown here also can be used to rig systems with three or more primary anchor points.

In industrial confined space entry or rescue, multipoint anchor systems should not be used to link weaker anchor points together in an attempt to form one strong anchor connection. Instead multipoint anchors are used in industrial operations as a way of adjusting

Figure 13.19 The attachment from the secondary anchor should be made directly to the knot that supports the rope system.

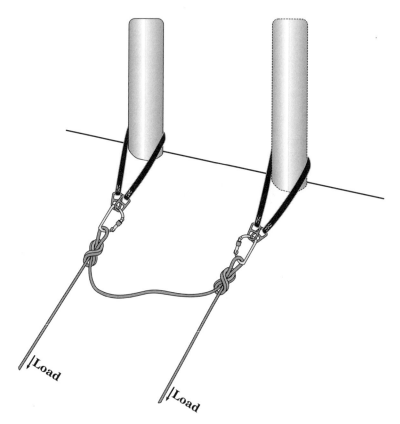

Figure 13.20 Two primary anchors back each other up in this example.

the direction of the pull any time the anchors are not in line with the configuration of the space (Fig. 13.23).

The options available to rig multipoint anchor systems are endless, limited only by the equipment and anchor points available and the experience and imagination of the riggers. Learning the basic designs of multipoint anchors allows team members to adjust to conditions in the field and to competently rig multipoint anchor systems.

Consider the following points when designing multipoint anchor systems:

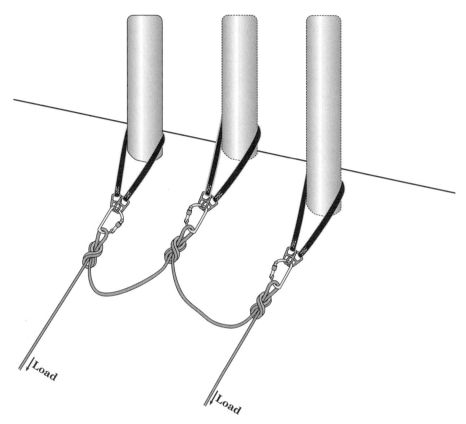

Figure 13.21 Two primary anchors share a common secondary anchor in this example.

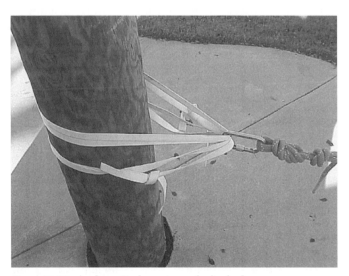

Figure 13.22 Anchor attachments can be backed up.

- The system should be built so that the weight of the load is distributed between the anchor points as equally as possible.

- Failure of one of the anchor points in the anchor system must not lead to the failure of the entire anchor system.

- If one of the anchor points fails and rope attached to the system does a pendulum swing along a sharp edge or abrasive surface, damage or failure of the rope could

Figure 13.23 A multipoint anchor system is used to adjust the direction of a lower in this example.

result (Fig. 13.17*b*). If failure is possible, pad any edges or rough surfaces that the rope could contact if this occurs.

Using multipoint anchor systems can add an extra margin of safety to the system since the weight of the load is distributed between two anchor points and each anchor point backs up the other. The tradeoff is the time that it takes to rig the system and the increased possibility of shock loading or pendulum swings if one of the anchor points fails. The anchor attachments can be backed up as discussed in the previous section. The system also can be configured to include a backup anchor as shown in Fig. 13.25. Rigging multipoint anchor systems can be time-consuming. In entry operations time is not as critical as in rescue operations, and multipoint anchor systems can be thought out in preplanning the space.

Multipoint anchor systems can be classified into two main categories: (1) load-sharing and (2) load-distributing or self-adjusting. Ideally each anchor point used in the system should see an equal proportion of the load. Remember that the key to anchoring is that the anchor points are stable and strong enough to hold the weight of the load on the system. Individual anchor points chosen for multipoint systems should each be capable of handling the entire load. As we will see later, compound forces must be considered anytime angular forces are generated with load-sharing multipoint anchor systems.

Load-sharing multipoint anchor systems. *Load-sharing anchor systems* are just what the term implies, an anchor system that distributes the weight of the load over two or more anchors. See Figs. 13.24 and 13.25 for examples of two-point load-sharing anchor systems and Fig. 13.26 for an example of a three-point load-sharing anchor system.

The amount of the load each anchor sees is dependent on the relative lengths of the anchor connectors and the direction of the pull. If one connector in an anchor bridle is shorter than the other, the anchor point attached to the shorter connector will see more of the load than the other anchor point. This will cause the distribution of forces between the anchor points to be uneven.

The direction of the pull affects how the force is distributed between the anchor points. A system is rigged for a pull in a certain direction. If this direction changes once the main line is loaded, significantly more of the load can be shifted to one or the other of the anchor points. If one anchor fails, the load will shift to the remaining anchor, causing a pendulum swing that could damage the rope (Fig. 13.17*b*). Pad all angular edges and rough surfaces that the rope could contact if this occurs.

When attaching to load-sharing anchor systems, do not use a single carabiner to make the connection if it will be subjected to a three-way pull. Triloading the carabiner significantly reduces its working strength since carabiners provide maximum strength only when loaded parallel to the long axis. Use a trilink in this position instead, as shown in Figs. 13.23 and 13.24. Other options include the use of one carabiner attaching each anchor connector to an anchor plate (Fig. 13.27) or a rigging ring (Fig. 13.28).

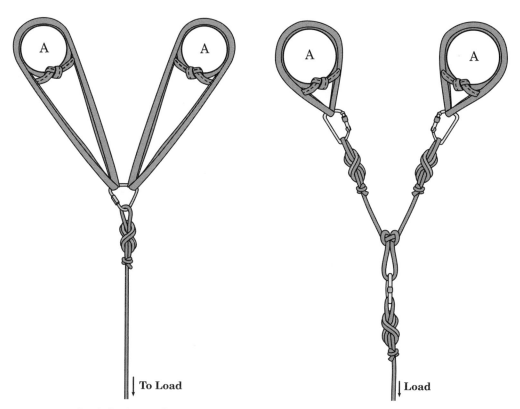

Figure 13.24 Load-sharing anchor systems are anchor systems that distribute the weight of the load over two or more anchors.

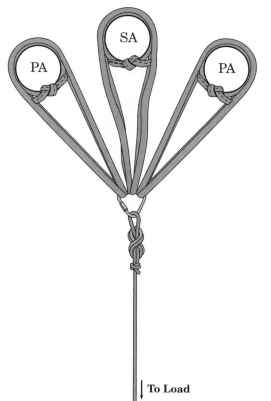

Figure 13.25 Two-point load-sharing anchor systems should be backed up if the anchor points are questionable.

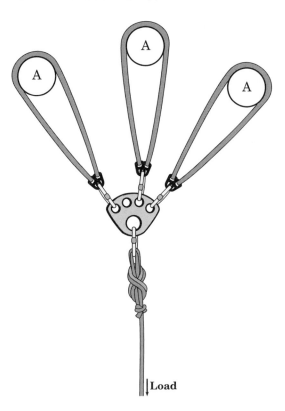

Figure 13.26 This load-sharing anchor system distributes the weight of the load over three anchor points.

Figure 13.27 Anchor plates can be used in load-sharing anchor systems.

Load-distributing or self-adjusting anchor systems. Load-distributing anchor systems are used when a change of direction in the main line could occur during the operation. A simple load-distributing anchor system can be made between two anchor points by placing a 180° twist in a webbing loop sling as shown in Fig. 13.29. This places the sling in a figure 8 configuration. Clipping a carabiner around the point where the two lengths of webbing cross, as shown in Fig. 13.29c, makes the point of attachment for the load-handling system. Connecting in this way prevents the carabiner from slipping off the loop if one of the anchor points fails. Despite its name, this system does not actually distribute the load evenly between the two anchor points unless the load is centered directly between the two.

Compound forces on multipoint anchor systems. Angular forces in multipoint anchor systems can significantly increase potential loads on anchor points and other system components. As an example, assume that we are rigging a multipoint anchor system using two equal-length webbing slings to form a load-sharing bridle. As the interior angle between the legs of the bridle increases, the amount of force placed on the anchor points and other

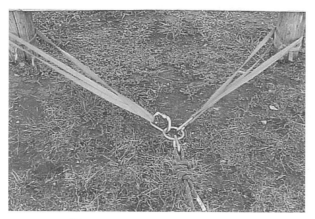

Figure 13.28 Rigging rings can be used in load-sharing anchor systems.

system components also increases. The following formula can be used to calculate the forces placed on each anchor as a function of the angle between the legs of the bridle:

$$\text{Force placed on each anchor} = \frac{\text{load}}{2 \cos (\text{interior angle}/2)}$$

For example, if we place a 100-lbf load on the two-point anchor system with the bridle legs parallel to each other (0° interior angle), each of the anchor points will see half the load, or 50 lbf each (Fig. 13.30). If the angle between the bridle legs is increased to 90°, then each anchor will see approximately 71 lbf (Fig. 13.31). If the angle between the bridle legs is increased to 120°, then each anchor point will see 100 lbf (Fig. 13.32). At angles greater than 120°, the forces placed on the anchors are compounded so that both the anchor points will actually see a force greater than the load placed on the system. As an extreme example, if the angle between the legs of the bridle is increased to 175° and a load of 100 lb is placed on the system, each anchor point sees approximately 1146 lbf (Fig. 13.33).

Compounding forces can cause anchor points and other system components to fail under loads that we might otherwise expect them to withstand. For this reason compound forces must be avoided. When using multipoint anchor systems, avoid angles greater than 120° between legs of bridles, slings, and other system components. Angles of 90° or less are preferred.

Extending and redirecting anchors. In some scenarios, the only suitable anchor point may not be in a position where it can be utilized to make the entry. Anchors can be extended or redirected in situations such as this.

Extending an anchor may be as easy as placing a double-loop figure 8 in the end of a suitable kernmantle rope, lowering or pulling the end of the rope to the location where an anchor point is needed, and tying another double-loop figure 8 in the rope for attachment to the available anchor point (Fig. 13.34). Be sure to pad any sharp edges or rough surfaces to protect the rope extension.

Anchors can be redirected by using a load-sharing system that spreads the weight of the load between two anchor points. They also can be redirected by passing the main-line rope through a pulley as shown in Fig. 13.35. Pulleys used in this configuration are stationary and are called "directionals." Anytime directional pulleys are used in a system, compound forces placed on the pulleys should be considered. This is discussed below in the section on rigging for change of direction.

General rules for rigging anchor systems

A number of general safety precautions should be taken when rigging anchor systems. Several of these precautions have already been discussed in this chapter. Examples of general rigging rules for anchor systems include the following:

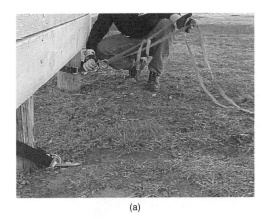

(a)

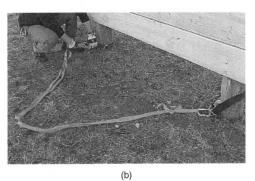

(b)

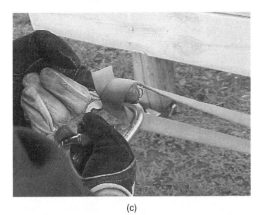

(c)

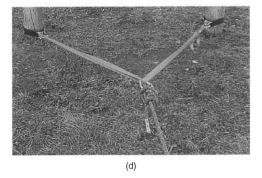

(d)

Figure 13.29 To configure a two-point load distributing anchor system, take these following steps. (a) Tie a webbing loop with a water knot. This loop should be long enough to extend between the two anchors and give an acceptable angle (less than 120°, preferably 90° or less) when loaded. Connect one end of the webbing loop to one anchor attachment. Then place an 180° twist in the webbing loop, placing it in a figure 8 configuration. (b) Attach the other end of the webbing loop to the other anchor attachment. (c) Clip a carabiner around the point where the webbing crosses at the midpoint of the figure 8. (d) This is a completed two-point load distributing anchor system.

■ Make sure that any questionable anchor points are backed up with an acceptable anchor point.

■ Pad all sharp edges and abrasive surfaces that soft components of the anchor system may contact.

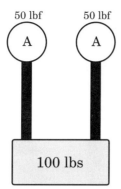

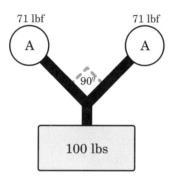

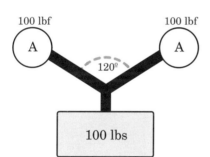

Figure 13.30 If a 100-lb load is suspended as shown with the legs of the anchor bridle parallel to each other, each anchor point will see half the load (50 lbf).

Figure 13.31 As the angle between the legs of the anchor bridle increases, so does the force that each anchor will see. At 90°, with the same 100-lb load, each anchor will see approximately 71 lbf.

Figure 13.32 With the same 100-lb load and 120° between the legs of the anchor bridle, each anchor will see approximately 100 lbf.

Figure 13.33 With the same 100-lb load and 175° between the legs of the anchor bridle, each anchor will see approximately 1146 lbf.

- Attach as close to the base of anchor points as possible.
- Seek out the strongest available part of a structural system for the point of attachment.
- If in doubt, load-test anchor points in advance.
- Avoid angles in anchor systems that create compound forces on anchor points and system components.
- Remember the KISS method. Keep anchor systems simple and safe.

Figure 13.34 Anchor points can be extended to locations where points of attachment are needed but no suitable anchor points exists.

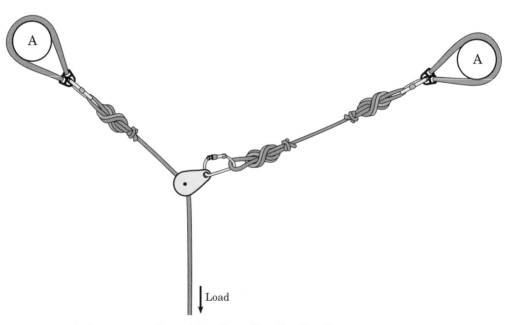

Figure 13.35   Anchors can be redirected by using a "directional" pulley.

Rigging Basic Systems for Hauling or Lowering a Load

Mechanical-advantage (MA) systems make the performance of work such as raising or lowering a heavy load easier by reducing the effort required to complete the task. Common examples of MA systems used in confined space operations include winch systems and block-and-tackle systems. Winch systems will be discussed later in this chapter in the section on tripod operations, because winches are most commonly used with tripods in confined space operations.

Block-and-tackle systems

Block-and-tackle systems can be simple or compound. We will initially focus on simple MA systems and turn our attention to compound systems later. Block-and-tackle MA systems reduce the amount of force needed to lift a load, but the tradeoff is that they increase the distance you have to pull the rope to move the load a given distance.

Block-and-tackle systems can be constructed easily with a couple of pulleys, rope, carabiners, a rope-grabbing device, and a preestablished anchor attachment. When used in raising or lowering operations, these systems should be used in conjunction with a safety belay line. Safety belays are discussed later in this chapter.

To understand their construction and to prevent confusion in building the systems in time-critical situations, let's look at how they work. Consider a 100-lb load suspended from a rope (Fig. 13.36). The load obviously is being pulled downward by gravity with a force of 100 lb, its weight. To keep the load suspended requires an equal but opposite upwardly directed force. The anchor system provides this force. If we want to lift the load 100 ft, then we must pull 100 ft of rope in the upward direction, which would be very difficult.

Now let's add a single pulley as shown in Fig. 13.37. The force required to keep the load suspended is still 100 lb, and to move the load 100 ft still requires 100 ft of rope to be pulled. This configuration is classified as a 1:1 system, meaning that it provides no mechanical advantage. The pulley in this system is considered a directional pulley because it changes the direction in which the force is applied to the system to lift the load. The line pulled to move the load, or *haul line,* now can be pulled in a direction that

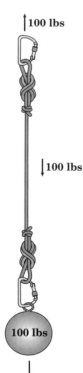

100 lbs

100 lbs

100 lbs

Figure 13.36 To keep this 100-lb load suspended, an equal but opposite force must be provided.

would make the haul team's job easier. Directional pulleys are the stationary pulleys in the block and tackle system. This means that they are anchored and nonmoving.

Simple mechanical advantage systems.   Now let's add a second pulley to the system as shown in Fig. 13.38. The second pulley is called a *traveling pulley*. It moves up and down with the load. The load is equally divided between the two lines so that each line supports 50 lb to keep the load suspended. This means that only 50 lb of force has to be applied to the haul line to keep the 100-lb load suspended. This is a 2:1-MA system. The amount of force needed to lift the load has been reduced by half. (*Note:* Frictional and angular forces that increase the amount of effort or force needed to lift the load are ignored in this example.) The tradeoff for the reduction in force required to move the load is that the distance the haul line would have to be pulled to move the load a given distance has doubled. For example, if we want to move the load 100 ft, we would have to pull 200 ft of haul line to do so. The more common configuration of the 2:1-MA system used in confined space operations is shown in Fig. 13.39. In this configuration the lines in the system run parallel to each other, increasing the efficiency of the pulleys by reducing the angular forces that increase the amount of effort needed to move the load.

Figures 13.38 and 13.39 are examples of 2:1-MA systems. An easy way to determine the mechanical advantage of a simple MA system is to count the number of lines running to the load. The knot in the end of the rope will be attached at the anchor in an even-numbered system (e.g., 2:1, 4:1, 6:1) and attached to the traveling pulley at the load in an odd-numbered system (e.g., 3:1, 5:1).

A 3:1-MA system adds another directional pulley to the system (Fig. 13.40), which means that you could actually have either a double-sheave pulley or two single-sheave pulleys at the anchor and a single sheave at the load. The knot is at the load since it is an odd system. There are three lines to the load that equally split the 100-lb load between them, so each only holds 33.3 lb to suspend the weight of the load. This means that only 33.3 lb of force have to be applied to the haul line to keep the load suspended. The amount of effort or force needed to lift the load has been reduced by two-thirds. Again the tradeoff is the distance that the haul line would have to be pulled to move the

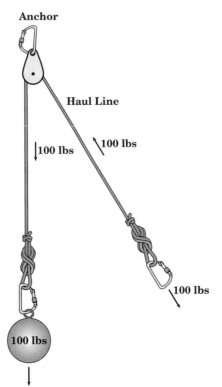

Figure 13.37 Adding a directional pulley provides no mechanical advantage. This is a 1:1 system.

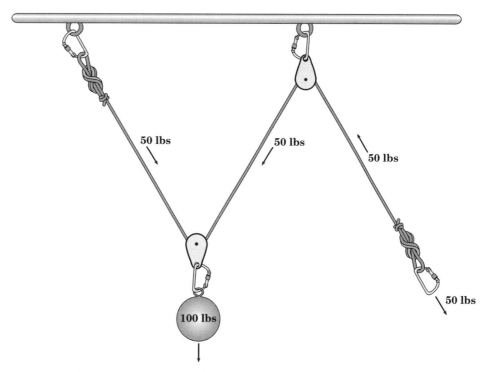

Figure 13.38 Adding a traveling pulley that moves up and down with the load provides mechanical advantage. This is a 2:1-MA system.

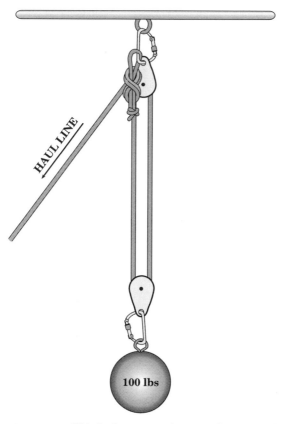

Figure 13.39 This is the more common configuration of the 2:1-MA system seen in confined space operations.

load a given distance. In fact, if we want to move the load 100 ft, we would have to pull 300 ft of haul line to do so.

A 4:1-MA system adds another traveling pulley to the system (Fig. 13.41). The knot is at the anchor since it is an even system and there will be four lines to the load that equally split the weight of the load. This means that only 25 lb of force is needed on the haul line to keep the load suspended. The amount of force needed to lift the load has been reduced by three-fourths; however, to move the load 100 ft will require pulling 400 ft of haul line.

In planning to use MA systems in confined space operations, keep in mind that as the mechanical advantage increases, the amount of effort needed to move the load decreases but the distance the haul line must be pulled increases. The amount or length of rope needed to rig the system increases with increasing mechanical advantage. The depth or distance of the pull also should be considered. For example, if a tank were 25 ft deep, we would need a minimum of 100 ft of rope plus the length needed for knots, the haul line, and extension to the anchor attachment in order to rig a 4:1-MA system.

General guidelines for rigging simple MA systems.   Whenever you build a simple mechanical-advantage system, some basic rules apply:

- The "key" number of the system indicates the total number of pulley sheaves in the system. For example, if you are building a 3:1 system, a total of three sheaves will be required. Obviously, for the 3:1, one double-sheave pulley and one single-sheave pulley will be most practical to use.

- In an odd-numbered setup, the pulley with the greater number of sheaves is typically placed at the anchor.

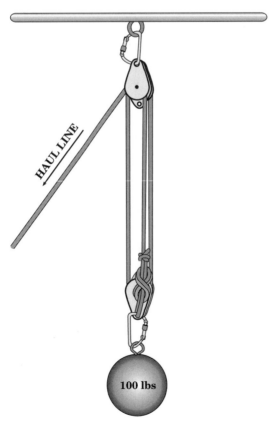

Figure 13.40   This is a simple 3:1-MA system.

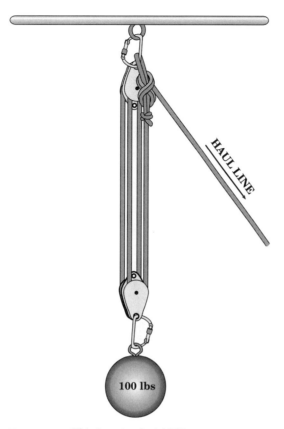

Figure 13.41   This is a simple 4:1-MA system.

- If you are building an even-numbered system (e.g., 2:1 or 4:1), attach the end of the rope at the anchor. On the other hand, if you are building an odd-numbered system (e.g., 3:1 or 5:1), attach the end of the rope at the load.

- Reeve the rope through the pulleys alternately. For example, to complete the 3:1 system, attach the end of the rope to the load, then thread the rope through one of the sheaves on the anchor pulley and through the sheave on the traveling pulley, then finish by threading it through the other sheave on the anchor pulley.

- In rigging the system, always reeve the rope through the pulley sheaves in such a way that adjacent lines don't cross or rub against each other.

- In most cases, it is easier to lay the pulleys out close together on the ground or some other flat surface to build the system rather than trying to build it in place suspended from the anchor point.

Using rope-grabbing devices as ratchets in MA systems.   When using MA systems in vertical applications, incorporate a rope-grabbing device that functions as a ratchet into the system on the line opposite the haul line as shown in Fig. 13.42, or on the haul line as shown in Fig. 13.43. The device acts as a ratchet to prevent the load from falling when the haul team releases the haul line. Some pulleys, such as the Haul Safe™ pulley by RSI, include a ratchet cam that is prefabricated to an extended middle plate. The use of a ratchet in the system reduces the chance that the haul team will lose control of the load. It is important to realize that the rope-grabbing device used in this application is not a belay device to be used during lowering operations. This is especially true when mechanical rope grabs such as the Gibbs ascender are used, as these devices can severely damage the rope under shock-load conditions.

The rope-grabbing device must be installed on the line opposite the haul line or on the haul line itself. The mechanical advantage of the system reduces the force that the rope

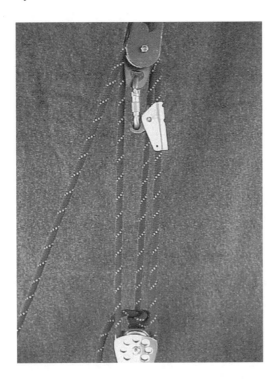

Figure 13.42 A Gibbs ascender is used as a ratchet on this MA system. It is attached to the line opposite the haul line.

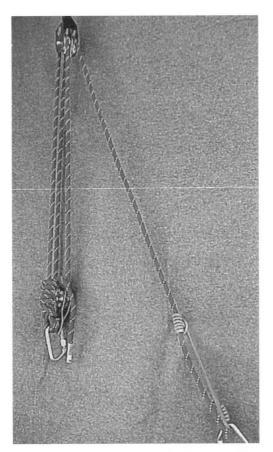

Figure 13.43 A Prusik hitched loop of accessory cord can also be used as a ratchet device. In this system it is shown attached to the haul line.

grab must apply to the rope to stop and hold the load. For example, in holding a 200-lb load suspended with a 4:1 system, the rope-grabbing device theoretically sees only 50 lbf.

Both soft rope-grabbing devices, such as Prusik hitched loops of accessory cord, and mechanical rope-grabbing devices, such as the Gibbs ascender or the MIO industrial rope grab, commonly are used as ratchet devices in MA systems. Mechanical rope grabs such as the Gibbs ascender can damage the rope significantly if a shock load occurs. While shock loads should always be avoided with rope systems, this is especially critical when devices such as the Gibbs ascender are used.

If you are using a mechanical rope-grabbing device, such as the Gibbs ascender or the MIO industrial rope grab, the engraved arrow on the device should point toward the load when installed in the system (Figs. 13.44 and 13.45). It is a good practice to test the system to ensure proper installation before putting the system into operation.

Hard linking sometimes causes problems when mechanical rope grabs are used as ratchets in hauling systems. *Hard linking* refers to the placement of hardware in a configuration that denies the system the flexibility needed for proper operation. This can occur when a Gibbs ascender is attached to the becket of a pulley for use as a ratchet. If

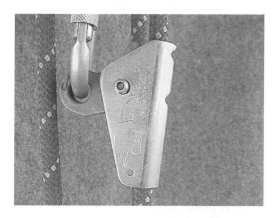

Figure 13.44 When using mechanical rope grabs such as the Gibbs ascender as a ratchet, the arrow should point toward the load when installed on the line opposite the haul line.

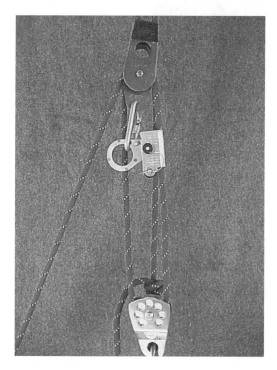

Figure 13.45 The MIO industrial rope grab can also be used as a ratchet as shown.

the Gibbs ascender is attached directly to the becket with a carabiner, the ascender sometimes binds on the carabiner. This can result in improper loading of the carabiner and can place the ascender in a bind so that it does not align well with the rope. Using a basket-hitched loop sling of one-inch webbing to attach the ascender to the becket prevents this. Because of this, manufacturers now offer some models of the Gibbs ascender with a webbing "soft link" installed at the factory.

Applications for simple MA systems in confined space operations.   Once an MA system is rigged, it can be used in a variety of confined space applications. For some routine work entries, MA systems are used to lower workers into a space and haul them back out of the space. To accomplish this, attach the anchor pulley to the head of a tripod or a structural anchor, attach the traveling pulley to the entrant's harness, and operate the MA system as described below (Fig. 13.46). In other routine entries, the MA system is attached to the entrant's harness even though a ladder or some other means is available for the entrant to descend into the space. If an emergency occurs involving the entrant, use the MA system for nonentry rescue by hauling the entrant back out of the space.

Simple MA systems have numerous uses in entry rescue operations. They are used to lower rescuers into vertical spaces, then used to haul the victim out. MA systems are also useful for performing rescues from horizontal spaces. Rescue teams commonly carry the traveling pulley of a simple MA system into a space, or else pull it into the space following entry, then attach it to the packaged victim so that team members outside the space can haul the victim out (Fig. 13.47).

For many applications it is convenient to use a prebuilt MA system, rather than rigging the system at the time when it is needed. This is especially true for confined space rescue. One thing to keep in mind is that rope smaller than $1/2$- in kernmantle can be used to rig a general-purpose MA system, since the load will be borne by all the lines in the system. For example, $3/8$-in (9.5-mm) static kernmantle rope can meet general-use requirements when used in a MA system. As discussed in Chap. 17, a small prerigged

Figure 13.46 MA systems can be used for both lowering and raising operations.

Figure 13.47 MA systems can be used for hauling packaged patients from horizontal spaces.

4:1-MA system is a very valuable tool for numerous confined space rescue applications. Several of those applications are described in Chap. 17.

Hauling operations with simple MA systems.   To use simple MA systems in hauling applications, attach the anchor pulley to the anchor system and extend the system by taking the traveling pulley to the location of the entrant or load. Once rescuers have entered the space, support personnel outside the space can tie a butterfly loop in the rescuer's tag line and connect it to the carabiner on the traveling pulley. The rescuer pulls the traveling pulley to the victim, thereby extending the system. Once the system is extended, the carabiner on the traveling pulley is attached to the victim.

A safety belay line should be attached to the entrant for all vertical hauling or lowering operations. The safety belay will catch the load if the MA system fails.

Personnel roles in hauling operations.   Like all aspects of confined space activities, hauling operations are a team activity. These operations must be carefully coordinated and controlled in order to be carried out safely. Good communication between everyone involved is vital. Personnel involved include the team leader, belayer, attendant, spotter, entrant or rescuer, and the haul team members.

The team leader is responsible for directing the operation for maximum efficiency and safety. The team leader must do a standard safety check to make sure that everything is properly rigged, all needed safety precautions have been taken, and the belayer and haul team members are in place and ready.

The belayer must be ready to use the belay system to catch a falling load throughout the hauling operation. Rigging and operating belay systems are discussed later in this chapter.

A person must be designated as spotter during hauling operations to make sure that the load does not get caught or tangled. This duty may be performed by the attendant for simple entries, but complex spaces may require an internal spotter to watch the load as it is being hauled and relay instructions to the attendant. The spotter's duty is critical to the safety of the operation. If the load hangs up and the haul team continues to haul, severe personal injury or equipment damage could result.

The entrant or rescuer has an active role in hauling operations. In some operations, the entrant or rescuer may be the load. In some cases a rescuer may have to guide a patient being hauled to prevent the patient from becoming caught or entangled. The team leader should be positioned to see the attendant or spotter and to be seen by the safety belayer and the haul team.

As the term implies, the *haul team* pulls on the haul line to move the load. Haul teams must be very sensitive to resistance in the haul system as this may indicate that the load is hung or the MA system is fouled. If resistance is felt, stop hauling immediately and find out what is causing it! This is especially true with higher-mechanical-advantage systems, because increasing the mechanical advantage increases the likelihood that the haul team may fail to notice the increased resistance. Some organizations do not use MA systems with more than 4:1 mechanical advantage for this reason.

Haul team staffing levels.   One way to estimate the number of members needed on the haul team is to divide the number 12 by the mechanical advantage of the system. For example, if we are using a 4:1-MA system, three haul team members would be estimated using this "rule of 12." This number may need to be modified according to the load, working conditions, and the duration of hauling. For example, if a 300-lb one-person load is hauled with a 4:1-MA system, the haul line will see a quarter of the load or 75 lbf. Each haul team member would be responsible for exerting 25 lb of force in order to haul the load. This doesn't sound like a lot of force, but haul team members may begin to tire during long hauls, especially under tough environmental conditions such as high heat. In that case, an additional haul team member may be desirable. More than four haulers seldom are recommended with a 4:1-MA system because too many haulers may overpower the system and fail to notice increased resistance caused by a hung load or a tangled system.

Haul team commands.   Verbal commands typically are used to direct hauling operations. In some cases, commands may be given by radio or intercom system. In high-noise environments, hand signals may be used. Examples of hand signals are shown in Fig. 17.16.

Examples of common verbal commands include "prepare to haul," "haul," "set," and "Stop!" The "prepare to haul" command is given by the team leader to notify everyone to get into position. On hearing this command, the haul team removes the slack from the system by pulling on the haul line until the system is taut. The belayer removes all slack from the safety belay system and replies "belay on" to let the team leader know she is in position and ready to arrest any fall that may occur. When the team leader issues the "haul" command, the haul team begins pulling the haul line to move the load. The team leader gives the "set" command to notify the haul team to stop hauling and slowly release the haul line, allowing the ratchet to set so that they can regroup for the next haul. The "Stop!" command is given when the haul team needs to stop hauling immediately. Any member of the team can issue this command and should do so immediately if an unsafe condition is recognized. Everyone involved in the operation should repeat the command "Stop!" to verify that they heard the initial command.

Lowering with simple MA systems.   The MA system can also be used as a lowering system. This is basically the reverse of the hauling operation. Commands and procedures differ slightly.

When MA systems are used to lower a load, it is critical that the haul team maintain control of the haul line and slowly feed rope into the MA system to carefully lower the load. One technique that can be used for this is an offset belay. The offset belay is performed by two or more haul team members who position themselves to perform body belays on the haul line, as shown in Fig. 13.48.

Initially the team leader gives the command "prepare to lower," which notifies everyone to get into position for the operation. The safety belayer should reply "belay on" when in position and ready to belay the load being lowered. If a ratchet is used in the MA system, it must be unloaded and unlocked before the lower can proceed. To do this, the team

Figure 13.48   Team members perform an offset belay to lower a load with a MA system.

leader may give a command such as "short haul" to notify the haul team to pull the haul line just enough to shift the weight of the load onto the MA system and off the ratchet. The team member designated as the ratchet tender either removes the ratchet from the rope or holds it open during the lower. Remember that the ratchet is not intended to be used as a fall-arrest device and can damage the rope severely if shock-loaded.

Next the team leader gives the command "lower." The haul team performing the offset belay slowly feeds rope from the haul line into the MA system in order to lower the load. The safety belayer slowly belays the safety line. The command "Stop!" notifies all team members to stop the lowering operation. This command should be given by anyone on the team who recognizes a dangerous situation. In response, all team members should repeat the stop command to acknowledge that they received it.

Rigging and using the Z rig system.   The *Z rig* is another type of simple MA system (Fig. 13.49). It provides a 3:1 mechanical advantage. As used in vertical applications, it incorporates a rope-grabbing device as a haul cam, a rope-grabbing device to serve as a ratchet, two single-sheave pulleys, and two carabiners. For strictly horizontal applications, the ratchet can be omitted from the system.

The Z rig can be used any time rope length limits the use of the more conventional block-and-tackle systems described previously. For rescue applications, it can be rigged using the tag line already attached to an entrant or victim in a space. This may allow nonentry rescue to be performed. Remember that the tag line must be life-safety-rated when this technique is used for rescue from vertical-type spaces. The use of the Z rig in rescue applications is discussed further in Chap. 17.

Rigging the Z rig. To build the Z rig, follow four simple steps:

1. Place the running end of the rope through a directional pulley and attach the pulley to the anchor.

2. If the system is to be used in a vertical application, attach a rope-grabbing device to the rope just to the load side of the anchor pulley to serve as a ratchet. If a mechanical rope grab is used, attach it to the rope with the arrow pointing toward the load and connect it to the carabiner above the pulley with a short basket-hitched webbing loop sling.

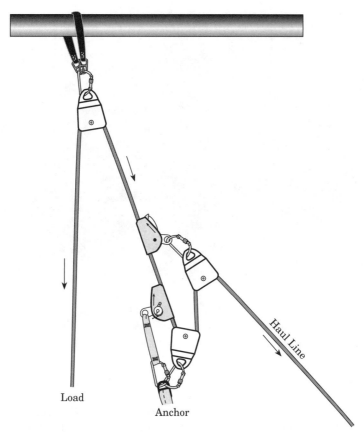

Load

Anchor

Haul Line

Figure 13.49   The Z rig is a 3:1-MA system and is used here with a high-point directional pulley.

3. Attach a second rope-grabbing device to the standing part of the rope between the anchor pulley and the load to serve as a haul cam. Install it with the arrow pointing toward the load.

4. Install a traveling pulley in the haul line and connect it to the haul cam.

Either a mechanical rope grab or a triple-wrapped Prusik loop of accessory cord can serve as the rope-grabbing device (Fig. 13.50). The precautions previously mentioned regarding potential damage to the rope due to shock loading of the ratchet apply here. Take great care to ensure that the system is not overstressed, for instance, due to a hung load and an overzealous haul team. In such a case the rope may be severely damaged, especially if a mechanical rope grab such as the Gibbs ascender is used for the haul cam. Because of this, some rescue organizations use only soft rope grabs, such as Prusik hitched loops of accessory cord, in this application.

The Z rig can be used with an added directional pulley. This directional pulley can be placed to change the direction of the force applied to lift the load. In this configuration the added pulley is referred to as a *high-point directional pulley* (Fig. 13.49). A directional pulley also can be placed in the haul line as shown in Fig. 13.51 to change the direction of the pull. Directional pulleys add no MA to the system but may allow a more convenient positioning for the haul team.

Overview of Z rig operation.   To use the Z rig, pull the haul cam along the main line as far as possible toward the load and release or set it, extending the Z rig as far out as possible. Use the Z rig to pull the load as far as possible toward the anchor. When the team leader gives the command "haul," the haul team pulls the haul line until told to "stop" or until the haul cam is next to the ratchet cam. Once the haul cam is close to the

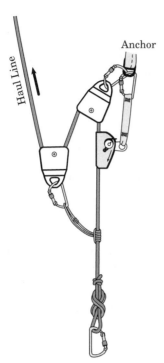

Anchor

Haul Line

Figure 13.50 A triple-wrapped Prusik hitched loop of accessory cord serves as the haul cam in this Z rig example.

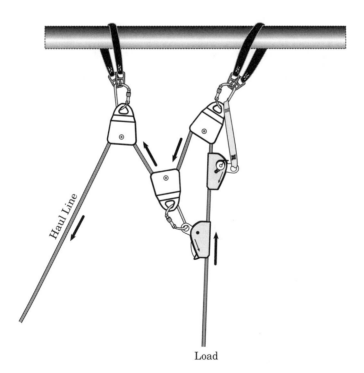

Haul Line

Load

Figure 13.51 A Z rig used vertically may require an additional directional pulley on the haul line to change the direction of the pull.

ratchet cam, the team leader gives the command "set" and the haul team slowly releases the haul line, allowing the ratchet to set. Now the ratchet cam holds the load. The team leader gives the command "slack," and the haul cam is again pulled as far as possible along the main line toward the load and released or "reset." This sets the stage for another haul. The process is repeated until the load has been hauled to the desired location. As always, any team member can give the command "stop" anytime an unsafe condition is noted.

Compound mechanical advantage systems.   A compound MA system is actually two or more simple MA systems that have been "stacked" so that one system applies a pulling force to another system's haul line.

Figure 13.52 shows a 2:1 (MA) system attached to the haul line of another 2:1 system. If the load is 100 lb, the first system reduces the force required to haul the load to 50 lbf. The second 2:1 system reduces the force further to 25 lbf. This provides a total mechanical advantage of 4:1. The mechanical advantage of any compound system can be calculated by multiplying the advantages of the component systems.

Compound systems are used in a variety of configurations to achieve mechanical advantages ranging from 4:1 to 9:1. It is important to use caution and limit the number of haul team members when using compound systems with a high mechanical advantage. Injury to the person being hauled or damage to system components could easily result if the load hangs up and the haul team continues to haul with these high-powered systems. As in all hauling operations, a spotter must be assigned to watch the load as it is being hauled and issue the order "stop" immediately if the load or rigging gets hung or tangled.

In this text we will focus mainly on using simple MA systems. Other texts offering more extensive coverage of compound MA systems are available (Frank 1998).

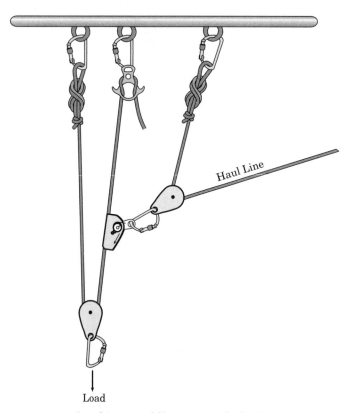

Figure 13.52   Attaching a 2:1-MA system to the haul line of another 2:1-MA system forms a compound 4:1-MA system.

Piggyback applications. In some situations where vertical block-and-tackle systems cannot be used, another option is to use a rope-grabbing device to "piggyback" a simple MA system to a main line attached to the load (see Fig. 17.17). Piggyback applications are limited only by the length of the main line, thereby providing the mechanical advantage of the conventional systems without the length limitations normally associated with the conventional use of simple MA systems.

The piggyback system is significantly more complicated to rig and operate than the simple MA systems and the Z rig. For this reason it is considered beyond the scope of basic rigging and is not included in this chapter. For routine confined space entries, the lead time available should allow arrangements to be made for simpler systems to be used; however, rescuers may find the piggyback system invaluable. The piggyback system is covered in detail in Chap. 17, "Rigging and Using Rescue Systems," and shown in Fig. 17.17.

Manufactured MA systems. So far all the MA systems we have covered are rigged from the basic rope, software, and hardware components described in Chap. 11. The rigged systems are used very commonly; however, manufactured MA systems are currently marketed as an alternative to rigged systems. Examples of manufactured systems include the Griptech BP1 and the Griptech SP2 by Grip Safety and Rescue Systems, Inc. (Fig. 13.53). The BP1 is marketed as a work/rescue system, while the SP2 is marketed as a rescue system. Both are designed to be used in vertical hauling applications. The BP1 system is available in a 3:1 or 4:1 mechanical advantage, and the SP2 system can be configured in a 2:1, 3:1, 4:1, or 5:1 configuration.

Both the BP1 and the SP2 systems eliminate the majority of rigging components needed to build a conventional MA system (Fig. 13.54). These systems are lightweight compared to winches and can be anchored anywhere a conventional vertical block-and-tackle system can. They have built-in drum friction control for lowering with the same system without having to change the rigging. The BP1 has a working load limit of 300 lb, while the SP2 has a working load limit of 600 lb. The BP1 can be purchased with an optional fall arrester that activates automatically and resets automatically after activation. It also has a built-in ball bearing swivel on the lower pulley to eliminate rope twisting. The SP2 is available with a haul lock or ratchet cam used to hold the load between hauls.

Figure 13.53 The Griptech SP1 is a manufactured MA system.

Figure 13.54 The Griptech is lightweight and can be anchored anyplace a conventional vertical block-and-tackle system can.

Rigging for Change of Direction

In rigging for confined space operations, you may find it necessary to redirect the pulling forces by changing the path of a rope within a rope system. Do this by placing directional pulleys in the system. Directional pulleys add no mechanical advantage to a system; they merely change the direction of the forces applied to move a load. Directional pulleys are always anchored and stationary.

Applications for directional pulleys

Directional pulleys have many potential applications in rigging basic systems for confined space operations. Examples of how you might use them follow here:

- Install an elevated or "high point" directional pulley above points where a rope will be hauled or lowered over edges (Fig. 13.55). This prevents excessive resistance in the system as the rope is pulled over the edge, and prevents possible damage to the rope.

- Anchor a high directional pulley over the entrance to vertical spaces to allow a Z rig system to haul an entrant above the entrance to the space (Fig. 13.49). This places the haul team away from the space in a safer, more convenient location.

- Use a directional pulley to redirect the haul line to make the haul team's job easier as shown in Fig. 13.51. It is easier for the haul team to raise a load by pulling from an upward angle down than to pull the same load from a downward angle up.

- Use directional pulleys in tripod operations to keep the force exerted by the haul line on the tripod within the "triangle of stability" as shown in Fig. 13.56. This keeps the tripod from being pulled over by the pulling force.

- Use directional pulleys when you need to redirect anchor points within an anchor system, as shown in Fig. 13.35.

Theoretically you can include as many directional pulleys as needed to make a rope system work when rigging the system; however, each directional pulley adds friction to the system and reduces the overall efficiency. Limit the number of directional pulleys in a given system to the minimum required to make the system work.

Rigging directional pulleys

Rigging directional pulleys is like any other type of basic rigging. Begin by selecting or establishing a suitable anchor point, rigging an anchor system, and attaching the pul-

Figure 13.55 High-point directional pulleys prevent excessive wear on a rope.

Figure 13.56 A directional pulley can be used in tripod operations to keep the force exerted by the haul line within the "triangle of stability."

ley. Follow the basic rules and procedures discussed earlier in this chapter. Avoid common problems related to hard linking and compound forces.

Hard-linking directional pulleys. *Hard linking* refers to the placement of hardware in such a way that the system lacks the flexibility required for proper operation. An example of this might be attaching a change-of-direction pulley directly to an eyebolt with a carabiner (Fig. 13.57). The pulley would be held rigidly and not afforded flexibility for side-to-side movement. This could prevent the pulley from being able to pivot to ensure that the sheave is fully aligned with the direction from which the rope is feeding into it. As a result, the pulley does not operate efficiently and damage results to the pulley or the rope. In the same example, if a basket-hitched webbing loop or some other type of flexible connection is placed in the system between the pulley and eyebolt, the pulley can pivot into any position required for proper operation (Fig. 13.58).

Figure 13.57 Beware of "hard linking" directional pulleys. The lack of flexibility can prevent the pulley from aligning properly with the rope.

Figure 13.58 Adding a flexible connection between the directional pulley and the anchor attachment will prevent "hard linking."

Compound forces on directional pulleys. Forces placed on directional pulleys can be compounded so that the force on the pulley and its anchor system is greater than we would expect on the basis of the weight of the load. The force on the pulley is a function of the interior angle of the lines feeding into and out of the pulley (Fig. 13.59) and can be estimated using the following equation:

Force on the pulley = 2 × weight of load × cos (interior angle/2)

As an example, assume that we have a 100-lb load on a line running through a directional pulley. According to the equation, if the angle between the lines is 180°, then no force is placed on the pulley. If we decrease the interior angle to 120°, then the calculated force on the pulley is 100 lbf, the full weight of the load. At interior angles less

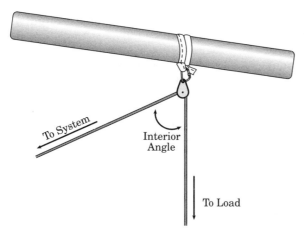

Figure 13.59   The forces placed on a pulley are a function of the interior angle of the lines that feed into and out of the pulley.

than 120°, the force on the pulley begins to be compounded. For example, at an interior angle of 90° the calculated force on the pulley is 141 lbf. If we reduce the interior angle to 0° so that the lines are parallel to each other, the calculated force on the pulley is 200 lbs or twice the load.

To deal with potential compound forces on directional pulleys, use directional pulleys that meet the strength requirements of NFPA 1983's general-use rating, which is 8093 lbf without failure, as discussed in Chap. 11. Be sure that all pulley anchor systems are attached to anchor points that are engineered or bombproof. Be sure that the entire anchor system is configured so that all components meet NFPA 1983 requirements for general use. If for any reason you must use a directional pulley that does not meet the general-use requirement, or you have doubts about a directional pulley for any reason, back up the pulley.

Backing up directional pulleys.   Directional pulleys can be backed up by attaching a separate connector to the anchor point above the pulley and locking a carabiner around both the anchor connector and the line running through the directional pulley (Fig. 13.55). The carabiner is there to catch the rope if the pulley fails and should hang just below the pulley. Rig the backup system with enough slack to keep the carabiner from interfering with the function of the pulley, but keep the slack to a minimum to prevent shock loading if the pulley fails. Remember that you can back up any pulley that you consider questionable for any reason.

Portable Anchor Systems and Winch Operations

Portable anchor systems such as tripods and hoisting frames are very handy for confined space operations. They allow us to establish a temporary anchor point and anchor a change-of-direction pulley or MA system directly above the entrance to a vertical space. Such devices may be required to allow a worker to be lowered into the space during routine entries. In some cases they provide fall protection or allow nonentry rescue to be performed. In other cases, rescue teams may deploy these devices in order to perform entry rescues.

Tripod operations

Tripods are the most common type of portable anchor system used in confined space operations. Although tripods are very useful, they must be selected carefully and used in ways that allow them to be safe and effective.

Tripod selection considerations. Tripods are available in different height ranges and usually have telescoping legs that can be adjusted to various lengths to accommodate different terrain and entry configurations. All tripods have working load limits established by the manufacturer. The working load limit of a tripod usually decreases with height, while the distance the legs must extend outward to maintain stability increases. Taller tripods that require larger leg extensions do not fit on top of some containers.

Some tripods have a maximum extension of 7 ft or less and may not provide enough lifting capability for rescue applications. Any tripod used in rescue applications must provide enough overhead clearance to allow a packaged patient to be pulled clear of the entrance to the space. In calculating lifting clearance, remember to allow for the overhead clearance that will be lost in rigging the retrieval system to the head of the tripod.

Using tripods safely. Tripods can be pulled over easily by rope mechanical-advantage systems if the load is hauled from outside the tripod's triangle of stability. The triangle of stability is the triangle on the ground formed by the three legs of the tripod. To prevent the tripod from being pulled over by the haul team, run the haul line through a directional pulley and anchor the pulley within the triangle as shown in Fig. 13.56. Never attach a directional pulley to the leg of a tripod, as the resulting force may cause the tripod to collapse. Attach the pulley to some other anchor point. If no suitable anchor point is available within the triangle of stability, locate the nearest suitable anchor and extend it to the desired pulley location.

Always place the legs of tripods in the locked position. Most manufacturers provide lengths of rope or chains to be fastened loosely around the bottom of the tripod legs as a backup to prevent the legs from slipping outward and to stabilize the base. Another alternative is to drive anchor posts or stakes and tie the legs off to them.

Other portable anchor systems

In addition to tripods, other portable anchor systems are used for confined space operations. These include a variety of portable booms, davits, and hoisting frame devices. These devices can be freestanding, as seen in Fig. 13.4, or mounted on existing supports, tank openings, or truck trailer hitches. Most of these devices are designed for vertical pulls; however, there are preassembled devices available that are configured for horizontal pulls. Hoisting frames generally are much sturdier than tripods, but tend to be heavier and bulky, and generally don't store as compactly as tripods. The general considerations and concerns previously mentioned in regard to tripods apply when selecting and using other portable anchor systems.

Using winches in confined space operations

Portable anchor systems can be used with rope block-and-tackle systems or with winches. Any of the block-and-tackle systems described earlier in this chapter can be anchored to a portable anchor device for use in confined space operations. In this section, we will focus on winch operations.

Rope- and cable-type winches suitable for confined space operations are available. The *rope-type winches* are typically of an open capstan design and are rigged by placing several turns of rope around the capstan. This design allows an unlimited length of rope to be used with the device. *Cable-type winches* use a steel cable to handle the load and are limited to the amount of cable that the winch contains. Cable winches suitable for confined space operations are readily available with 50 to 200 ft (15 to 60 m) of cable. The cable-type winches are far more commonly used in confined space operations and will be the major focus of this section (Fig. 13.60).

Like block-and-tackle systems, winches can be used to lower and retrieve entrants from a space only if they are designed and rated for human loads. Some winches are intended only for utility use and should never be used to handle human loads. Winches have a working load limit established by the manufacturer. To avoid overloading the winch, never exceed this limit.

Figure 13.60 A tripod with a cable-type winch is commonly used to raise and lower personnel during vertical confined space operations.

Some winches are equipped with a fall-arrest feature. The fall-arrest mode allows the entrant to connect to the winch cable and climb down into the space using a ladder or some other means while the cable feeds freely from the winch. If the entrant falls, the winch automatically brakes to arrest the fall (see Fig. 14.1). The use of tripods and fall-arrest equipment is discussed further in Chap. 14.

Never use powered winches to lower or haul human loads in confined space operations. If the entrant or patient becomes entangled he can be severely injured before the operator realizes it. Use manual winches instead. In using these winches to move human loads, follow all precautions previously discussed for using MA systems to avoid damage to equipment and injury to the person being moved. Use, store, inspect, and maintain all winches in accordance with the manufacturer's instructions.

Rigging and Using Safety Belay Systems

Any system created and operated by human beings can fail as a result of either equipment failure or operator error. Basic systems used in confined space operations are no exception. The failure of systems used to move human loads vertically can result in catastrophic falls. To address this concern, the use of safety belay or safety line systems has long been recommended to provide backup fall protection during all vertical operations.

The basic idea behind the concept of fall protection is that anyone exposed to a fall hazard is protected from the fall by at least two systems, a primary system and a secondary system, at all times. A common practice is to provide protection from any potential fall of 6 ft or higher. In some cases this is a regulatory requirement.

Fall protection is addressed further in Chap. 14, "Making Entry Safely." The focus there is mainly on equipment such as fall-arrest lanyards and engineered fall-arrest systems that are triggered by the inertia of a fall. In this part of the text we will examine belay systems operated by a team member designated as the belayer during vertical operations.

Any of the various MA systems discussed earlier in this chapter could serve as the primary fall-protection system during a vertical operation. For entries requiring that ladders be descended or ascended, the entrant's hands and feet can be considered the primary system. In this section, we will examine several options for rigging belay or safety line systems to serve as the secondary or backup system. Assume that the load to be belayed in our examples is a single entrant wearing a full-body harness.

Basic rigging for safety belays

Rigging safety belay systems is similar to rigging any other type of system discussed in this chapter. Begin by identifying suitable anchor points and establishing an anchor system. If feasible, use different anchor points and separate anchor systems for rigging the primary load-handling system and the belay system. If both systems are attached to a single anchor, it must be engineered, bombproof, or backed up to another suitable anchor. Remember the basic rigging practices we covered previously. For example, rough or sharp edges that the belay line might contact must be padded. Once the anchor system is established, attach a suitable belay device, such as one of those described below.

Examples of safety belay systems and components are shown in Figs. 13.61 through 13.70b. Systems such as these can be rigged quickly and easily. For the belay line, select a static kernmantle rope that is long enough and has the proper life-safety rating for the load to be belayed, as described in Chap. 11. Tie a double-loop figure 8 on a bight in the working end of the belay line and attach it to a D ring on the entrant's harness. It is a good idea to install a shock absorber between the end of the belay line and the harness (Fig. 13.61). If possible, attach the belay system to a D ring different from the one to which the primary system is attached, so that failure of a single D ring cannot result in the loss of both systems. Attach the carabiners so that the locking rings screw downward into the locked position and the gates point in toward the entrant when the carabiner is in the normal-use position. These precautions are intended to reduce the chance that locking rings will work into the unlocked position and gates will be pushed open during the operation.

Install the running end of the belay line in the belay device at the anchor system. From this point on, the process will differ depending on the belay device being used. Never use mechanical rope-grabbing devices such as the Gibbs ascender in a belay system. These devices may severely damage the rope if shock-loaded.

General belay procedures

In all cases, keep good belay principles in mind when acting as belayer. Keep slack in the belay line to a minimum. While some slack is needed to avoid interfering with the operation of the lowering system, excessive slack can cause a shock load on the belay system if the primary system fails. Shock loading can severely injure personnel on the system. It also can damage system components or cause the belayer to loose control of the load. It is recommended that no more than 18 in (45 cm) of slack be allowed in the belay line at any time. Less slack is better. The use of shock absorbers is also a good recommendation to prevent shock loads, especially during training. Always use standard safety gear and leather gloves for handling loaded ropes.

When acting as belayer, proceed as directed by the team leader. Remain alert and ready to act immediately to stop a falling load at all times while on belay.

Using the Munter hitch belay

The *Munter safety belay* is commonly used for fall protection during lowering and hauling operations. It is a quick and simple system to rig and operate and can effectively

Figure 13.61 A shock absorber should be placed between the end of the safety belay line and the person who is being raised or lowered. The shock absorber will prevent shock loading of the system that could cause severe personal injury and damage or failure of anchors, ropes, and other system components.

belay a falling single person load if used properly, but controversy exists regarding its use for belaying two-person loads. The Munter hitch belay technique requires positive action on the part of the belayer to arrest a fall. It is completely operator-dependent, as opposed to an automatic or "sniperproof" system like the tandem Prusik belay system described below.

Rigging the Munter safety belay.   To use the Munter hitch belay technique, attach an extralarge steel carabiner to the belay anchor and form a Munter hitch through the carabiner with the running end of the belay line (Fig. 13.62). Stand facing the belay anchor with the belay line running to the load in one hand and the idle line in the other hand. Keep both hands at least 1 ft (30 cm) back from the carabiner to prevent your fingers from being jerked into the carabiner if the line is suddenly loaded.

Operating the Munter belay.   To feed slack into the standing part of the belay line as a load is lowered, feed rope into the Munter hitch from the idle side while pulling rope through the hitch from the side toward the load. Maintain a minimal amount of slack in the safety line throughout the lower. One way to maintain an appropriate amount of slack is to use the following procedure:

1. Pull enough slack through the Munter hitch to allow the hand pulling the slack to twist a Z into the belay line.

2. As the lower progresses, the weight of the load will tend to pull the Z into a straight line.

3. Before the Z is pulled completely straight, release the rope and pull enough additional slack to allow another Z to be twisted into the belay line.

4. Repeat this process throughout the lowering operation.

To pull slack out of the belay line during hauling operations, feed rope into the Munter hitch from the side toward the load while pulling rope through the hitch from the idle side. The extralarge carabiner allows the Munter hitch to flip back and forth through the carabiner to change from a raising belay to a lowering belay. If a fall occurs, immediately lock the Munter hitch by tensioning the line on the idle side of the carabiner. This locks the Munter hitch to stop the falling load.

Figure 13.62 The Munter hitch is a belay technique that can be used on the safety belay line during lowering and hauling operations.

Concerns for belaying two-person loads with the Munter hitch. When properly used, the Munter hitch technique is effective for stopping a single-person load, especially when slack in the belay line is kept to a minimum. Controversy exists regarding its use to belay two-person loads, based on testing that indicated the Munter hitch would not dependably arrest the fall of a two person load dropped a distance of approximately 3 ft (1 m) (Roop et al. 1998, p. 112). For this reason the Munter hitch belay is not recommended for two-person loads. One alternative is to rig a separate Munter belay system for each person making up the load, provided adequate equipment and personnel are available. Another alternative is to use a different belay technique or device, such as one of the ones described below.

If a single Munter belay system is used to belay a two-person load, the following precautions will increase its effectiveness:

- Run the belay line over a properly padded 90° edge. The extra friction will help the belayer to stop the falling load.

- Install a shock absorber between the load and the end of the belay line. The shock absorber is intended to reduce any potential shock load at the Munter hitch. This reduces the chance that the belayer will lose control of the load.

- Keep slack in the belay line to a minimum. This can significantly reduce the degree of shock load that the belayer must contend with.

- Remain fully alert at all times when operating the Munter safety. Act immediately to stop a falling load before it gains significant momentum.

- Assign a second team member to perform body belay on the belay line as shown in Fig. 13.63. The body belayer will assist the Munter hitch belayer in arresting a falling load.

Resuming normal operations after arresting a fall with the Munter belay. After a fall has been arrested with the Munter belay, the load can be shifted back to the primary system easily. Prepare to do so by correcting the problem that caused the load to drop, whether related to equipment failure or operator error.

If basic hauling/lowering systems like those described in this chapter are being used as the primary system, use them in the hauling mode to lift the weight of the load, thus unloading the Munter hitch. Normal system operations can then be resumed.

For some rescue lowering operations, a device such as a brake bar rack or figure 8 descender is used in the primary system. These lowering systems are beyond the scope

Figure 13.63 A second team member can be assigned to perform body belay on the belay line to help the Munter hitch belayer control heavy loads.

of this chapter, but are discussed at length in Chaps. 17 and 18. These devices have the capability to lower but not to haul. To reestablish normal operations after a fall has been arrested, release the tension on the Munter hitch to shift the load back onto the lowering system. After control of the lowering system is reestablished, remove all slack from the lowering system and lock off or tie off the lowering device. The Munter belay operator then slowly releases the tension on the idle line running to the Munter hitch. This lowers the load a very short distance to shift its weight back onto the lowering system so that normal lowering operations can be resumed. The team leader must carefully coordinate these actions for a safe recovery.

Lowering with the Munter hitch.   Situations may arise in which it is desirable to use the Munter hitch to lower a person a significant distance to a safe location after a fall has been arrested. One example of this would be a situation in which an equipment failure that cannot be corrected readily occurs in the primary system.

The use of the Munter hitch for lowering a load is controversial (Roop et al. 1998, p. 113). Some authorities contend that it is an acceptable practice, while others contend that it should be avoided. Concerns have been raised because the rope runs against itself under load in passing through the Munter hitch when used for lowering. It is believed that this can generate significant amounts of heat that may weaken or damage the rope if the Munter hitch is used to lower the load for a significant distance.

The best recommendation is to avoid using the Munter hitch to lower a load except as a last resort; however, it may be necessary to do so in order to get the person hanging on the end of the belay line to a place of safety in an emergency. After this is done, the safest recommendation is to remove the rope used in the belay line from service.

Using the tandem Prusik belay.

The *tandem Prusik belay* (TPB) is another technique commonly used for fall protection during both lowering and hauling operations. When properly rigged and used, it can effectively belay both one-person and two-person loads.

Unlike the Munter hitch, the TPB system requires no positive action on the part of the operator to arrest a fall. If the belay operator stops tending the TPB at any time, the Prusik loops automatically set and lock the belay line. This has been referred to as a "deadman" control or "sniperproof" system, with the idea that the system will function to arrest a fall even if the belayer is rendered incapable.

Rigging the tandem Prusik belay.   To rig the TPB, attach two loops of accessory cord in tandem to the belay line with triple-wrapped Prusik hitches. Attach both Prusik loops to the belay anchor system (Fig. 13.64).

It is recommended that 8-mm-diameter accessory cord be used with a belay line of $^{7}/_{16}$- or $^{1}/_{2}$-in kernmantle rope in this application. Smaller cordage may be too weak to catch a falling load, or may even sever the line if shock-loaded. Larger cordage may not grip the rope well enough and could allow the belay line to slip through the Prusik hitch. The 8-mm accessory cord used in this application must be adequately flexible. If the cord is too stiff, the Prusik hitch may not grip the belay rope. If the cord is too soft, the Prusik hitches may be hard to unfasten and the cord may wear quickly. To test cordage suppleness, pinch a bend in the cord. The inside diameter of the bend should be about the same as the outside diameter of the cord (Fig. 13.65). A larger or smaller bend diameter may suggest that the accessory cord is either too stiff or too soft (Frank 1998, p. 25).

The two pieces of accessory cord used in the TPB must be of unequal lengths, with one measuring 54 to 56 in (137 to 142 cm) and the other measuring 66 to 70 in (168 to 178 cm). Tie each length of cord into a loop by connecting the ends with a double fisherman's knot and attach each loop to the belay line with a triple-wrapped Prusik hitch. The Prusik hitches must be tight and remain tight throughout the operation. If they become loose, they will not grip the belay rope properly to arrest a fall. When properly rigged, the Prusik hitch in the shorter loop should be positioned closer to the anchor with the Prusik hitch in the longer loop about the width of a fist farther away from the anchor.

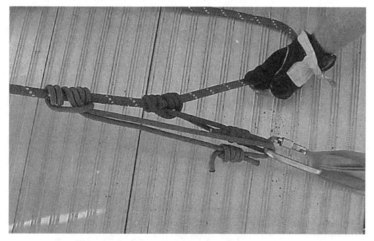

Figure 13.64   The tandem Prusik belay is another technique commonly used for fall protection during both lowering and hauling operations.

Figure 13.65 Pinch a bend in accessory cord to test its suppleness. The inside diameter of the bend should ideally be the same as the outside diameter of the cord.

The Prusik loops are intentionally short to minimize the potential shock load in arresting a fall.

The Prusik minding pulley is a very handy item to use in the TPB system (Fig. 13.67). Although the pulley is not helpful for lowering belays, it does not hamper operation of the TPB during lowers. When the operation changes from lowering to hauling, the Prusik minding pulley makes the hauling belay significantly easier to operate.

When rigging for some rescue operations, it may be desirable to install a load-releasing device such as a mariner's knot or a load-release hitch between the TPB loops and the anchor. The use of the load-releasing devices in lowering belay systems is described below. Note that these devices are not needed when belaying the basic lowering and hauling systems described here and are therefore beyond the scope of this chapter. Load-releasing devices are very important items in rigging and operating rescue systems and are covered in Chap. 17.

Operating the TPB during lowering operations. When using the TPB to belay a load that is being lowered, hold the Prusik hitch farthest from the anchor in one hand while pulling the belay line through the hitches with the other hand in order to feed slack into the belay line (Fig. 13.66). Maintain a slight amount of slack in the Prusik loops to keep the Prusik hitches from locking onto the belay line. Pay attention to be sure that the hitches do not accidentally lock, as this can be very annoying.

As the lower progresses, limit the slack in the belay system between the Prusik hitches and the load. Ideally there should be just enough slack to allow unhindered operation of the lowering system. The method of twisting a Z into the belay line to measure the slack, as described above for Munter belay, can also used with the TPB. If the lowering operation stops, pull all slack out of the Prusik loops and the belay line and stand by.

If the primary system fails during the lower, immediately release the Prusik hitches. The hitches will lock onto the belay line as the slack in the loops between the hitches and the belay anchor runs out. This arrests the falling load.

Operating the TPB during hauling operations. When using the TPB to belay a load being hauled, slack must be pulled out of the standing part of the belay line as the hauling operation progresses. Accomplish this by pulling the rope on the idle side of the Prusik minding pulley. The tandem Prusik hitches will move with the belay line until they contact the sideplates of the Prusik minding pulley. The extended sideplates of the pulley are designed to prevent the Prusik hitches from being pulled into the pulley and jamming

Figure 13.66 To belay a lowering operation with a TPB, hold the Prusik hitch farthest from the anchor in one hand while pulling the belay line through the hitches with the other.

between the rope and the pulley sheave. This allows the belay line to be pulled through the Prusik hitches with minimal effort.

One recommendation for enhancing the effectiveness of the Prusik minding pulley in this application is to attach a small personal carabiner around both lines exiting the Prusik minding pulley. Position the carabiner just below the pulley and above the Prusik hitches (Fig. 13.67). The carabiner is intended to force the belay line to enter the pulley from the bottom, eliminating any tendency for the Prusik hitches to slip to the side of the pulley and bind or jam the pulley.

Do not allow slack to accumulate in the belay line during the hauling operation. Slack increases the potential for shock loading of the belay system if a hauling system failure occurs. If such a failure does occur, immediately release the belay line. This allows the Prusik hitches to set on the belay line and stop the falling load.

Resuming normal operations after arresting a fall with the TPB. After a fall has been arrested with the TPB, or if the Prusik hitches accidentally lock, the entire weight being handled is loaded onto the TPB system. That weight must be removed before the Prusik hitches on the belay line can be unlocked. The load usually is shifted back onto the primary system to allow the Prusik hitches to be unlocked. This should not be a problem if the primary system is one of the basic MA systems such as those described in this chapter because all those systems have a hauling capability. Simply use the MA system in the hauling mode to remove the weight of the load from the Prusik hitches and unlock them. Normal system operations can then be resumed.

For some rescue lowering operations, a device such as a brake bar rack or figure 8 descender is used in the primary system. These lowering systems are beyond the scope of this chapter but are discussed at length in Chaps. 17 and 18. These devices have the capability to lower but not to haul. To reestablish normal operations after a fall has been arrested, some means is needed to shift the load back onto the lowering system. This can be done fairly easily provided a load-releasing device such as a mariner's knot or a load-release hitch was placed between the TPB and the belay system anchor. Once con-

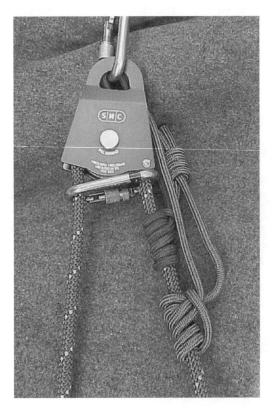

Figure 13.67 To enhance the effectiveness of the Prusik minding pulley for hauling operations, position a personal carabiner just below the pulley and above the Prusik hitches.

trol of the lowering system is reestablished, all slack is removed from the lowering line and the system is secured. The load-releasing device is unfastened and slowly extended. This has the effect of lengthening the anchor system, thereby lowering the load a short distance to shift its weight back onto the lowering system (Fig. 13.68). The load-releasing device then is retied and the TPB system rerigged so that normal lowering operations can be resumed. As always, the team leader must carefully coordinate these actions for a safe recovery.

Since all the basic MA systems described in this chapter can haul as well as lower, load-releasing devices should not be required when TPBs are used with these systems. For that reason the use of load-releasing devices is actually beyond the scope of this chapter. Load-releasing devices have some very important uses in some of the more advanced systems used in rescue operations, and rescuers will receive specific information on rigging and using these devices in Chap. 17.

Manufactured safety belay devices

So far we have focused on belay systems rigged with various components in the field. As an alternative to this approach, manufactured belay devices can also be used.

One example of such a device is the 540°™ Rescue Belay by Traverse Rescue (Fig. 13.69). Two models are offered by Traverse Rescue, the large and the small 540° Rescue Belay. The larger model is designed for use with kernmantle ropes from 11.5 to 13 mm in diameter and consists of a clutch-mounted oval spool fitted between two side plates, with a built-in release lever. To rig the device, remove the front plate, wrap the belay rope around the spool, and replace the front plate. Reported test results indicate that the device can belay a falling load of over 600 lb (280 kg). It is a self-locking device. According to marketing information, the built-in release lever releases the tension on the belay line, eliminating the need for a load-releasing device in the belay system.

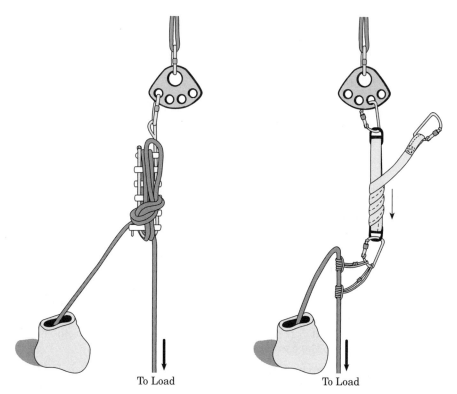

To Load   To Load

Figure 13.68   The use of a mariner's knot or load-release hitch allows the load to be shifted back onto the main line so that the Prusik hitches can be unlocked after a fall has been arrested.

Figure 13.69   The 540°™ Rescue Belay by Traverse Rescue is a manufactured belay device.

Other options for using safety belay systems

In this section we have focused on rigging and using safety belays for fall protection during hauling and lowering operations. These are by no means the only uses for safety belays. For example, safety belays can be rigged and used to provide fall protection when entrants must climb down ladders to enter spaces. While the entrant's hands and feet serve as the primary system, any of the belay techniques discussed previously can serve as the secondary fall-protection system.

As another example of alternative uses for a safety line system, consider a belay system used for fall protection during a ladder climb as shown in Fig. 13.70. The system shown can be rigged as follows, assuming that the reader is the climber and is wearing a full-body harness.

1. Prepare the belay line by tying a double-loop figure 8 on a bight in the end of a suitable life-safety-rated static kernmantle rope. Attach a shock absorber to the end of the belay line and to the dorsal D ring of the climber's harness.

2. Basket-hitch a 12-in (30-cm) presewn loop sling of one-inch webbing around the ladder beam below the bottom rung on the climber's strong side. Be sure to fasten the sling around the beam, because ladder beams are much stronger than rungs. Attach a figure 8 descender (Fig. 13.70) to use as the belay device. A Munter hitch belay could also be used.

3. The belayer ropes the belay line into the belay device. Use of the figure 8 descender is covered in Chap. 17, and other belay techniques in this chapter. The belayer feeds rope through the belay device as you climb.

4. Climb a short distance up the ladder, stopping with your feet about 5 ft above the ground, and establish an anchor. To do so, basket-hitch a 12-in (30-cm) presewn webbing loop sling around the ladder beam just above a rung at a convenient location about 5 ft above your feet. Attach a carabiner to the sling and lock the belay line into the carabiner.

5. Continue climbing up the ladder while the belayer feeds rope through the belay device at the bottom of the ladder and remains ready to react quickly in the event of a fall.

6. Move up the ladder and stop with your feet no higher than the location of the anchor you just established. Establish another anchor at a convenient location about 5 ft

(a)

(b)

Figure 13.70 (*a*) This is a belay system used for fall protection during a ladder climb. The climber is belayed with the friction control device anchored at the base of the ladder. (*b*) The climber establishes anchor points for the belay line at intervals all the way up the ladder.

above the last anchor you set. Lock the belay line into the carabiner and continue climbing (Fig. 13.70*b*).

7. Repeat the process, anchoring the belay line at intervals of about 5 ft all the way up the ladder.

The idea behind this system is that the maximum free fall that can occur is limited to less than 6 ft. Of course, a free fall of several feet could impart a severe shock load to the climber and the system. For this reason the use of the shock absorber between the climber and the belay line is critical. Setting the anchor points closer together would limit the free-fall distance and therefore would be safer, but would take more time and equipment. The ladder, like all other system components, must be sufficiently strong to function as a part of the belay system.

The use of the figure 8 descender as the belay device allows an unconscious climber to be lowered all the way to the ground after a fall has been arrested, assuming that the belay line is long enough. This would also be possible with a Munter hitch belay, although damage to the rope might result. It would not be possible if the tandem Prusik belay technique were used.

If personnel are making a significant number of trips up and down the ladder, a more convenient fall-protection system could be rigged by anchoring a fixed line at the top and bottom of the ladder and using a device such as the MIO industrial rope grab (see Fig. 14.8). Fall protection is discussed further in Chap. 14.

General Rules of Rigging

A number of general rules should be followed in rigging and using the systems described in this chapter for confined space operations. Most of the rules have been discussed previously in this chapter.

- Avoid rigging that produces compound forces on anchor points and other system components.

- Beware of the hard linking of hardware components. *Hard linking* is the placement of hardware components in a system in such a way that the system lacks the flexibility needed to operate properly.

- Avoid placing three-way loads on carabiners. In any situation where it is possible to have a three-way pull, such as the point of attachment of a rope system to a load-sharing anchor bridle, use a trilink, rigging ring, or some other method to prevent triloading the carabiner.

- Be sure that all carabiners are locked and properly oriented before a system is loaded. Avoid side-loading carabiners.

- Avoid contact between sharp edges or abrasive surfaces and software system components. Make sure that all edges are padded or edge rollers are in place to prevent the contact of the rope with rough or sharp edges. Any other points that soft components of a system might contact should be padded also.

- Avoid pendulum swings of ropes, especially along unpadded edges or abrasive surfaces.

- Avoid shock loads. Keep slack in all systems to a minimum. If shock loads are possible, use shock absorbers.

- Remove from service and destroy or dispose of any equipment that is shock-loaded, overloaded, or otherwise subjected to damaging conditions.

- Avoid using knives or other sharp objects around loaded ropes. While a rope is loaded, it can easily and effortlessly be cut into with the slightest contact with any type of sharp edge, such as a knife, scissors, or even unpadded abrasive or squared edges.

- Consider the working load limits of all system components and be sure not to exceed them. Remember that some configurations can compound the forces involved. MA systems can produce enough force to overload system components, such as when a load hangs up during a haul.

- Use a separate safety belay system during all vertical operations.

- If in doubt, back up anchor points and other system components.

- Remember the KISS pneumonic and keep all systems as simple and safe as possible.

Summary

In this chapter we covered rigging and using basic systems suitable for handling human loads during confined space operations. They can be used in lowering, hauling, and belaying operations. The systems we covered are simple systems that are relatively simple to rig and use. These systems should be adequate for conducting most routine confined space entry operations. Rescue operations may require rescuers to rig and use more complex systems. The basic systems covered in this chapter will serve as the basis for rigging and using the more advanced systems described in the part of this text devoted specifically to rescue.

Making Entry Safely

Two workers were moving electrical lighting and painting equipment in the hold of a 300-ft barge under construction. Painting operations had been under way for 10 hours using paint mixed with methyl ethyl ketone (MEK), a flammable liquid. No mechanical ventilation was performed, and one of the hatches to the hold was not open. No atmospheric testing was conducted before entry. The electrical equipment did not have the proper approvals for use in a potentially flammable atmosphere. While the workers were moving the equipment, flammable vapors within the hold of the barge ignited. Both workers died from the resulting burns.

Basic Considerations for Safe Entry

Safety doesn't happen by accident. A proactive approach, involving planning for personnel safety prior to entry, goes a long way toward ensuring that rescues are not required. Good planning ensures that work will be carried out efficiently during permitted entries (see Fig. 14.1). Likewise, a proactive approach to rescue can ensure that timely, effective rescue can be carried out if needed. These are basic principles underlying the requirements of the PRCS standard and all good standard operating guidelines.

In this chapter we will highlight considerations for making entry and operating safely in confined spaces. In doing so, we will review some topics from previous chapters, add some things that didn't quite fit in before, and make reference to information expanded on in later chapters. All the considerations included here apply in making permitted entries for work in confined spaces, and most also apply to entries made for rescue. They should be considered in developing standard operation procedures (SOPs) for either type of operation.

The Role of the Entry Permit

The focal point of a safe permitted entry is the entry permit described in Chap. 5. This impressive document identifies the work location and work to be done; provides for accountability of personnel; identifies hazards present, monitoring procedures, and safety precautions required; and provides for rescue should it be needed.

The information contained in the entry permit is critical to the safety of all personnel involved in the permitted entry, for obvious reasons. That same information will be one of the first sources utilized by rescuers in sizing up any emergency that occurs during the entry. For this reason it is of the utmost importance that the permit be filled out fully and accurately. Shortcuts or a "rubber stamp" approach could have disastrous results.

All that having been said, the permit will not necessarily identify all the hazards you need to be concerned about for all entry operations. In some cases hazards may lurk both inside and outside the confined space.

Figure 14.1 Planning prior to entry promotes efficient confined space operations.

Interaction of Personnel Involved

Confined space entry is a team activity, with each team member dependent on the other members of the team for a safe, effective completion of the mission. Personnel roles involved include entrant, attendant, supervisor, and rescuer. These roles were discussed in Chap. 2.

The entry supervisor

The supervisor has the responsibility for preparing the entry permit and signing it to authorize the entry. This requires identifying all the hazards present, making sure that they are addressed, and ensuring that rescue is either available or on standby. This is a huge responsibility, since all the considerations mentioned in this chapter and elsewhere within this text have to be considered, and addressed if need be, in completing the entry permit. The supervisor is ultimately responsible for seeing that the provisions of the permit are followed by personnel involved in the entry, and for the overall safety of the operation.

The attendant

The attendant is the kingpin of safety during confined space entries. The attendant's primary responsibility is to oversee the safety and well-being of the entrant throughout the entry. This includes monitoring conditions within the space to check for any prohibited conditions and monitoring the entrant's condition by checking for any signs that the entrant may be in trouble. This is important, because some chemical exposures may reduce the entrant's judgment and ability to realize that he is in danger. The attendant must also monitor conditions outside the space to ensure that all life-support systems are functional and that no hazards to the entrant arise outside the space. This includes keeping unauthorized personnel away from the area of operations.

If the attendant identifies a prohibited condition, a threat to the entrant, or any other indication of impending trouble, she must order the entrant to exit the space immediately. If an emergency occurs and the entrant is not able to perform self-rescue by exiting the space under his own power, the attendant must immediately summon help from the designated rescuers. If rescuers are not on standby during the operation, the attendant should attempt nonentry rescue (as described in Chap. 15) if feasible, while waiting for available rescuers to arrive.

In addition to performing a critical safety role, the attendant is also in a position to enhance the effectiveness of operations. Examples of important ancillary duties include

- Acting as a communications link or a liaison between the attendant and support personnel outside the space

- Feeding out and taking up slack in lifelines and hoses as needed to prevent tangling and damage

- Ordering up tools and equipment and relaying them to the entrant

- Helping the entrant into and out of the space

In performing secondary duties, it is important that the attendant not be distracted from performing the primary duty to the safety of the entrant. Also, in performing the secondary duties such as passing tools, it is wise to keep in mind that entry is considered to have been made into a space if any part of a person's body breaks the plane of the entryway. The attendant might technically "enter" while passing tools into a space, and thereby be in violation of the PRCS standard.

The entrant

Entrants are typically the people at greatest risk from the hazards of the permit space. Most of the provisions of the entry permit are intended to keep them from getting into situations requiring rescue; therefore, the entrant has a major personal responsibility to follow those provisions. The entrant must be

- Adequately trained to perform the assigned task safely and effectively

- Aware of the hazards present and precautions required

- Provided with appropriate protective gear and safety equipment and trained in its use

- Able to recognize signs and symptoms of exposure

- Trained to detect prohibited conditions within the space

In the event that a prohibited condition develops or the entrant experiences signs or symptoms of exposure, he must notify the attendant immediately and then attempt self-rescue. If ordered out of the space by the attendant for any reason, the entrant must exit immediately.

The rescuer

The intent of OSHA's permit spaces standard is that the rescue team or service be an integral part of every entry. In some cases, such as entry into IDLH atmospheres, rescue personnel are required to be on standby at the location of operations during entry. Even in less hazardous situations where rescuers are not required to be on standby, the permit form must document the availability of the rescue service on call. The specific procedures the rescue team or service may be called on to perform are detailed in Chaps. 15 through 18. For now, keep in mind that rescue is one function that must be on line during all permit space entries.

Assessing the Hazards

As detailed in Chap. 3, a wide variety of hazards may be encountered during confined space operations. Don't restrict your concerns to the hazards indicated by the permit form. Keep your mind and eyes open, and avoid a tunnel-vision approach that focuses on only the most obvious hazards. Don't restrict your attention to the hazards within the space. Hazards originating outside the space may affect the safety of personnel involved in the operation, both inside and outside the space.

Consider the activities to be carried out in the space and how they could alter conditions there. Some activities, such as application of coatings or removal of sludge, may cause an atmospheric hazard level much higher than that present when operations began.

Hazardous materials are involved in many confined space operations. Special concerns such as selection of chemical protective clothing and decontamination may need to be considered, as addressed in Chaps. 8 and 10. Permitted entries into such spaces may require cross-training in hazardous-waste operations. Responding to emergencies in such spaces may require cross-training in hazardous-materials emergency response. Hazardous-waste and emergency response training and equipment are described in 29 CFR 1910.120(e), (q).

The Role of Air Monitoring Equipment

Confined space operations are somewhat unusual in the degree to which one is required to use air monitoring devices to stay safe, as explained in Chap. 4. The importance of these devices in assessing conditions inside the space is obvious, and they may need to be used extensively outside the space as well. For example, are support personnel outside the space possibly at risk from contaminants being emitted from the space, especially as purging of the space is conducted? Is the air being used to ventilate the space of good quality? Questions such as these can be answered only by using air monitoring equipment.

Remember that air monitoring devices may do nothing more than mislead dangerously if they are not properly calibrated and used. Personnel using them must be adequately trained to interpret the results, or they can overestimate or underestimate the hazards significantly.

Air monitoring devices are worthless if they are not used. One common mistake is to use an instrument initially to verify acceptable entry conditions, then put it away for the rest of the operation. Remember that conditions can change for the worse for any number of reasons as confined space operations progress.

Controlling the Hazards

To the maximum extent practical, hazards should be eliminated or controlled. Safe work practices and personal protective equipment should be the last line of defense, instead of the only line of defense, in keeping personnel safe. Hazards can be controlled using practices such as purging, ventilation, isolating the space from sources of energy and material, zoning the scene, and other procedures described in Chap. 6. Remember to control hazards originating outside the space as well.

Using Personal Protective Equipment and Related Gear

In many confined space operations, it is not possible to eliminate or control all the hazards prior to entry. In those cases, personal protective equipment may be the last line of defense in the safety of personnel.

Respirators

Respirators frequently are required for confined space entry, as explained in Chap. 7. IDLH atmospheres always required atmosphere-supplying equipment. Whenever that type of equipment is used, major concerns exist regarding air supply.

When SCBA is used, the air supply must be monitored throughout the entry, and the entrant must leave the space at some predetermined interval. If the entrant waits until the alarm sounds to begin the trip out of the space, he may find that he is past the point of no return. This is the point beyond which the reserve air supply remaining in the cylinder will be used up before the entrant reaches the exit point.

The same concern applies when SARs are used. Most SARs are equipped with a "5-min" escape cylinder, with the actual breathing time available from the cylinder varying from user to user and situation to situation. If the primary air supply is lost when the entrant is beyond the point of no return, the amount of air in the cylinder will not be sufficient to reach a safe atmosphere. In some cases, larger escape-air cylinders should be considered. Training in dealing with SAR or SCBA emergencies (described in Chap. 17) can help entrants avoid panic and maximize the reserve air supply for valuable additional escape time.

Impairment of mobility is another important concern when atmosphere-supplying respiratory equipment is used. As discussed in Chap. 7, the type of equipment used will have a direct bearing on the degree of difficulty involved in working in a tight space, since different types have different degrees of bulkiness (see Fig. 14.2).

Chemical protective clothing

For entries involving hazardous materials, chemical protective clothing, gloves, boots, and various accessory items may be vital links in the protective ensemble. As noted in Chaps. 8 and 10, it is critical that these items be properly selected for the hazards present, properly donned, and properly used. Decontamination may be required for entrants leaving the space. Personnel dealing with such hazards should have appropriate training in hazardous-materials emergency response.

Personal safety equipment

In many cases minor physical hazards are the main threat to confined space entrants. In such cases, basic safety gear such as hardhats, eye protection, gloves, and protective footwear maybe very important to the safety of personnel. Items of this type were discussed in Chap. 9.

Other safety-related gear

PPE that is rarely needed for other operations may be very important for confined space operations. For example, harnesses and tag lines will be required (see Fig. 14.3), unless it can be proved that they will pose a hazard to the entrant (such as through entanglement) or will not help significantly if nonentry rescue is attempted. Kneepads and elbow pads may be critical protective items as entrants crawl and maneuver in tight enclosures during some entries. Small-profile rescue helmets may be preferred to regular hardhats and are preferred to bulky fire helmets for most entries (see Fig. 14.4).

Personal alarm safety systems

A significant piece of equipment for confined space operations is a personal alert safety system (PASS) device (see Fig. 14.5). This is a portable alarm that activates automatically

Figure 14.2 Design of respiratory protective equipment directly affects the degree to which it impairs mobility in tight places. Both these SCBA units have 30-min-rated air cylinders, but the high-pressure cylinder on the left is significantly less bulky than the low-pressure cylinder on the right.

Figure 14.3 Harnesses and tag lines are standard confined space equipment, and kneepads and elbow pads can provide abrasion protection in tight spaces.

Figure 14.4 Size and resulting mobility restrictions vary significantly between fire helmets, hardhats, and rescue helmets for confined space operations.

if the wearer becomes immobile. The unresponsive person can be found much faster and more easily by rescuers if he is wearing a PASS. PASS devices should be part of the standard ensemble worn by all rescue personnel, and are highly recommended for personnel involved in permitted entries as well.

Personal lights

Since confined spaces are usually dark places, all entrants should carry lights. Since a person using a hand light is limited to working with one arm, the use of helmet lights, or other hands-free lighting, is recommended (see Fig. 14.6). Even if the work area is well lighted, every entrant should carry a personal light as a backup in the event of failure of the lighting system. Using the Braille method to find your way out of a space is not recommended. If a helmet light or hand light is your primary light supply, then carry at least a small light as a backup. All lights should have appropriate safety approvals (as discussed in Chap. 4) if flammable atmospheres may be encountered.

Communication equipment

Another important aspect of safe and effective confined space operations is the ability to communicate. As a minimum, you must be able to communicate with others on your

Figure 14.5 The personal alert safety system (PASS) device activates automatically if the wearer remains immobile for a specific length of time, and makes lost or incapacitated entrants much easier to locate.

Figure 14.6   A helmet light frees up both hands for use in dark areas.

entry team and the attendant. If you will be outside the range of voice contact, communication equipment can provide a vital communication link. Specific considerations for communication are discussed below.

General Safe Work Practices

Ironically, many of the accidents that occur in permit spaces don't involve exotic chemical hazards. Instead, they involve common physical hazards such as those that we find in many construction, manufacturing, or industrial workplaces. Here again, a proactive approach is bound to be more effective than a reactive approach in dealing with workplace hazards.

One such proactive approach is *job safety analysis* (JSA), described in Chap. 3. This process involves dividing a task into steps required to complete it, identifying the types of accidents that could happen at each step, and devising measures to prevent the accidents from occurring.

To the greatest extent possible, the conditions that make accidents possible should be eliminated rather than relying on the precautions of personnel to prevent accidents. In Chap. 6 we covered using procedures such as purging, ventilation, lockout/tagout, and other isolation procedures to eliminate hazards. Hazards can be eliminated in other ways. For example, slippery surfaces can be covered with slip-resistant mats or grates. Open entryways or holes should be properly covered or guarded. Ladders and handrails should be designed and used in accordance with OSHA standards and maintained in good condition.

Ideally, safe behaviors by human beings should be the last line of defense against injury; however, no workplace is totally safe, especially permit-required confined spaces! Everyone involved should be trained and encouraged to work safely throughout all confined space operations. Developing a safety first attitude can help us avoid rushing, taking shortcuts, and knowingly violating safety rules.

Some general guidelines for safe entry follow:

- Plan all entry activities in advance, including protective gear, tools and equipment, and safety precautions required.

- Have a brief "hole side" safety meeting before entry to be sure that everyone involved understands the work to be done, hazards involved, safety precautions, and emergency procedures.

- Know the layout of the space prior to entry to avoid surprises such as unexpected vertical drops.

- Know the exact pathway to take to the work location and back.

- Identify any alternate escape routes out of the space in case the primary route becomes blocked.

Musculoskeletal injuries represent a threat to our safety during confined space operations. Examples include injuries due to heavy lifting or repetitive motions. Some general guidelines to avoid musculoskeletal injuries include

- Work smarter instead of harder by using powered equipment or a mechanical-advantage system as an alternative to the "Armstrong" method for moving heavy loads.

- Plan heavy or repetitive work so that adequate help is available.

- Use good posture in the workplace, using the legs instead of the back to lift loads manually.

- Avoid bending or twisting at the waist when handling loads.

- Avoid long-duration operations that require the same motion to be performed over and over, or arrange for workers to alternate in performing the task.

- Arrange the workplace and the task to accommodate your comfort and safety to the maximum extent practical.

Considerations for Tool and Equipment Use

Tools and equipment used in confined space operations sometimes are a source of danger for personnel. For example, electrical tools and equipment may provide a source of ignition for any flammable atmospheres encountered or a source of electrical shock for personnel. Electrical equipment used in confined spaces should be operated on low-voltage direct current or outfitted with ground-fault circuit interrupters. Under no circumstances should gasoline-powered equipment be used inside a confined space.

Examples of measures commonly used to prevent ignition include

- Use of electrical equipment such as radios, intercoms, hand lights or helmet lights, and area lighting systems that have the proper safety approvals (as described in Chap. 4)

- Using nonsparking tools such as bronze or beryllium alloy wrenches, hammers, and screwdrivers

- The use of special clothing and footwear to prevent the development of static electrical charges on workers (as described in Chap. 9)

- Using grounding and bonding systems as needed to prevent the accumulation of static electrical charges, for example, during some ventilation operations (as described in Chap. 6)

Remember that none of these precautions make working in a flammable atmosphere safe. They merely reduce the probability of ignition if one is encountered. Never intentionally enter a flammable atmosphere, no matter what precautions you are prepared to use.

Remember also that hot work such as welding, cutting, and brazing operations within confined spaces requires a separate hot-work permit in addition to the entry permit. The hot-work permits are required to address the additional hazards related to using compressed gases and/or arc welding inside permit spaces, as well as the constant ignition source involved. Compressed-gas cylinders and welding machines should never be taken inside confined spaces.

Emergency Retrieval and Fall Protection

Some confined space operations require entrants to travel downward into vertical spaces through the top. Other operations may require entrants to climb upward after entering through the bottom of a space. Workers sometimes have to climb ladders or stairs and walk along catwalks just to reach the space to be entered. Activities such as these can expose workers to significant fall hazards. The topic of fall protection is separate from but related to confined space operations. As a minimum, all workers should be protected from potential falls of 6 ft or more in vertical distance, unless more protective requirements are in place. A full coverage of fall protection is not possible here, but comprehensive references are available (Ellis 1993). Fall protection consists of two parts: fall prevention and fall arrest.

Fall prevention

Fall-prevention measures are used to prevent falls from occurring. These include measures such as the use of handrails on stairs and guardrails along catwalks and around openings or dropoffs. Good housekeeping can be an important aspect of fall prevention by eliminating conditions that could cause slips, trips, or falls to occur. Fixed-positioning systems, such as a harness connected directly to a suitable anchorage in the work area, can also prevent falls but allow no worker mobility.

Fall arrest

Fall-arrest measures, such as personal fall-arrest systems, are used to terminate falls if they occur. A simple example of a personal fall-arrest system is a harness connected by a fall-arrest lanyard and shock absorber to a suitable anchorage in the work area (see Fig. 14.7). Such a system allows worker mobility but still provides fall protection. Various equipment and options for establishing fall-arrest systems are available (e.g., see the fall-arrest system shown in Fig. 14.8). Much of the equipment described in Chap. 11 and the tying and rigging techniques described in Chaps. 12 and 13 can be applied in establishing personal fall-arrest systems. Specific instructions for rigging fall-arrest systems are included in Chap. 13. Areas where workers routinely are exposed to significant fall hazards, such as ladders, should be equipped with engineered systems, such as the rail and inertial brake device described below, as a better alternative to rigged systems (see Fig. 14.9).

Special considerations for vertical ascents and descents

Falls can occur when going up or down ladders or stairs, especially steep stairways such as ship ladders. In negotiating ladders and stairs, a safe practice is to maintain three points of contact at all times. In climbing a ladder this means that three of the four limbs must be in contact with a rung or beam at all times. When a hand releases its grip

Figure 14.7 A simple fall-protection system can be rigged with a class III harness, fall-arrest lanyard with shock absorber, and an anchor.

Figure 14.8 In this fall-arrest system the rope grab moves up and down the rope beside the ladder, but locks to the rope if the climber falls.

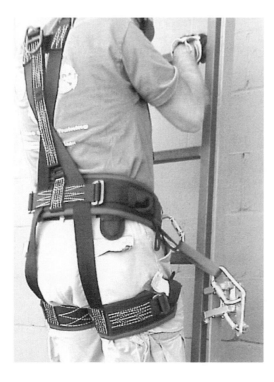

Figure 14.9 This inertial brake moves up or down a rail mounted to a ladder, but locks against the rail for fall-arrest if the climber falls.

to go for a higher hold, the other hand and both feet must be in contact with the ladder. As a foot is raised from one rung to the next, the other foot and both hands must be in contact with the ladder. As an alternative, the hands can slide up the beams for constant contact during ladder climbs. Note that these techniques don't leave a free hand for carrying tools and equipment. The best bet for getting them up and down ladders is to hoist them up after you using a rope, or a mechanical advantage system or winch for heavy loads. Light and small tools may be worn up or down stairs or ladders in tool pouches or packs.

Backup fall protection

Another important concept in fall protection is *backup fall arrest.* This refers to the use of two systems, a primary system and a secondary or backup system, to protect personnel from falls.

Let us use the case of a climber going up or down a ladder as an example. The climber's hands and feet can be considered one system. For a backup system the worker might wear a full-body harness connected to a rope grab running along a suitable rope or cable fixed vertically beside the ladder (see Fig. 14.8). If the worker falls, inertia will cause the grab to grip the rope or cable, arresting the fall. A similar system utilizes an inertial locking device that rides a metal rail attached to the ladder to arrest the fall (see Fig. 14.9). Retractable lifelines and webbing lanyards that feed off spools fitted with inertial locks may be used in vertical applications (see Fig. 14.10). Various belay techniques, as described in Chap. 13, can be used to provide backup fall protection when better techniques aren't available for ladder climbs and other vertical operations. Under OSHA regulations, only personnel designated by their employers as qualified climbers are allowed to "free climb" without backup fall protection, and then only to reach a work area where adequate fall protection will be used.

As noted in Chap. 13, backup fall protection is also important in raising and lowering operations, such as when personnel are lowered using a winch or mechanical-advantage system. It is advisable for the backup system to have an anchor system different from that of the primary system. Some winches are equipped with a fall-protection feature. Such winches have a setting that allows cable to feed out freely as the person descends or to retract under spring pressure as the person ascends. If a fall occurs, an inertial

Figure 14.10 This retractable lanyard has an inertial lock for fall-arrest.

brake locks the cable spool to arrest the fall. As an example, assume that such a device is mounted to a tripod placed over a manhole equipped with a ladder (see Fig. 14.11). The cable is attached to the entrant's harness, and the winch is set in the fall-arrest mode. The worker climbs the ladder using the winch's fall protection feature as a back-up. If nonentry rescue is required, the entrant can be hoisted out of the hole with the winch. For vertical entries in which the entrant must be lowered into the space with a winch, a separate fall protection system should be used to back up the winch. A retractable lifeline or web lanyard, or one of the safety belay techniques described in Chap. 13, could be used for this. The best recommendation is that the safety system have a separate anchor system rather than also being anchored to the tripod.

Communication Procedures

One dangerous aspect of confined space entry is that the entrant is effectively isolated from the outside world. This is the reason that such a strong emphasis is placed on the role of the attendant in constantly monitoring the safety of the attendant. This may be easy to achieve for simple spaces that allow the attendant a clear line of sight to the entrant throughout the entry. In some entries, constant voice contact between entrant and attendant may be possible. It may be possible to extend the range of voice contact by installing voice amplifiers in respirator facepieces, as seen in Fig. 14.12; however, in more complex spaces neither line of sight nor voice contact with the entrant may be possible. In all cases, remember that constant contact with the entrant is the only way to be sure that the entrant is safe at all times. Some methods and considerations for achieving this are discussed below.

Role of internal attendant

In some cases it is desirable to designate an entrant to remain at a position within the space from which she can observe, or remain in voice contact with, personnel who are working outside the view and voice realm of the attendant outside the space. The "internal attendant" should have no other duties that might distract her from acting as communications link.

Figure 14.11 The winch on this tripod has a fall-arrest setting that allows cable to feed out smoothly under normal conditions but locks if a fall occurs.

Figure 14.12 Communication equipment may include respirator facepiece voice amplifiers, two-way radios, and intercom systems.

Hand signals

In situations where personnel can see each other but voice contact is not possible, simple hand signals may be used for basic communication. There is no universally used set of signals, but any signals can be used as long as their meanings are clearly understood by everyone on the team. A few examples of simple hand signals are shown in Fig. 14.13.

OATH system

Confined space entrants sometimes use the OATH system to communicate with the attendant using tugs in the lifeline or tag line. OATH is a pneumonic device used to help personnel remember the following signals:

O = OK. One tug on the rope indicates that the entrant is okay.

A = advancing. Two tugs indicate that the entrant is advancing and needs rope fed into the space.

T = take up. Three tugs indicate that the entrant is retreating and needs rope taken up.

H = help. Four tugs indicates that the entrant needs help.

Concerns have been expressed about the effectiveness of the OATH system, namely, that it is inconvenient to maintain during operations, that the number of tugs may be muddled or misinterpreted, and that an entrant in trouble may not be able to tug the rope to call for help. Obviously the OATH method should be used only when more effective communication techniques are unavailable.

Radios

Two-way radios are commonly used for communications during work activities and rescue operations (see Fig. 14.12); however, radios simply don't transmit well into or out of many confined spaces. The radio waves rarely can pass through materials making up or

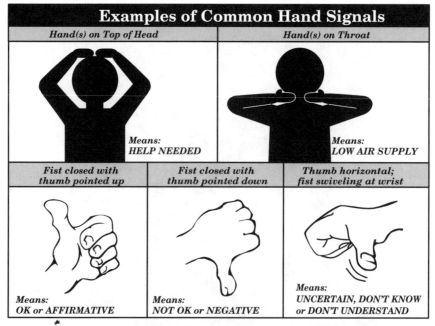

Figure 14.13 Simple hand signals may be used to communicate with other team members.

surrounding the space. While radios may allow adequate communication during some simple entries, they may fail to receive or transmit during operations involving more complex spaces. Remember that only radios with the proper safety approvals (as described in Chap. 4) should be used if flammable atmospheres may be encountered.

Intercom systems

Because of the problems associated with the use of radios, intercom systems are sometimes used in confined space operations. *Intercoms* use a hard-wired system to transmit communications. Some intercom systems have been designed and marketed specifically for confined space operations. Use approved electrical equipment if flammable atmospheres may be encountered.

Dealing with loss of communication during entry

One problem that may be encountered during confined space operations is a failure of communication between entrant and attendant. Entry standard operating procedures (SOPs) should provide guidance in dealing with this. In a situation involving voice contact, if the attendant cannot get a response from the entrant the only sound assumption is that an emergency exists inside the space. The same thing applies in situations involving loss of communication by radio or intercom. Backup communication equipment and procedures are recommended.

SOPs must address how a failure of communication equipment is to be dealt with. This includes the actions an entrant should take if unable to communicate with the attendant. The safest alternative would be for the entrant to prevent a false alarm by exiting the area of operations and letting the attendant know that no emergency existed.

Emergency alarms

An entrant involved in a confined space emergency needs some way to inform the attendant immediately. Alarms should be given verbally by voice, radio, or intercom if possible so that specific information about the problem can be related. Lacking that, some prearranged signals should be available such as rope tugs, sounding signal devices, or sending a designated signal by rapping heavy tools on solid objects. One reason that PASS devices (Fig. 14.5) are recommended for all entries is that they have a "panic button" that can be activated by the wearer as a way of notifying the attendant that an emergency is in progress. The continual piercing tone from the device also allows rescuers to find an unresponsive entrant much faster and easier.

Predesignated alarm signals must also be available for personnel outside the space to use in warning entrants to exit the space immediately in the event that communication equipment is unavailable. Signaling devices might include air horns, sirens, or banging on a vessel with a solid object.

Dealing with Claustrophobia

The new college edition of the *American Heritage Dictionary* defines *claustrophobia* as a "pathological fear of confined spaces." It is derived from two Latin words, *claustrum,* meaning "an enclosed space," and *phobia,* meaning "a persistent, abnormal, or illogical fear of a specific thing or situation." Claustrophobia is a common hazard of confined space entry (see Fig. 14.14).

Signs and symptoms

Signs and symptoms of claustrophobia may include the following:

- Feelings of anxiety or panic

Figure 14.14 Claustrophobia is one hazard of confined space entry.

- Faster breathing and heart rate
- Sensation of shortness of breath or suffocation
- Being in a cold sweat
- Feeling disoriented and trapped, as if "the walls are closing in"

It is important to realize that the symptoms are episodic in nature. This means that when you experience an episode, it is likely to diminish or disappear, only to recur later. If you have an episode, you may very well have several more in the course of a single entry. The symptoms may cycle through increasing then decreasing levels of intensity.

Triggering events

Certain conditions or events are likely to trigger a claustrophobic episode. These triggers may include

- Initial entry into a confined space
- Entering an especially small passage
- Feeling restrained or having difficulty moving
- Working in bulky gear
- Restricted visibility, especially total darkness
- Becoming disoriented or lost

Coping with claustrophobia

One way to cope with claustrophobia is to anticipate that it may happen, especially during a triggering event, so that it doesn't catch you off guard. If you should have an episode, realize that there is nothing wrong with you. It is a common reaction many people have in response to being confined. Even people with lots of previous confined space experience can experience an episode. You may not be able to keep it from occurring, but you can control your reaction to it and keep it from escalating.

Above all else, don't overreact, panic, and loose control. Ripping your respirator facepiece off is not an appropriate response. When you experience an episode, realize that it will probably lessen in severity momentarily but possibly occur again later.

Try to control your breathing by taking slow, deep breaths and exhaling fully each time. This may prevent you from hyperventilating. If you are wearing a respirator, it may prevent a buildup of carbon dioxide in the facepiece. Hyperventilation and overexposure to carbon dioxide both are known to cause feelings of anxiety.

Try to remain focused on your mission as a whole instead of the way you feel at any given moment during the entry. Remain in steady communication with the other members of your team. Trust in your training, equipment, and team.

Finally, if you know that you have a tendency to be claustrophobic, use training sessions to become more familiar with your reaction and limitations. You may be able to overcome it completely by exposing yourself to gradually intensifying triggers using a "baby steps" approach.

Emergency Provisions

One of the most important aspects of the PRCS standard and the permitting process is the intention that effective, timely rescue be available if needed during an entry. If entries are made into IDLH atmospheres, rescuers are required by OSHA's respiratory protection standard to be on standby at the scene during the entry. For entries into spaces where immediate rescue is not warranted by the hazards, merely having rescuers available on call during the entry may be sufficient for compliance.

The rescue service must be properly selected according to their ability to handle any emergency that could reasonably be expected to occur. Selection criteria for rescue team evaluation are included in Appendix F to the PRCS standard, found in App. III to this book. The attendant must have a means of summoning the rescue service promptly if they are not on standby at the time of entry. The topic of rescue will be addressed in detail beginning with the next chapter.

Concluding the Entry

All entry personnel must exit the space as soon as operations are concluded. An immediate head count should be done to verify that everyone is out. It is important that all personnel exit in an orderly fashion, bringing out all tools and equipment that were taken into the space. Any items remaining in the space could adulterate products or cause major damage to process equipment. Once the entry has been concluded and the permit invalidated, reentering the space to retrieve a forgotten tool will technically need to be repermitted. Once everyone and everything are clear of the space, all isolation procedures can be systematically undone and preparations made to return the space to service.

Summary

This chapter attempted to pull together information from various places and combine it around the theme of safe confined space entry. Safe and efficient confined space operations, whether for routine work or for rescue, don't happen by accident. They require planning, preparation, due regard for safety considerations, and a team approach. This is the philosophy underlying the requirements of the PRCS standard. Following those requirements should ensure a safe entry and a safe exit for people who are required to enter permit spaces.

Rescue

15

Overview of Confined Space Rescue

A 280-lb pound worker in a sandblasting operation fell roughly 60 ft through a hole in the top of a water tank to the bottom of the tank, suffering femur and wrist fractures. The water tank was elevated 120 ft above the ground and connected to the ground by a 6-ft-diameter standpipe. A temporary handrail and toeboard prevented the worker from falling the additional 120 ft down the standpipe. To reach the victim, rescuers were required to climb up a fixed ladder running 180 ft to the top of the tank. The rescuers anchored kernmantle ropes to the top of the tank and rappelled down through the hole in the top of the tank 60 ft to reach the injured worker. The two paramedic rescuers conducted an assessment, provided intravenous (IV) fluids, stabilized the fractured femur, and immobilized the patient on a backboard. Assistance was requested from an industrial confined space rescue team in a nearby town. With the arrival of the additional rescuers, the backboarded patient was secured to a plastic basket litter rigged to be lowered in the vertical position. The rescuers anchored a brake bar rack and established a lowering line attached to the basket. A separate safety belay line was established using a Munter hitch at a separate anchor point. The rescuers lowered the patient 120 ft to the bottom of the standpipe, where he was removed through an opening in the base of the standpipe. Support personnel on the ground took charge of the patient and transported him to a nearby hospital. Rescuers involved in the operation noted that a recent practice session in basket lowering techniques had given them a big advantage in getting the large patient over the edge to begin the lower.

Basic Considerations for Rescue

Rescue operations are time-critical, stressful activities (see Fig. 15.1). This is especially true of confined space rescue. These rescue operations may require us to make decisions and take actions in a very short timespan in order to save the lives of patients. Confined space rescue operations also require us to operate in unfamiliar settings under difficult conditions, so high levels of physical and mental stress may be involved.

The same characteristics of confined spaces that make them dangerous to work in also make rescuing people from them difficult. The restricted entry/exit pathway of confined spaces may make accessing the patient difficult, and removing the patient even more difficult. Time-consuming procedures may be required to isolate the space from hazards and prepare the space for a safe entry. In some cases we must work in SCBA or SAR and chemical protective clothing, making the process even more difficult. Confined space rescue can be uniquely challenging.

Confined spaces are unique in the degree to which they have been deadly to rescuers. NIOSH estimated that 60 percent of the fatalities in confined spaces were attempting to rescue others when they became part of the problem. Some of the rescuer fatalities

Figure 15.1 Confined space rescues are time-critical, high-stress operations.

were workers attempting impromptu rescues of coworkers, but some were trained rescuers.

Considering all the factors working against a successful confined space rescue, it becomes clear that a well-organized, methodical process is required. As in all types of emergency response operations, this process should begin long before the call is received. In this chapter we will use preemergency considerations as a point of beginning, progressing from there through the actions to take on arrival on the scene.

Rescue Psychology 101

Before starting into the nuts and bolts of how to perform rescues, we need to establish some basic priorities or "get our minds right." Keeping our priorities straight is important because it can keep us from substituting emotion for reason in the decision-making processes. The role of emotion in making decisions is one of the main reasons for the high percentage of rescuer fatalities in confined spaces. When the average person sees another person in trouble, there is an almost overpowering urge to go and help, even in a permit space. This is especially true in cases involving coworkers, friends, and family members.

Prioritizing things in a rational way will help us avoid placing ourselves unduly at risk. Basic concepts we will use to establish our priorities are the pyramid of survivability, the pyramid of priority, and risk-versus-benefit analysis.

Pyramid of survivability

According to the pyramid of survivability (Fig. 15.2), you should give the highest priority to your personal safety during a rescue operation. If you become involved in an accident during rescue operations, other members of your team may be endangered by the accident or in attempting to rescue you, and rescue and treatment of the original patient will be slowed. The first step in making the situation better is to stay out of trouble. Don't become part of the problem.

The safety of your team should be your second priority. Teamwork is required for effective rescue. If all team members are actively monitoring the safety of team activities, then safe, effective rescue operations should result.

The safety and well-being of your patient should be your third priority. Never place a higher priority on the patient's well-being than that of yourself or your team. If you or your team become part of the problem, you will not be able to help the patient.

Pyramid of priority

According to the pyramid of priority (Fig. 15.3), once the safety of rescuers is assured, rescuers should give the highest priority to safeguarding the lives and safety of people who are involved in the emergency, or who may become involved in it if it progresses. Only after all life-safety issues have been addressed can responders turn their attention to the second priority, which is stabilizing the actual incident. Only after the incident is stabilized can responders turn to the third priority, which is the conservation of property.

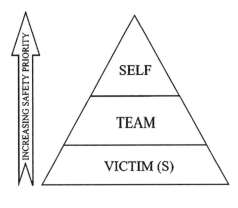

Figure 15.2 The pyramid of survivability establishes safety priorities for the individual rescuer, rescue team, and victim during rescue operations.

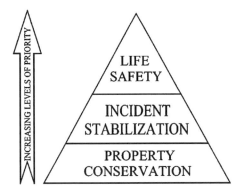

Figure 15.3 The pyramid of priority establishes relative priorities for life safety, incident stabilization, and property conservation during emergency operations.

As an example, assume that a worker is unconscious due to a chemical release from a damaged valve within a space. According to the pyramid of priority, it makes no sense to delay removing the patient from the space while we try to repair the valve (unless the repair is required to make it safe to remove the patient). It would make no sense to try to clean up the material released by the damaged valve (property conservation) before the valve is repaired (incident stabilization).

Risk-versus-benefit analysis

Risk is exposure to the chance of injury or loss. Some degree of risk is inherent in all emergency operations. People who are unwilling to assume any level of risk at all typically avoid serving as rescuers. With the exception of a few certifiable lunatics, all emergency response personnel place limits on the level of risk we are willing to accept. For most of us, the level of risk we consider acceptable for a given action is directly related to the level of benefit we perceive as resulting from the action. This is the basis of risk-versus-benefit analysis, as shown in Fig. 15.4.

Most emergency responders routinely perform operations that involve a low level of risk with a high probability of saving a life. Most EMS calls probably fall in this category. If we could quantify both benefit and risk and plot them along the axes in Fig. 15.4, these low-risk, high-benefit operations would fall in the lower right quadrant.

Most responders are willing to carry out, or ask others to carry out, operations that involve a high level of risk if it is likely that a life may be saved by their actions. A prime example of this from the fire service is search-and-rescue missions conducted by firefighters during structure fires. High-risk, high-benefit operations would plot in the upper right quadrant in Fig. 15.4.

Conversely, most responders are unwilling to assume, or ask others to assume, a high level of risk if there is little or no probability of saving a life. Most responders place significant limits on the level of risk they are willing to assume to achieve objectives that don't involve life safety, such as property conservation or body recovery. High-risk, low-benefit operations would plot in the upper left quadrant in Fig. 15.4. At some point this

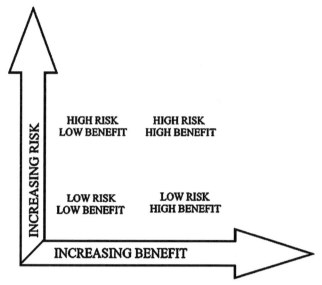

Figure 15.4 Risk-versus-benefit analysis relates the level of risk to the potential benefit of a rescue operation.

becomes difficult to generalize about because neither risk nor benefit is well quantified and human judgment is required.

Problems result when emotion takes the place of reason and no clear risk-benefit analysis is evident. In such cases, rescuers may undertake highly risky operations with very little or no potential benefit. An example of this phenomenon is the Phoenix tank explosion incident described in the introduction to Chap. 1. In that incident, rescuers placed a very high priority on immediately removing the victim from the toluene tank. In order to do so, the rescuers undertook a high-risk procedure in attempting to cut a hole in the side of the tank with a rotary saw. Autopsy results later indicated that the victim was already dead before the tank explosion occurred. One firefighter died, and 14 others were injured in what was actually a body recovery operation.

Types of Rescue

Confined space rescue operations can be classified into three types: self-rescue, nonentry rescue, and entry or internal rescue. These three types or levels of rescue differ significantly in the hazards to rescuers, the probability of survival of the patient, and the degree to which external rescuers are involved.

Self-rescue

Self-rescue is performed by the victim rather than external rescuers. The victim recognizes a forbidden condition and exits the space without assistance from outside.

Assume that a worker in a permit space begins to be exposed to organic vapors due to a failure of ventilation of the space. The worker continues to work in the space until she begins to experience dizziness and feelings of nausea, both of which she has been trained to recognize as symptoms of exposure. She notifies the attendant of the emergency and exits the space before the symptoms become acute enough to prevent her from being able to do so. As a variation of this example, assume that the attendant notices that the entrant is exhibiting behavioral effects that he has been trained to recognize as signs of exposure. In this example, the attendant might order the entrant to exit the space as a form of self-rescue.

Self-rescue is the preferred type of rescue. It involves the least hazard to rescuers, since their only patient contact occurs after the patient is out of the space. The fact that the patient is able to exit unassisted usually indicates a good probability of surviving the incident without serious harm.

Nonentry rescue

Nonentry rescue is used in situations in which the entrant has to be rescued by others, but rescue procedures can be performed from outside the space. Nonentry rescue may be performed by the attendant or others trained to do so, while awaiting the arrival of the designated rescue team or service. In some cases rescue team members are able to perform nonentry rescue procedures.

Assume that an entrant is working inside a storage tank after having been lowered using a tripod and winch through a manway on top of the tank. The entrant suddenly collapses and the attendant immediately notifies the designated rescuers. The attendant may be able to use the tripod-winch setup to hoist the entrant up out of the space while rescuers are en route.

Nonentry rescue is the second order of preference for confined space rescue. Here again, rescuers are not exposed to the internal hazards of the space. The victim will almost certainly be removed from the space more quickly than if internal rescue is required. In many confined space emergencies nonentry rescue is not a feasible option.

Entry or internal rescue

Entry or internal rescue requires that rescuers enter a confined space to access, stabilize, package, and remove the patient from the space. It requires the greatest assumption of risk on the part of rescuers and has the worst prognosis for the patient. In many instances factors such as space configuration or the nature of the incident require that entry rescue be performed if an entrant can't perform self-rescue.

Internal rescue operations in permit spaces are some of the most challenging technical rescues that responders can be called on to perform. We will devote the rest of this chapter, and most of the rest of this book, to meeting that challenge.

Putting It All Together: The Rescue Quad

In order to maximize the effectiveness of a confined space rescue operation, four factors must be considered: time, the victim or patient, the space, and the rescuer. These factors are distinct but interrelated, as represented by the rescue quad (Fig. 15.5).

Time

Time is a common nemesis of anyone who provides rescue or emergency medical services. This is especially true in the arena of confined space rescue.

Assume that a worker without respiratory protection becomes unconscious in an extremely oxygen-deficient space, or becomes pulseless and breathless for some reason. The clock is ticking for that worker. We know that biological death, or permanent death of brain cells, can begin in 4 min and is very likely in 6 to 10 min. The patient's chance of survival drops roughly 10% for every minute that he is in full arrest. Unless rescuers are on standby ready to initiate rescue immediately during the entry, there may be little chance of saving the victim.

As another example, assume that a worker has suffered major trauma due to a fall within a confined space where no chemical or atmospheric hazards are present. Again time is a critical factor. If the patient can be accessed, properly packaged, removed from the space, and transported to an appropriate medical facility within the "golden hour" following the injury, his chances of survival are enhanced. If the process takes longer, the patient's chances for survival may be significantly reduced.

The time required for initiating rescue is an important aspect of the time factor. A common source of controversy has been the question of what constitutes "timely" rescue. According to OSHA, the answer depends on the hazards present within the space and the likely effects on the victim. OSHA requires rescuers on standby outside the space during entries into potentially IDLH atmospheres. This is because the workers can suffer brain damage and death in a few minutes should respiratory protective or ventilation equipment fail. On the other hand, if the only hazards present are physical hazards that might result in lacerations or broken bones, a response time of 10 to 15 min may

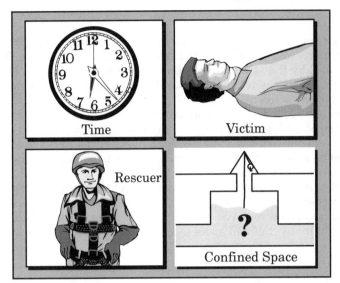

Figure 15.5   The rescue quad relates four critical factors involved in confined space rescue. (*Adapted from Wieder et al.* (1992), Fig. 5.97.)

be adequate. In this case, merely having a rescue team available to be called if needed may be adequate.

Response time has several individual components, including the time required for the rescue team or service to be notified of the emergency, the travel time required for the rescuers to arrive at the location, and the setup time required for rescuers to be ready to enter the space. Additional time is required to enter the space and access the patient. These factors must be considered by a facility in deciding whether to rely on an off-site rescue service or develop an on-site rescue team.

The time factor is impacted by whether the operation is actually a rescue or a body recovery. Rescue operations are time-critical. Recovery operations are not.

The victim

The victim or patient is the focal point of the rescue operation (Fig. 15.5). To begin with, how many victims are involved? What are their likely locations within the space? Information from the entry permit and attendant will help in establishing this.

Another important consideration is the likely condition of the victim. Attempting immediate voice contact may reveal whether the patient is conscious, unconscious, injured, or trapped. Voice contact may also allow assessment of the victim's mental state and an opportunity to assure the victim that help is on the way.

The mechanism of injury involved relates to a number of important considerations. The mechanism of injury should indicate whether spinal injury is possible. This, of course, will impact other considerations, such as the equipment and techniques required to package the patient for removal from the space. Some mechanisms of injury may be so severe as to make it obvious that the patient is dead and the operation should shift down into the recovery mode.

In cases involving a reportedly "trapped" entrant, the specific nature of the entrapment must be determined. In some cases, the entrant may be uninjured but trapped by the configuration of the space, such as when a worker slides into the bottom of a hopper and cannot climb back up because the sides are too steep (Fig. 15.6). In other cases, the entrant's limbs may need to be extricated from machinery in order to complete the rescue. In these cases, whether limbs are crushed or merely pinned is a major size-up consideration (Fig. 15.6). If the limb is merely pinned, the resulting injuries may not be life-threatening, or may even be minor, depending on the force involved. In contrast, any crushed limb must be treated as a major, life-threatening injury because of the possibility that crush syndrome may result. Crush syndrome occurs when muscles are crushed with a resulting release of myoglobin, potassium, and other substances. These substances produce cardiac arrythmias, renal failure, and various organ system failures. Without timely, aggressive medical treatment, the death of the patient from crush syndrome is likely.

The victim's condition has a direct bearing on the time factor. For example, CPR cannot be performed on a breathless and pulseless patient in an anoxic atmosphere. In such a situation, a "grab and go" approach may be justified. In contrast, a patient who has injuries from a fall inside a space with no atmospheric hazards may warrant very careful packaging for cervical spine (C-spine) stabilization before being removed.

The space

The space is the location where all the other factors come together (Fig. 15.5). It is the location of the highest hazard level on scene. In some cases it is equivalent to the hot zone of a hazardous materials incident. The hazards of the space must be identified and assessed carefully. This dictates the actions rescuers must take to establish a safe area of operations prior to entry and to improve the patient's chances of survival. The hazards also dictate the type and level of protective equipment that rescuers must use when entering the space. This, in turn, impacts other factors such as rescuer stress, impairment of mobility and dexterity, and time required for rescue operations.

The size, configuration, and other features of the space involved are also significant (see Fig. 15.7). For example, the size and shape of a space determine the amount of room available to access, treat, and package a patient. The size, shape, location, and orientation of entryways have a direct bearing on the time required to access the patient and remove the patient from the space. The availability of anchor points must also be considered.

Simple spaces may make it possible to remove a packaged patient through a single simple vertical or horizontal operation. Complex spaces may require multiple horizontal to vertical transitions, with complex rigging and difficult maneuvering required to complete the removal. Once a patient is removed from elevated spaces such as vertical storage tanks or baghouses, high-angle rope rescue techniques may then be required to complete the operation. In an attempt to come up with a concise way of classifying spaces, ROCO Rescue has developed an alphanumeric system for typing spaces on the basis of factors such as these (Roop et al. 1998, p. 312).

The rescuer

During confined space rescue operations, the rescuers must deal with the complex interactions of all four factors making up the rescue quad (Fig. 15.5). This requires working with other rescuers as a team in order to remove the victim from the space in a timely fashion. Rescuers must do a complete hazard assessment; take the necessary precautions; use safe entry procedures; access, treat, and remove the patient from the space; and safely terminate the rescue operation, all in a timely fashion.

Rescue operations are highly stressful. They demand a high level of physical and mental fitness on the part of the rescuer. In order to meet this challenge, rescuers must

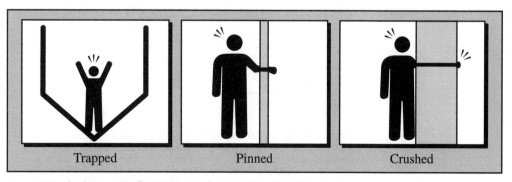

Figure 15.6 An important determination for situations involving entrapment is whether the patient is trapped and/or has body parts that are either pinned or crushed.

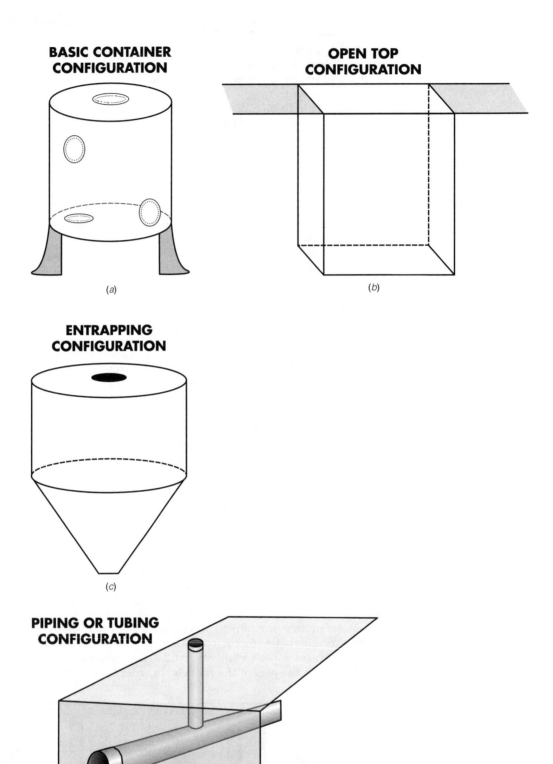

Figure 15.7 (*a*) Difficulty in performing rescues from tanks, vessels, or vaults varies with factors such as size; shape; and the size, location, and orientation of entryways. (*b*) Conventional portable anchors such as tripods and hoisting frames may not be usable for open-top spaces such as pits and vats. (*c*) Rescue may be difficult from spaces with entrapping configurations. (*d*) Patient packaging may be difficult or impossible within small-diameter piping or tubing systems.

**COMPLEX
CONFIGURATION**

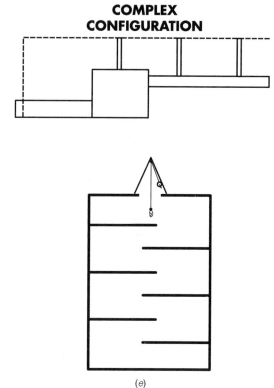

(e)

Figure 15.7 (*Continued*) (*e*) Confined spaces with complex configurations require complicated rescue procedures such as horizontal-to-vertical rigging transitions.

be adequately trained and equipped to carry out anything they may be expected to do. The level of success of most rescue operations is predicated on the level of training and the mental readiness and tenacity of the rescuer.

In the course of this, as in all human undertakings, communication between all the players involved is critical. Everyone involved must remember that rescue team members work together as a team to maximize effectiveness.

Steps in the Confined Space Rescue Process

Given the challenging nature of confined space rescue, it is obvious that a careful, methodical approach is needed. This section of the chapter will walk through a step-by-step approach to confined space rescue. This approach begins with things that should be done long before a call for confined space rescue is received and progresses through steps required to perform an actual rescue and terminate the operation. Information generated through the entire process is fed back into the preplanning step after terminating a rescue operation. The process can be viewed as a cycle, as seen in Fig. 15.8, rather than a series of distinct steps.

Step 1: preemergency preparation and planning

Confined space rescue, like most special operations in the emergency services, requires a significant amount of advance preparation in order to be done effectively. To begin with, a confined space rescue team or service capability must be developed. This requires that team members be designated and provided with adequate training before being called on to perform a rescue (see Fig. 15.9).

As a minimum, the team or rescue service as a whole and all team members individually, should comply with all requirements of Paragraph K and Appendix F of *OSHA's Permit-Required Confined Spaces Standard* (29 CFR 1910.146). In addition, NFPA 1670 *Standard on Operations and Training for Technical Rescue Incidents* is strongly recommended as a minimum standard, even for non-fire-service organizations. This will require the rescue

Figure 15.8 Steps in the rescue process can be viewed as a cycle.

Figure 15.9 The time for training is before the call is received.

organization to establish functional capability at either the operations or technician level for confined space entry rescue. Individual rescuers should be competent at the awareness, operations, or technician level for confined space rescue depending on their assigned rescue role. For organizations with a technician-level confined space rescue capability, it is strongly recommended that all technician level rescuers meet the competency requirements for confined space rescue listed in NFPA 1006 *Standard for Rescue Technician Professional Qualifications*. The standards identified here were described in Chap. 2.

The team must be well organized. The use of the incident management system (IMS) is urged for organizing and managing all rescue team operations (see Chap. 16). Written SOPs for confined space rescue should be developed and implemented during preemergency preparation. The SOPs should be adequate for any rescue operations that the organization could reasonably expect to be called on to perform.

Develop preplans for individual spaces or types of spaces during preemergency preparation. Remember that the PRCS standard gives rescuers the right to be informed of hazards, to preplan spaces, and to practice rescues from actual or representative spaces. A variety of forms or formats, either in hard copy or on computer, can be used to organize the information. An example preemergency worksheet for planning confined space rescues is included as Fig. 15.10. Having basic information about a space gathered ahead of time is invaluable in the event that an actual rescue is required from the space. Once again the old adage that "size-up begins with preplanning" holds true.

Obtain appropriate rescue equipment and supplies for use in confined space rescue operations. The preplanning process will make clear what equipment and supplies are needed. All equipment should meet applicable provisions of NFPA 1983 *Standard for Life Safety Rope and System Components*. Organize and store the equipment for rapid, orderly deployment, which may require storage areas and vehicles dedicated for the purpose. Address issues related to inventory, inspection, evaluation, and retirement of equipment, including appropriate documentation.

Step 2: Size-up

Once a confined space emergency is under way, size-up of the incident should begin as soon as the rescue team or service arrives at the scene. The person in charge of the first group of responders establishes initial command, makes an initial report, and directs the placement of any additional units en route.

From this point onward, the incident commander may wish to use a tactical worksheet to organize the emergency operation. A sample tactical worksheet for confined space rescue is included as Fig. 15.11. The worksheet incorporates the four elements of the rescue quad, and includes various considerations for responding to confined space emergencies. It includes a generic command structure for use with the incident management system, which is covered in Chap. 16. Note that in using the command structure, only the roles required to complete the emergency operation need to be filled. The others can be left blank. The tactical worksheet also includes a space for sketching the space or incident scene. Note that the form provides for geographic division or "sectoring" of the scene. By tradition, the main entrance or approach to the space is designated sector A and the other quadrants proceeding clockwise are designated sectors B, C, and D (see Fig. 15.11). Always include a north-pointing arrow to orient the worksheet to the cardinal directions.

Rescuers should attempt to establish communication with the victim early in the rescue, unless someone else, such as the attendant, is already maintaining contact. Someone must guard the entryway and act as "hole watch," attempting to maintain communication throughout the operation.

Establish control of the incident scene quickly. Zoning for site control during permitted entry operations was discussed in Chap. 6. The zoning concept can also be applied in a confined space rescue operation, as shown in Fig. 15.12.

Conduct an initial size-up by gathering relevant information, as follows:

1. Obtain the entry permit, if available.

2. Question the attendant, supervisor, other workers, or bystanders.

3. Make visual observations of site conditions, weather conditions, and any obvious hazards to personnel.

4. Determine relevant information such as characteristics of the space, number of victims, probable location, time of last contact, and mechanism of injury.

At this point a preplan for the space should be utilized, if one is available. Use the information gathered during the initial size-up to verify and supplement the information in the preplan. Assess available resources. If additional resources are needed, such as through mutual aid agreements, call for them as soon as possible to allow for lag time in their arrival.

CONFINED SPACE RESCUE
PREEMERGENCY PLANNING WORKSHEET

Date:

Location/ID of Space:

Point of Contact: Contact Info:

PART 1: HAZARD IDENTIFICATION

<u>**Hazardous Substances**</u> (attach MSDS if available)

Name of Substance:			
Physical State	Sol Liq Gas	Sol Liq Gas	Sol Liq Gas
Toxic? Y/N Routes:	Inh Ing Ctc Abs	Inh Ing Ctc Abs	Inh Ing Ctc Abs
PEL (if applic.):			
STEL (if applic.):			
IDLH (if applic.):			
Skin Hazard?	Yes No	Yes No	Yes No
Acute Symptoms:			
Flam. Gas/Vapor?	Yes No	Yes No	Yes No
Flash Point:			
LEL:			
UEL:			
Typical Concentration			
Vapor/Gas Density:			
Combust Dust?	Yes No	Yes No	Yes No
Corrosive?	Yes No	Yes No	Yes No
Reactivity?	Yes No	Yes No	Yes No
Other Hazard Information:			
Hazard Detection Equipment Needed:			

Typical Oxygen Content	Deficient (<19.5%):	Enriched (>23.5):

<u>**Biological Hazards:**</u>

<u>**Radiological Hazards:**</u>

<u>**Physical Hazards**</u>

Energy Sources:
Engulfment Hazards:
Pressure:
Temperature Extremes:
Falls:
Noise:
Other:

Figure 15.10 A preemergency worksheet is helpful for planning confined space rescues ahead of time.

PART 2:   TYPE/LAYOUT OF SPACE (Make scale drawing or attach drawing, including scale, north arrow, and landmarks for referencing location. Identify suitable anchor points.)

PART 3: RESCUE OPERATIONS AND LOGISTICS

Probable Emergencies	Required Rescue Operations	Equipment Needed	Personnel Needed

Additional Rescue Resources Available:

Figure 15.10   (*Continued*)

PART 4: HAZARD CONTROL PROCEDURES

Procedures	Operation / Location	Equipment Required
Ventilation		
Isolation		
-Lockout/Tagout		
-Blanking		
-Line Breaking		
-Dbl. Block & Bleed		
-Mech. Stabilization		
Scene Security		

Additional Hazard Control Requirements:

PART 5: PERSONAL PROTECTIVE EQUIPMENT

<u>Respiratory Protective Equipment:</u>

<u>Chemical Protective Clothing:</u>

Level of Protection: A   B   C   D

Chemical Protective Clothing

<u>Type:</u>   <u>Material:</u>

Suit

Gloves:

Boots:

Other:

<u>Other Protective Ensemble Items:</u>

Figure 15.10 (*Continued*)

CONFINED SPACE RESCUE TACTICAL WORKSHEET

Location		Time on Scene	Weather Conditions

ACTION CHECKLIST

Establish Command On Scene	
Contact Attendant or Supervisor	
Obtain Entry Permit	
Attempt Victim Contact	
Is Space Preplanned? Yes / No	
Request Technical Assistance	
Abate External Hazards	
Abate Internal Hazards	

Time Reported:

C

NORTH

B   D

A

Name of Patient	Time of Contact	Condition of Patient
1		
2		
3		
4		

Hazards Present	Y/N	Describe
Atmospheric		(log on back)
Toxic		
Corrosive		
Engulfment		
Energy		
Temperature		
Biological		
Other		

(OVER)

Figure 15.11 A tactical worksheet is helpful for organizing confined space rescue operations.

ATMOSPHERIC MONITORING					
	Time/Result	Time/Result	Time/Result	Time/Result	Time/Result
Oxygen Content					
Flammability					
Toxic 1:					
Toxic 2:					
Other:					

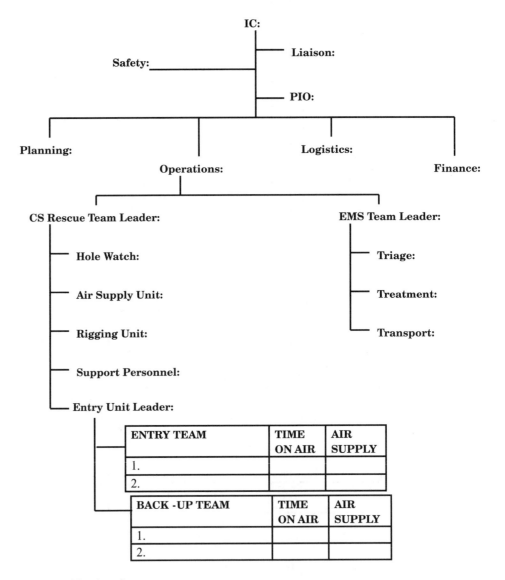

Figure 15.11 (*Continued*)

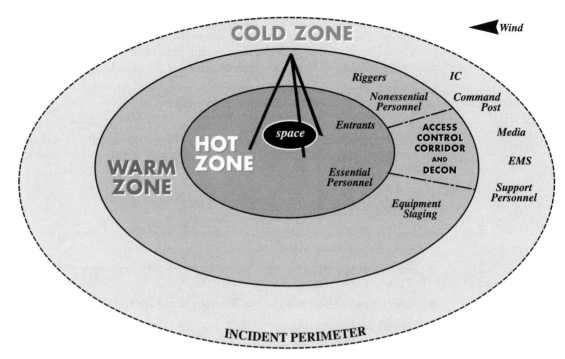

Figure 15.12 The concept of site zoning can be used to establish and maintain control at the scene of a confined space rescue.

As soon as possible, make a decision about whether entry is likely to be required for rescue. In some cases, nonentry rescue may be possible.

As part of the size-up process, do a hazard assessment using all available information. Conduct air monitoring to assess atmospheric hazards. Identify any hazardous materials within the space. Identify hazards related to energy and/or material that could impinge on the space. Identify any other hazards, such as physical hazards of the space itself. Assess any hazards to responders operating outside of the space at the incident scene.

As early in the process as feasible, make a determination regarding whether the operation should continue in the rescue mode or downgrade to the recovery mode. This is an important consideration, as previously noted in discussing risk-versus-benefit analysis and the rescue quad. In some cases, definitive signs of death may be visually evident from outside the space, such as in cases where a dramatic mechanism of injury is involved. In other cases, the victim will have to be accessed and determined to be dead in accordance with applicable emergency medical protocols before the operation can be shifted into the recovery mode. Rescuers should avoid declaring a victim to be dead based solely on atmospheric conditions within a space and the patient's duration of exposure, unless there is complete certainty that the conditions within the space and the duration of exposure involved offer no chance of survival.

If the information gathered during size-up indicates that entry is actually required and that it can be performed without undue risk to responders, then proceed to the next step. Otherwise continue gathering information and consider other options.

Step 3: developing a plan of action for entry and rescue

Before any actions are taken toward entry and rescue, it is important to develop a specific action plan. Design this plan to achieve the strategic goals established by the incident commander using tactical options developed by the operations officer, as discussed in Chap. 16. In accordance with the plan, assign specific tasks to individual team members.

Having SOPs to follow and a preplan available for the space will make this step much quicker and easier than starting from scratch. In such cases, the general preplan may simply be fine-tuned to develop the specific plan of action. It is good operating practice to always have a backup plan, in case the primary plan fails.

Step 4: initiating hazard control and protective provisions

Entry into a confined space cannot be made until all hazards have been controlled or adequate protective provisions are in place. These were well covered in previous chapters and will not be elaborated on here. Examples of procedures that may be required include

- Ventilation procedures as required to control flammable, oxygen-deficient, oxygen-enriched, or toxic atmospheres
- Isolation of the space through procedure such as lockout/tagout, blinding, line breaking, or double block and bleed
- Selection of appropriate personal protective equipment such as SCBA or SAR, chemical protective clothing, and accessory items
- Other procedures as required to control hazards of the space and incident scene
- Ongoing air monitoring, as needed

Step 5: deploying and rigging rescue equipment

Identify rescue equipment that may be needed in order to safely enter the space, package the patient, and remove the patient from the space. Such equipment may include tag lines, mechanical-advantage systems, tripods, personal fall-protection systems, and SKED stretchers.

In some cases rescue systems are assembled on scene from available components. In other cases, prerigged systems are used. Deploying these systems requires that suitable anchor points be identified.

Step 6: performing rescue

Rescuers are exposed to the highest level of hazard during this step, as they enter the space in order to locate and access the patient. Once the patient is reached, conduct patient assessment and treat life-threatening injuries, if practical. Package the patient for stabilization of injuries and for easier removal from the space, as described in Chap. 18. Remove the patient from the confined space, provide further emergency medical treatment as needed, and transport him to an appropriate medical facility for further treatment.

Step 7: terminating the rescue operation

Once the patient has been transported off the scene, termination of the rescue operation can begin. At this point, make an accounting of the well-being of all rescue personnel. Identify any injuries that occurred, any potential exposures, and any exposure-related signs or symptoms. Verify that all personnel and equipment have been removed from the space, and that it is safe to remove all lockout/tagout items and undo other isolation procedures in preparation for returning the space to service.

After the rescue is completed, inventory, inspect, clean (and decontaminate if needed), and properly store all equipment used in the operation. Repair or replace any damaged equipment promptly.

Follow up on the condition of rescuers. This may require that medical monitoring be performed following operations involving potential exposure of personnel to hazardous substances. Document and follow up on exposures and injuries. Evaluate rescuer stress and provide critical-incident stress debriefing (CISD) if needed.

Critiquing the rescue operation is also an important part of terminating the incident (see Fig. 15.13). Hold the critique soon after the operation is concluded and offer everyone involved a chance to ask questions and provide input into the process. Use the lessons learned in the operation, as evident in the conclusions from the critique, as a basis for making needed modifications to SOPs and preplans. This completes the cycle by feeding back into the first step of the rescue process (see Fig. 15.8).

Summary

Confined space rescues can be some of the most challenging and physically and mentally demanding operations that rescuers will face. Performing them well requires keeping priorities straight and using a careful, methodical approach. This approach should begin well in advance of the first confined space rescue call, and extend through every step in the process in order to maximize the safety and effectiveness of team operations.

Figure 15.13 Lessons learned through critiquing an operation allow us to avoid repeating mistakes.

Rescue Team Organization and Management

Rescue operations require a high degree of organization in order to be effective, and confined space rescue operations are no exception. Rescue personnel may be called on to overcome resource limitations in a short timeframe under high-stress conditions in order to save lives. At the same time, the safety of everyone involved in the operation must be assured.

To complete a rescue successfully, team members and support personnel may be required to carry out different operations at different locations simultaneously. Rescue operations may require the coordination of people, equipment, and supplies from the public, private, and civilian sectors, and these operations may involve multiple responding agencies.

Given the potential dangers and challenges of confined space rescue, it is obvious that some type of formal managerial system is needed to organize the actions of all the personnel involved. One system that has been used with a high degree of success in managing all types of emergency response operations is the incident management system (IMS), also known as the incident command system (ICS).

Origins of the Incident Management System

The origin of the IMS concept can be traced to California in the 1970s. At that time an organization known as FIRESCOPE was formed through a joint effort between local, state, and federal officials involved in wildland firefighting operations. The system, originally referred to as the *incident command system,* was developed to meet the huge managerial and logistical challenges of firefighting operations for large wildland and wildland/urban interface fires. ICS, or IMS as we now know it, worked so well that it is widely used today by both public-sector and private-sector organizations in responding to all kinds of emergencies.

Incident Management System Overview

IMS is a managerial concept intended to control and coordinate resources within five major functional areas during an emergency operation. The five major functions of IMS are command, planning, operations, logistics, and finance. They are related as shown in Fig. 16.1. In all cases one person, the incident commander, is ultimately responsible for managing the entire response operation.

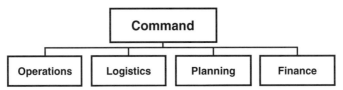

Figure 16.1   The IMS controls and coordinates resources within five major functional areas during emergency operations.

IMS is a very versatile, highly adaptable system applicable to both minor and major incidents. It can be used to manage single jurisdiction/single-agency responses, single-jurisdiction/multiagency response, and multijurisdiction/multiagency responses.

The main reason for the success of IMS is that it can provide a high level of control over the actions of a large number of people. This control is made possible by the following operating principals, characteristics, and components of IMS.

Unity of command

Unity of command simply means that everyone working within the system receives orders from and reports back to one designated supervisor. This avoids conflicting multiple directives and prevents multiple redundant report backs. Those who have worked in a position where they answered directly to more than one boss should see the wisdom in this.

Span of control

It is generally believed that a person in a managerial role in an emergency response operation can effectively supervise three to seven other people. As a general rule, a span of control of five is considered reasonable for someone supervising the activities of others within the IMS.

If a responder's span of control is exceeded, her ability to coordinate and control the actions of subordinates can be overwhelmed. This in turn can lead to a breakdown of the system. To avoid this, managerial responsibilities must be delegated according to an established chain of command as the system expands and more personnel become involved in it. This allows centralized command and control of the IMS to be maintained throughout the system no matter how many people become involved in it.

Division of labor

Few people are adequately trained to do all the tasks that may be required to complete a rescue operation. A logical division of labor assures that personnel are assigned tasks that are appropriate to their training, experience, and abilities. This allows for the most efficient utilization of everyone in the system.

Accountability

The IMS provides accountability for all resources operating within the system. This means that the location and function of any given person, piece of equipment, or other resource is accounted for at all times through the system. Because of the importance of accountability, freelancing is not allowed. *Freelancing* refers to an individual acting without orders, on his own initiative, outside the controls of the chain of command, and is considered a cardinal sin within the IMS.

Common terminology

Within the IMS, major organizational functions such as command, operations, and logistics are predesignated and named. Common terms are used for personnel roles and equipment used in tactical operations. Common identifiers are used for facilities in and around the incident scene, such as the command post and staging areas.

Modular organization

The IMS utilizes a modular format that allows the organizational structure to build or unfold from the top down as needed. We can think of these modules as boxes or slots on an organizational chart that can be activated or filled if they are needed.

Control of all five functional areas of the system rests initially with command. If required by the incident, the commander may appoint someone to manage any or all of the other four functional areas. As the command structure unfolds, still others may be designated to oversee activities within the functional areas. The ability to delegate responsibility as needed makes the system adaptable to both major and minor incidents.

Integrated communications

Clear and concise communications are vital to the success of any response operation (see Fig. 16.2). To provide this, IMS requires the use of an integrated and coordinated communication plan. Such a communication plan identifies communication procedures and protocols, allocates radiofrequencies and their uses, and specifies procedures to receive, acknowledge, and record all communications.

It is strongly recommended that all communications be conducted in plain language, rather than using "10 codes." This prevents confusion in multiagency or multijurisdictional operations, since the meanings of codes vary from place to place.

Unified command structure

In multiagency and/or multijurisdictional incidents, command may need to be a shared responsibility. Unified command allows representatives of all affected agencies or jurisdictions to be involved in formulating objectives and making strategic decisions. In such a case, one person should be the designated IC for purposes of conveying orders down the chain of command and receiving information back.

Good Teamwork requires
Good Communication.

Figure 16.2 Poor communication is a common source of problems during emergency operations.

Consolidated action plans

Any emergency operation requires a plan of action. A mental plan may be adequate for simple incidents. For complex or long-term operations a formal, written plan probably will be needed. The plan should address strategic goals, tactical objectives, and logistical requirements. The planning function is further discussed below.

Predesignated incident facilities

Several types of facilities are commonly used in areas where emergencies are being managed using the IMS. Examples include the following:

- The *command post* (CP) is the location from which all operations are directed. There should be only one CP per incident. In multiagency operations, representatives from all agencies involved should all be located at the command post.

- The *incident base* is the location on-scene where support activities are carried out. This will be the primary location of the logistics section of the IMS on scene.

- *Staging areas* are locations designated for resources that are not immediately assigned but must be available on short notice. Personnel and equipment are held at staging areas until ordered to deploy onto the incident scene.

Comprehensive resource management

Examples of resources utilized during emergency response operations include personnel, vehicles, equipment, and supplies. Under the IMS, all major resources are managed comprehensively. This means that the status and location of resources are monitored throughout the incident. Three typical status conditions are assigned, available, and out of service.

Functional Areas of the IMS

The IMS is designed to provide comprehensive coordination and control of all personnel and equipment operating within five functional areas (see Fig. 16.3). Each of the five functional areas is described below.

Command and command staff roles

Command is responsible for administering the entire incident management system. The incident commander (IC) is ultimately responsible for the entire response operation.

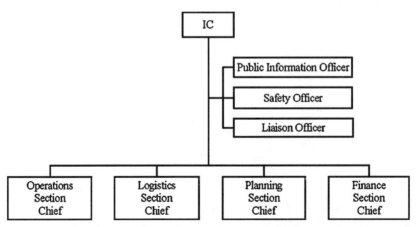

Figure 16.3 This diagram shows the organization of the functional sections and command staff roles within the IMS.

This includes responsibility for overseeing the assessment of the incident, formulating strategic goals for the response, and developing a plan of action. The IC must also ensure that the plan is successfully carried out. In addition to the command function, the IC is responsible for managing the other four functional areas of the IMS unless he has officially designated someone else to manage those areas.

In major response operations, the commander may require direct assistance from command staff personnel (see Fig. 16.3). Command staff roles include the following:

Public information officer.   The public information officer (PIO) is responsible for gathering accurate information about the incident and making that information available to the public. The PIO must be the only source through which such information is released. Important aspects of the PIO's role are providing a safe location for the media to gather, providing timely information about the incident, and arranging interviews and photo and film opportunities. This will keep members of the media from roving the scene and interfering with the response operation.

Safety officer.   The safety officer is responsible for the safety of the response operation. This includes ensuring that all hazards are identified, that the hazards are controlled or personnel are protected from them, and that all operations are carried out safely. The safety officer is responsible for advising the IC regarding safety issues and has the obligation to order the operation altered, suspended, or terminated if personnel involved are unduly at risk. For large or complex operations, the safety officer may require an entire staff to carry out his duties. A separate safety officer may be designated to operate within any branch, group, or division within the system.

Liaison officer.   The liaison officer is responsible for providing a point of contact for representatives of various agencies that may become involved in a response operation. In a major event, the liaison officer could be required to coordinate the involvement of agencies such as fire service, law enforcement, emergency medical service, private industry, and public works, to name just a few.

Planning

The planning section gathers, evaluates, and disseminates information about the incident and develops action plans for the response. If a preplan is available, it provides a basic information and planning framework that can be fine-tuned to develop the action plan. In some cases planning requires the expertise of outside experts such as engineers, chemists, or toxicologists.

Logistics

The logistics section provides the equipment, supplies, and services required for the rescue operation to function. Everything from ropes to cranes to communication equipment to breathing air to food, water, and toilet facilities for responders may be required to complete a rescue operation. The logistics function is the part of IMS that might be called on to provide items such as these on short notice at 2:00 on a Sunday morning.

Finance

The finance section is responsible for dealing with monetary matters related to the rescue operation. For example, completing a rescue operation may involve expenses ranging from rental fees for heavy equipment needed on short notice to providing food, facilities, and overtime pay for rescue personnel. All this adds up, and can be overwhelming for some response organizations. In many cases reimbursement is possible, as when a public-sector response agency performs a rescue for a private industry. Careful documentation of expenditures is required in all cases.

Operations

The operations section is responsible for all tactical operations at the rescue scene. The tactical operations are carried out in order to meet the strategic goals established by command. Examples of tactical operations include ventilating and isolating a space, making rescue entries, packaging and extricating a patient, and initiating patient transport. The operations section is the part of the IMS for which most of the information in this section of the book will be relevant. Operations will be discussed in more detail later.

Deployment and Operation of the IMS

The senior member of the first group of responders on the scene should establish command and assume the role of incident commander. That person will remain in the role of IC unless relieved by someone with more seniority who assumes command later as the response progresses. A command post (CP) location must be established as the IC's base of operations. The command post could be a special vehicle designed and equipped to be the nerve center of an emergency operation or any other response vehicle designated as the CP (see Fig. 16.4).

The IMS allows a response operation to be divided into functional or geographic subparts. This allows responsibilities to be delegated smoothly and enhances the overall efficiency of the system. The term *group* is used in reference to functional divisions, such as the rescue group of a response operation. The term *division* is used in reference to geographic divisions, such as the north division of a response operation. The term *sector* is used interchangeably to refer to both functional and geographic divisions. Geographic division or *sectoring* was incorporated into the tactical worksheet (Fig. 15.11) included and discussed in Chap. 15.

Any managerial responsibility that is not delegated to someone else remains with command. In responding to a very simple incident, the initial IC may be able to handle all supervisory duties related to operations, planning, logistics, and finance and act as safety officer at the scene. In major operations the IC's span of control will very quickly be overwhelmed unless some of her responsibilities are delegated to others early in the response operation.

Delegation of authority begins with the designation by the IC of one or more section chiefs who are assigned to assume control of the functional sections of the IMS (see Fig. 16.3). In turn, the functional sections may be split into branches, with a branch director appointed by the section chief to supervise activities within each branch (see Fig. 16.8). Likewise, each branch can be split into several groups, with a group supervisor, appointed by the branch director, in charge of each. Each group can be divided into units, with a unit leader appointed by the group supervisor to head up each unit. With a span of control of five, each section of the IMS can contain up to five branches, 25 groups, and 125 units.

Figure 16.4 A command post should be established at all incident scenes, and may be simply a designated response vehicle.

Whenever managerial authority for a part of the system is delegated or changes hands, an opportunity exists for information to be lost or a break in the chain of command to occur. To avoid this, anyone assuming a managerial role within the system should obtain a briefing from whomever authority is assumed, then maintain a log of activities and remain in the position until properly relieved through the system. All personnel will have specific duties to perform according to their assigned roles within the system.

To avoid confusion and ensure that all required duties are carried out, a strongly recommended practice is to provide checklists and worksheets to guide personnel in performing their required duties. Key players in the IMS should be identified clearly through the use of vests or other identifiers that indicate the wearer's role in the IMS (see Fig. 16.5). The locations of important incident facilities, such as the command post, staging areas, and incident zone boundaries, must be identified clearly.

Incident Management System in Confined Space Rescue

In this part of the chapter we will use some hypothetical scenarios to illustrate how the IMS might operate in confined space rescue operations. The scenarios vary from simple, single-agency operations to complex, multiagency operations. Our main focus is on the operations function of IMS, since that is the part of the system to which the information in this book mainly applies. Note that a simple, generic command structure is incorporated into the confined space rescue tactical worksheet (Fig. 15.11) included and discussed in Chap. 15.

Scenario 1: a simple single-agency response

Assume that a fire service engine company responds to a report of an unresponsive worker inside a municipal utility vault. The engine company consists of the company officer and a crew of four firefighters. A private ambulance is en route. The company officer and two of the crew members are paramedics and function at the operations level for confined space rescue. The other two crew members are EMT basics that function at the awareness level for confined space rescue.

On arrival, the company officer learns that the incident occurred during a permitted entry into the vault. The attendant attempted nonentry rescue after calling for help. He succeeded in cranking the entrant up out of the space using a tripod and winch with the cable attached to the entrant's harness; however, he ran out of lifting capability before

Figure 16.5 Basic equipment such as tactical worksheets, clipboards, and vests for identification of roles within the system enhance the effectiveness of the IMS.

Figure 16.6 The command structure for a single-agency rescue operation may be very simple.

getting the victim's legs clear of the space. The victim is now responsive but complaining of nausea and headache.

How would the incident management system be implemented for this incident? Figure 16.6 illustrates a likely system. The company officer establishes command on arrival on the scene and assumes the role of incident commander. A quick size-up reveals that the incident can be handled with the personnel already on the scene.

Evidence indicates that the entrant fell victim to an oxygen-deficient atmosphere, but revived after being pulled up into fresh air by the attendant. To bring the incident under control simply requires that the crew complete the process of removing the entrant from the space, provide appropriate emergency medical treatment, and transfer the patient to the private ambulance for transport to a medical facility. All equipment and supplies needed for the operation are on the scene.

Because of the simple nature of this incident and the limited number of personnel required to handle it, the IC is able to fill all managerial roles required to operate the IMS. This requires the IC to develop a simple mental action plan, oversee safety of the operation, and command the operations conducted by the engine company. No other functions or support staff roles other than command, planning, operations, and safety are required. In commanding the operation, the IC's span of control is never exceeded, so delegation of authority is never required.

Scenario 2: a complex single-agency response

The next scenario is similar to the first, except that on arrival the company officer learns that a flash fire has occurred in the utility vault. There are three victims in the vault who appear to be badly burned and are lying motionless 15 ft below the entryway. The company officer immediately establishes command and develops a strategy based on the first priority—life safety—as outlined in the pyramid of priority discussed in Chap. 15. A quick size-up indicates that additional assistance is needed, since the victims may die inside the vault if they do not receive help quickly.

In order to stabilize the incident—the second level in the pyramid of priority—and attempt a safe and successful rescue, the incident management system must be expanded. One member of the engine company is assigned to the safety officer position and begins sizing up the hazards of the confined space, the area around the space, and any additional hazards that rescue personnel will face. Additional resources are obviously needed, so the IC calls for help and prepares to organize and manage the incident. This includes requesting deployment of the special operations confined space rescue team, all of whom are paramedics who function at the technician level for confined space rescue.

In the meantime, size-up indicates that the atmosphere within the space is oxygen-deficient and still contains flammable gas. The remaining engine company crew members are assigned to begin ventilating the space to make it safe for entry and to increase the victims' chances for survival.

As the additional resources arrive, the IC directs their actions according to the strategy he has developed. As these critical resources and additional responders arrive, authority is delegated to maintain a reasonable span of control as tasks are assigned (see Fig. 16.7).

The operations position of the IMS is assigned at this time. The operations officer maintains a span of control when assigning the tactical groups for the incident. This allows the division of labor to be broken down into manageable parts.

From the time command is established at the beginning of the incident, the IMS can be expanded using its modular organization to meet the demands of the incident. As the command structure expands, personnel are placed in charge of other functions or roles

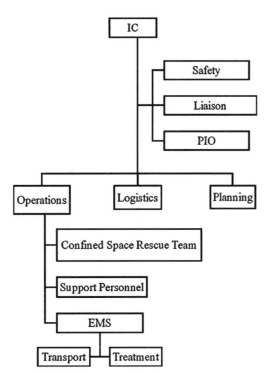

Figure 16.7   A complex single-agency rescue operation requires delegation of authority to main-tain a reasonable span of control throughout the IMS.

such as logistics, planning, public information officer, and liaison officer in order to maintain a reasonable span of control.

Command may change hands several times in the course of a response operation. For example, early in this scenario the fire department shift officer might arrive and assume the role of IC from the company officer. The new IC could then place the company officer in charge of the operations section of the IMS. As personnel continue to arrive and the IMS continues to expand, someone with more seniority or incident management experience may be placed in charge of operations. At that time the company officer could be placed in charge of the support personnel unit, which consists largely of his engine company and is involved in maintaining ventilation and other operations that support the confined space rescue team. The flexibility of the IMS allows command to be transferred and personnel to be reassigned from one role to another as required for completion of the operation.

Scenario 3: a complex multiagency response

For this scenario, assume that a major explosion has occurred at an industrial facility during a shutdown for annual refurbishing. A number of confined space entries by contractor employees were simultaneously under way at the time of the explosion. The spaces are interconnected and all were involved in the explosion to varying degrees. Initial response was by the facility's fire brigade, whose members also function at the technician level for confined space and rope rescue.

Command of the incident.   Initially the senior member of the onsite rescue team, who is also the plant safety director, establishes command and begins size-up. It is immediately apparent that additional help is needed, as 24 entrants are involved and most are seriously injured with some fatalities likely. Fires are burning in some areas and hazardous materials are being released. The plant IC immediately calls for help from the local fire department in accordance with the facility emergency response plan (ERP).

With the arrival of fire department personnel, the senior official in charge of the first arriving response unit assumes the role of acting IC in accordance with the ERP and begins to coordinate the involvement of off-site resources. The facility IC functions jointly with the acting IC under a unified command structure. Early in this scenario the acting IC requests mutual aid assistance from a neighboring municipality. A designated representative of the assisting agency is dispatched to the command post and works with the acting IC under the unified command structure. While command is shared between all three representatives, only the acting IC issues orders and receives reports back through the chain of command. This preserves unity of command and prevents needless confusion and duplication. The role of acting IC may change hands several additional times as personnel with greater seniority in the preestablished chain of command arrive on scene.

Command staff and major functional sections.   As more personnel arrive and the IMS unfolds, a command structure as shown in Fig. 16.8 evolves. In this scenario, it is important to have all command functions and support staff positions activated. This requires that the roles of safety officer, liaison officer, and public information officer be filled. Section chiefs are designated and placed in charge of the operations, logistics, planning, and finance sections. The section chiefs answer directly to the IC. Keep in mind that we

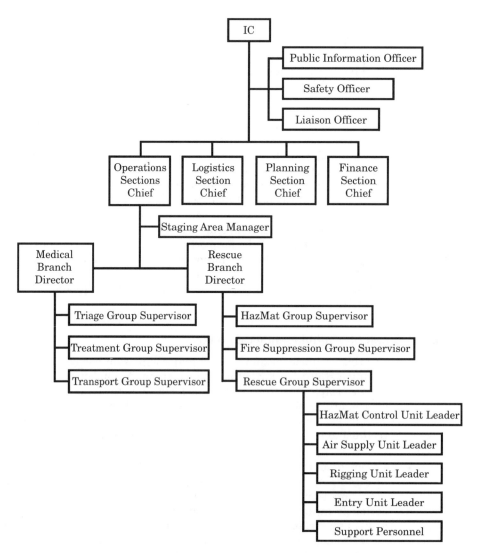

Figure 16.8 A complex multiagency rescue operation may require a complex command structure.

are focusing mainly on the operations section, and specifically the rescue group within that section because that is where the focus of this book mainly lies. In actually dealing with this scenario, the other functional sections also would need to be expanded significantly.

The operations section.   In this scenario the operations section is split into two major branches, the rescue branch and the medical branch, with a branch director placed in charge of each branch (Fig. 16.8). The branch directors answer directly to the operations section chief. A staging area manager is needed to control staged resources. This individual answers directly to the operations section chief.

The rescue branch.   In order to bring this incident under control, the rescue branch is divided into three groups as shown in Fig. 16.8: the hazardous-materials group, the fire suppression group, and the rescue group. In order to create a safe field of operations for the rescue group, the hazardous-materials group must bring released hazardous materials under control, and the fire suppression group must extinguish fires still burning in the area. A group supervisor is placed in charge of each group, and answers directly to the branch director. In an incident such as this it may be advisable to have a safety officer attached to the rescue branch to directly monitor and address safety issues affecting the rescue branch. The rescue branch safety officer would work directly with the rescue branch director and coordinate with the incident safety officer.

The rescue group.   The rescue group can be divided into several units as shown in Fig. 16.8. These units perform duties as described below:

- The hazard control unit performs ventilation, isolation, and other hazard control procedures as required to make the confined spaces and the area of operations safe to occupy.

- The air-supply unit ensures that adequate supplies of breathing air are available for rescue team entries.

- The rigging unit rigs up any rescue systems or other equipment required to transfer rescuers or patients.

- The entry unit makes entries into the spaces to access, assess, treat, package, and remove patients from the space.

- The extrication unit performs difficult extrication operations to free patients.

- The support personnel unit performs various duties as required to support the operations of the rescue group.

A unit leader is designated when each unit is activated. The unit leaders answer directly to the group supervisor.

The medical branch.   The medical branch is charged with handling patients involved in the emergency. Obviously this is an important aspect of bringing the scenario under control. The medical branch director oversees the medical branch and answers directly to the operations section chief. The medical branch is divided into three groups: triage, treatment, and transport. The triage group is responsible for assessing each patient's injuries and determining the order in which they should be treated and transported. The treatment group is responsible for providing emergency medical treatment at the incident scene. The transport group is responsible for providing or arranging transport to an appropriate medical facility for definitive care. A group supervisor is placed in charge of each group and answers directly to the medical branch director.

Rigging and Using Rescue Systems

Chapter 13 covered rigging and using basic systems, such as simple mechanical-advantage systems, that might be required for safe entry or for nonentry rescue during confined space operations. All the techniques described in Chap. 13 can also be used in performing entry rescue operations. Rescue team members may need to employ more complex systems in order to safely remove victims from confined spaces. This chapter covers procedures for rigging and using several rescue systems intended specifically for use in performing rescues.

There are a number of ways to accomplish the necessary ends in confined space rescue. This chapter describes some of the ways for rigging and using rescue systems; however, no two rescue situations are exactly alike. The specifics of each incident must be considered to determine the best way to rig and use rescue systems to solve the problem at hand. Training, experience, and professional judgment are required. A good approach is to remember the KISS mnemonic—keep it simple and safe—and select the simplest systems that will do the job safely.

Establishing Anchor Points for Rescue Systems

As noted earlier, before any system can be deployed for use in a confined space operation suitable anchor points must be established. All the information on anchors provided in Chap. 13 applies in rigging rescue systems. The information in that chapter dealt mainly with using structural anchors and engineered anchor points. Conventional anchors such as life-safety-rated-tripods, structural members, or certified eyebolts provide excellent anchor points for rescue rigging; however, rescuers may find that such anchors are unavailable in the timeframe required or simply will not work in the situation at hand. In order to perform a rescue under those circumstances, rescuers may be forced to use unconventional or improvised anchor points.

Aerial apparatus

One option that may be available to fire service personnel is the use of aerial platforms or aerial ladders to establish high anchor points. Some aerial units have the added advantage of being equipped with large breathing air cylinders that can be utilized in supplied-air systems if the rescuers are so equipped (Fig. 17.1). A vehicle is limited because it can provide a suitable anchor point only if it can be properly positioned in relation to the space to be entered.

In using fire service apparatus as an anchor, all the normal safety factors such as the angle of operation of boom or ladder, safe maximum load capacities, and proximity to power lines must be observed. It is important to realize that the apparatus itself is not used to move rescuers or patients into or out of the space. It merely provides a convenient point of attachment for the mechanical advantage or other systems used to move

Figure 17.1 Some aerial apparatus can provide a convenient high anchor point for rescue operations.

them. Take care to ensure that the apparatus is completely stable for as long as it serves as an anchor. Set brakes and chock wheels securely. Shut the apparatus down with warning lights and siren on so that they will provide an immediate warning if anyone attempts to crank the unit, or use some other tactic to prevent it from being activated.

Cranes and other powered industrial equipment

In some cases powered industrial equipment such as cranes or boom trucks can provide anchor points if other suitable anchors are not an option. It is critical to realize that the powered equipment is not used to move human loads into or out of a space. Patients or rescue personnel could easily be severely injured or killed by a relatively minor control error. Once the equipment is in position, shut it down and be sure that it remains stable. Chock the wheels. Disconnect the energy source to the starting system or lock and tag it out, or else post a guard to ensure that it is not tampered with during the rescue.

A-framed ladders

Anchor points can be improvised readily using ladders that are commonly available to both fire service and industry personnel. For example, a pair of ladders can be easily A-framed to provide an anchor point over a space. The A-framed ladders are erected and held upright by ropes running from the top of the A configuration to four anchors on the ground (see Fig. 17.2). This type of setup is highly portable and can be deployed in situations where other anchor options cannot. To set up an A frame quickly requires preplanning for equipment selection, good team organization, division of labor, and, most importantly, practice. At least seven team members are required to erect an A frame safely and efficiently.

A frames can provide a very high anchor point. This makes them useful in situations where a tripod or davit frame is too short, such as over manhole cones or well curbs. They may be used over open-top pits that are too wide to allow placement of a tripod. A-framing can be accomplished through the following steps.

Step 1: selecting and positioning ladders and guyline anchors. Select two ladders of adequate and similar length. Although 16- to 20-ft single ladders work well, any available length can be used as needed. Ideally the beams of one ladder should be slightly narrower than the other so that they will nest inside the beams of the other ladder as shown in Fig. 17.2a. If this is not possible the beams are offset instead of nested. Fire service ladders should meet applicable NFPA requirements. Other ladders used should meet all relevant OSHA and ANSI requirements, have a type 1A duty rating, and have a rated load capacity of at least 300 lb (1.33 kN).

Place the ladders on their beams and position them with the butts on opposite sides of the space. Ideally each butt is about one-fourth of the ladder length back from the center of the entryway to the space. Stated another way, the butts of the ladders are

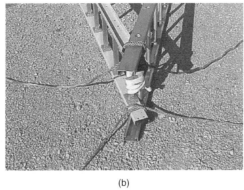

(a)

(b)

(c)

(d)

Figure 17.2 (*a*) To begin the A-framing procedure, place two ladders on the beams in the A configuration; (*b*) to attach the guylines to the top of the A configuration, form a clove hitch at the midpoint of a rope, place it over the ladder beams, and tighten it around the point where the beams cross; (*c*) to raise the A frame, two team members heel the ladder butts while one or two others lift the tip of the A configuration into an elevated position and two team members at the far guy anchors hoist the frame upright with the guylines; (*d*) as the A frame is pulled fully upright by the two team members at the far guy anchors, two team members at the near guy anchors tension their guylines so that the frame is held erect by the four lines under tension.

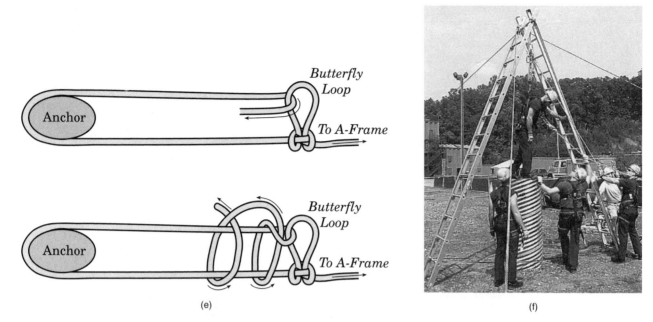

(e)   (f)

Figure 17.2 (*Continued*) (*e*) The trucker's hitch, or variations of it, can be used to tension and tie off the guylines to hold the A frame in the vertical position; (*f*) a high anchor point can be established by girth-hitching webbing loops around both ladder beams on each side of the A frame to form a load-sharing anchor system.

about one-half the ladder length apart with the entryway to the space centered between them. Cross the tips of the ladders and tie two rungs together with a piece of webbing or rope at the point where the beams cross (see Fig. 17.2*a*). If roof ladders are used, fold out the hooks and place them so that the hooks of one ladder hook over the top of the other ladder.

Establish four anchors for the guylines. In some cases it may be possible to use existing items such as utility poles (see Fig. 17.2*c, d*), fence posts, or structural members as anchors. Temporary anchors can be established by driving items such as metal tubing, large wooden stakes, or steel fence posts or signposts (see Fig. 17.2*c, d*) into the ground with a sledgehammer. They should lean well back away from the space to keep the guy ropes from sliding up the anchor posts when tensioned. These anchors don't have to be bombproof, because the guy ropes don't have an excessive amount of tension on them. All they have to do is keep the ladders balanced. The force of the load is on the ladder beams, not the guy ropes.

Position the guy anchors roughly at the corners of a square with the space near the center of the square. Ideally each post is positioned roughly one ladder length back from the center of the entryway, with the distance along each side of the square running about two ladder lengths.

Step 2: attaching the guy ropes to the A frame.   Two guy ropes are required to erect the A frame. Utility ropes are fine because the guy ropes don't see a lot of tension, as discussed above; however, they must be of adequate length for the ladders being used. For example, at least 120 ft of rope is recommended for A-framing 20-ft ladders. It is much better to have extra rope than not enough.

Figure 17.2*b* shows the steps in attaching the guy ropes to the A frame. To rig the rope to the ladder, make a clove hitch at the midpoint of one guy rope. Place the clove hitch over the upper tips of both the upper ladder beams. Tighten the hitch around the ladder beams at the point where the two beams cross. Orient the clove hitch so that the side of the hitch on which the lines cross is positioned on the inside of the ladder beams so that tension on the guy ropes will keep the clove hitch tight. Two team members each take one line from the clove hitch and assume positions near the anchor points on the far side of the entryway from the ladder tips. They pull the slack out of the guy ropes and then stand by.

Next make a clove hitch at the midpoint of the other guy rope. Lift the lower ladder beams just clear of the ground and slip the clove hitch over the tips of both lower ladder beams. Tighten the hitch around the beams at the point where they cross. Orient the clove hitch so that the side of the hitch on which the lines cross is positioned on the inside of the beams. Two team members each take one line from the clove hitch and assume positions near the anchor points on the near side of the ladder tips. They pull the slack out of the guy ropes and then stand by.

Step 3: raising the A frame, tying off the guy ropes, and stabilizing the bases. When the ladder is ready to be raised, have a team member heel the butt of each ladder (see Fig. 17.2c). To begin the raise, a team member lifts the tip of the A frame up above head height and begins moving toward the team members at the heel position, raising the A frame higher. Preferably, two team members raise the tip of the A frame, especially if heavy ladders are used. As the tip is being raised, the two team members positioned near the guy anchors on the far side of the entryway from the tip begin hauling on their guy ropes and continue to haul, thereby pulling the A frame upright. While this is going on the team members at the guy anchors on the near side of the tip feedout line as the A frame is raised, but begin applying tension to the lines as the A frame nears a vertical position. This prevents the A frame from being pulled all the way over by the team members pulling from the opposite side.

Once the A frame has been pulled into a vertical position (Fig. 17.2d), the four team members on the guy ropes maintain equal tension to keep the A frame upright. While doing so they tie off all four guylines to the anchors. Use the trucker's hitch (Fig. 17.2e) to tension and tie off the guylines to the anchors. Before tensioning and tying off the guylines, check the positions and angles of the ladders. Incline both ladders toward each other at an angle of roughly 75° to the ground. At this angle the ladders provide the best stability and strength. This angle also allows for easy climbing if required to attach system components to the A frame.

Once the A frame is properly positioned and the guylines are tight, secure the base of each ladder. If the only major force on the A frame will be directed straight down from the tip, it may be adequate to simply tie the ladder bases together with rope or webbing to prevent them from "kicking out" under load (see Fig. 17.2f). The best practice is to secure the ladder butts against movement in any direction. This is accomplished by driving suitable stakes or posts into the ground and tying the bases of the ladders to them with webbing or ropes.

Step 4: rigging systems to the A frame. Once the A frame is erected, it is easy to climb up either side to attach a mechanical-advantage system, change-of-direction pulley, or other components as needed. To begin with, establish a suitable point of attachment. One way to do this is to girth-hitch a doubled webbing sling around both ladder beams on one side of the frame so that a doubled webbing loop of about 18 in hangs down into the center of the frame. Repeat the same process with an equal-size sling of webbing on the opposite beams of the A frame. Hook a carabiner through the webbing slings hanging down into the center of the frame to create a load-sharing anchor system (see Fig. 17.2f). Each webbing sling must be configured to provide adequate strength for either light use or general use, depending on the intended load, as discussed in Chap. 13.

Note that this setup places the load mainly on all four of the ladder beams rather than any of the rungs. This is important because the beams are significantly stronger than the rungs. Other attachment points as needed for directional pulleys or rope-grabbing devices can be rigged wherever needed, using webbing slings that are girth-hitched or basket-hitched around the beams. Whenever hoisting operations are carried out with an A frame, avoid applying lateral pulling forces to the top of the frame, as these may have the effect of trying to pull the frame over. A good practice is to try keep all pulling forces directed between the ladders.

As an alternative to hanging a system on the A frame after it is raised, the system can be attached before the A frame is raised (see Fig. 17.2c). If life-safety ropes are used as guylines, they can be attached to the ladder tips in such a way as to leave large butterfly loops hanging down when the A frame is erected. In that case the two rope loops are used as a load-sharing anchor system instead of the webbing loops used in our example.

Gin ladders

Another type of improvised elevated anchor is a gin pole rigged using a ground ladder, as shown in Fig. 17.3. The gin ladder configuration is not as strong as the A frame, but it has the advantage of being useful when only one ladder is available. It can be used when rescue must be made from a space that is too wide to allow an A frame to be used, such as a large open-top pit.

Most of the general guidelines for selection of equipment as discussed for A framing apply to gin ladders, but the guy ropes and anchors used in the gin ladder setup will see significantly more of the load than those used with the A frame. Life-safety-rated rope is recommended, and strong anchors should be utilized. A gin ladder is rigged as shown in Fig. 17.3 and described below. At least four team members are required to erect the gin ladder. Preplanning and practice are required for safe and fast gin ladder deployment.

Step 1: positioning ladder and guyline anchors.   Start the process by placing the ladder flat on the ground, positioned so that the ladder tips point away from the space and the butt is located a safe working distance back from the space. Establish two anchor points for the guy ropes. From the butt of the ladder, locate the anchors roughly one ladder length back and one-half a ladder length over (see Fig. 17.3). Angle driven anchors well away from the butt location.

Step 2: attaching the guy ropes to the ladder.   Attach the guylines by tying two butterfly knots roughly the width of the ladder beams apart near the midpoint of the guy rope. Slip one butterfly knot over each ladder tip.

Step 3: raising the gin ladder, tying off the guy ropes, and stabilizing the base.   To raise the gin ladder at least one team member heels the ladder while another lifts the ladder tip over her head and begins walking toward the butt, raising the ladder progressively higher as she goes. At the same time team members located at the guy anchors feed rope out to allow the ladder to be raised but keep slack out of the ropes. The team member raising

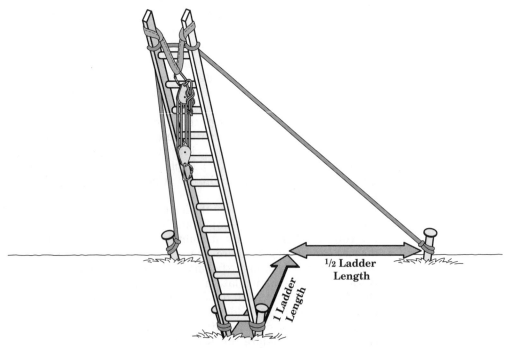

Figure 17.3   A gin ladder can be rigged to provide a high anchor point over open-top spaces too wide to allow the use of A frames.

the ladder continues until the ladder is vertical, then pushes it past the plumb point so that the weight shifts onto the guylines. The team members at the two anchor points then lower the ladder with the guylines until the tip is positioned over the entrance to the space. Once the ladder is at the proper angle, tie off the guylines at the anchors. Use a clove hitch with a safety knot. Drive two vertical anchor posts adjacent to the beams at the base of the ladder and tie them to the beams with rope or webbing to stabilize the base of the ladder.

Step 4: attaching systems to the gin ladder.   Mechanical-advantage systems or other components can be anchored to the top of the ladder using loops of webbing as shown in Fig. 17.3, much as described above for the A frame. From a safety standpoint the best practice is to attach components at the top of the ladder before it is raised.

The angle of inclination of the gin ladder is an important consideration. The flatter the angle (meaning the smaller the angle between the ladder and the ground), the more tension the guylines and anchors will be subjected to in handling a given load. The angle must be small enough to provide enough "reach" to position the top of the ladder over the space, while maintaining a safe working distance between the space and team members working around the base of the ladder. Never apply lateral pulling forces to the top of the gin ladder because it has little lateral stability. All pulling forces on the top of the ladder should be directed downward.

Rappelling

Rappelling is a technique for descending a fixed line in which the rappeller uses a friction control device, such as a brake bar rack or a figure 8 descender, attached to the rappeller's harness to control the rate of descent down the line. Rappelling is not a very commonly used technique in confined space rescue, although in some cases it may be required to access the patient. If a rescuer must be lowered during confined space operations, it is generally preferred that his descent be controlled by another team member from a secure location using a lowering system, as described below.

Rappelling is a very valuable technique for training in the use and operation of descent control devices. Skill in operating these devices is critical in operating lowering systems, as we will see later in this chapter. Rappelling is also a good way to build general confidence in working with rope systems. It has the added advantage of being fun! The following discussion assumes that four or more team members involved in a training session will rotate through all the roles described below.

Rigging a rappel station

All team members should wear class III harnesses equipped with front waist D rings. Provide fall protection as needed for all team members, especially if working around unguarded edges. All team members must wear helmets and suitable rope gloves.

Use life-safety-rated static kernmantle rope for rappel lines and safety belay lines. Use light-use-rated ropes for one-person loads only. Use general-use-rated ropes for either one-person or two-person loads, as discussed in Chap. 11.

Like any other operations involving rescue systems, rappelling requires that good anchor points be established—in this case for both the rappel line and a top safety belay line. Pad any rough or angular edges contacted by ropes.

Establish an anchor or anchor system for the rappel line as described in Chap. 13. If possible, use a high-point anchor so that the rappel line is attached above the rappeller's head, as this makes loading onto the rappel line easier. A long-standing practice is to put a simple figure 8 stopper knot in the end of the rappel line. This keeps the rappeller from being able to rappel off the end of a short rappel line, as has happened in the past. Feed out the rappel rope over the edge until the end of the rope reaches the ground below, with an additional 15 ft (4.5 m) of rope fed out to allow bottom belay to be performed. Attach the rappel line to the anchor with a suitable anchor knot.

If it is feasible, use a separate anchor or anchor system for the top safety belay. Any of the safety belay techniques described in Chap. 13 can be used.

Personnel involved in rappelling operations

In order to operate smoothly and safely, rappelling operations must be well organized. This requires a specific division of labor. Ideally the following personnel roles should be involved in all rappelling operations:

- The rappel team leader is responsible for directing and overseeing the safety of all rappel operations.

- The rappeller is the person who actually descends the rappel line.

- The top safety belayer is responsible for tending the safety belay line attached to the rappeller and using it to arrest any fall that should occur.

- The bottom belayer guards the bottom of the rappel line and uses tension on the line to arrest a fall or lower the rappeller to the ground, if needed.

Preparing to rappel

Before the rappeller approaches the edge, the top safety belayer attaches the safety line to the dorsal D ring of the rappeller's harness. A good practice is to install a shock absorber between the rappeller and the end of the safety belay. The safety belayer leaves enough slack in the line to allow the rappeller to go over the edge, and holds the safety line in the locked-off position line. Rigging and operation of the safety belay line were covered in Chap. 13.

Throughout the entire rappelling operation, the rappel leader oversees the safety of everyone involved. If an unsafe condition occurs, he yells "Stop!" and all personnel immediately cease operations until the unsafe condition is eliminated. Likewise, anyone on the team who observes a safety hazard immediately yells "Stop!" to bring the operation to a halt. In reply, all team members should repeat the exclamation "Stop!" to verify that they have heard the command.

In the event that tools, equipment, or other objects fall, the term "rock" is used to warn personnel at lower levels. This use of the term is borrowed from wilderness rescuers, who use it to warn others when loose rocks are dislodged. If you hear someone yell "rock" above you, remain under any available cover and above all else, don't look up!

Rappelling using the brake bar rack for descent control

As discussed in Chap. 11, the brake bar rack offers definite advantages over the figure 8 descender as a descent control device. The figure 8 is also commonly used, and will perform adequately in many situations. In this subsection we will describe the rappelling procedure using the brake bar rack for descent control. In the next subsection we will cover use of the figure 8 descender. Most of the following information applies regardless of which type of descent control device is used.

Roping in the brake bar rack. All instructions given here assume that the rappeller is right-handed and that the rack is set up for a right-handed rappeller, as seen in the accompanying figures. For left-handed rappellers, reverse the orientation of the bars on the rack and reverse the following instructions regarding use of the right hand versus the left hand.

Attach the rappel rack to the front waist D ring of the rappeller's harness, as shown in Fig. 17.4a. Position the carabiner attaching the rack to the harness so that the gate is pointed toward the rappeller and the locking ring screws downward into the locked position. This puts the gate and locking ring in the most sheltered position available. The team leader must pay special attention to ensure that the carabiner remains in that position. Any time the rappel line is unloaded, it is very easy for the carabiner to flip around so that it is side-loaded or the gate is pointed out when it is loaded again.

When the rappeller is ready to rope in the brake bar rack, he notifies the rest of the team. This may be accomplished by shouting a command, such as "on rappel." In a high-noise environment a prearranged hand signal, such as raising the hand to the head in the fashion of a salute, can be used. Before proceeding, the rappeller must receive verbal replies and/or hand signals from the other team members indicating that they are in position and that it is safe to proceed. The top safety belayer issues a reply, such as "safety on," indicating that she is set up to use the safety to arrest any fall that may occur from that point onward. Likewise, the bottom belayer replies "belay on," indicating that he is guarding the bottom of the rappel line.

Next the rappeller ropes all six bars into the brake bar rack, as shown in Fig. 17.4. If a high-point anchor is used, the rappeller begins by moving the rack as far up the rappel rope toward the anchor as possible before roping in the bars (see Fig. 17.4b). If a low point anchor is used, the rappeller pulls all slack out of the rappel line and pinches the line between the left index finger and thumb at a point just beyond the edge to be negotiated (see Fig. 17.4c). He holds the "index point" at the end of the brake bar rack to begin roping in (see Fig. 17.4d). With this technique, the end of the rack should fall just beyond the edge when the rappeller loads onto the rappel line. In some cases, the index

(a)

(b)

Figure 17.4 (a) Before the brake bar rack is roped into the rappel line, attach the rack to the harness in the proper orientation (b) to begin roping in for a high-anchor-point rappel, move the brake bar rack as far as possible up the rappel line toward the anchor.

Figure 17.4 (*Continued*)   (*c*) Before roping in for a low-anchor-point rappel, index the rappel line to the edge to be negotiated; (*d*) place the index point on the rappel line at the end of the brake bar rack prior to roping in for a low anchor point rappel; (*e*) to rope in the brake bar rack, rotate each brake bar into the rack and use the rappel rope to hold each bar in place; (*f*) rope all six bars into the rack initially.

point may need to be "fudged" slightly to the anchor side of the edge to compensate for the rappeller's weight and stretch in the system.

The friction provided with all six bars roped in is ample to keep the rappeller stationary after loading onto the rappel line. For training purposes it is still a good practice to lock off the rack and then tie it off before going over the edge. Figure 17.5 shows how to lock and tie off the newer-style brake bar rack. Figure 17.6 shows how to lock and tie off the older-style brake bar rack.

Going over the edge.   After the brake bar rack is properly roped in and tied off, the rappeller again uses a voice command such as "on rappel" or a hand signal to notify the

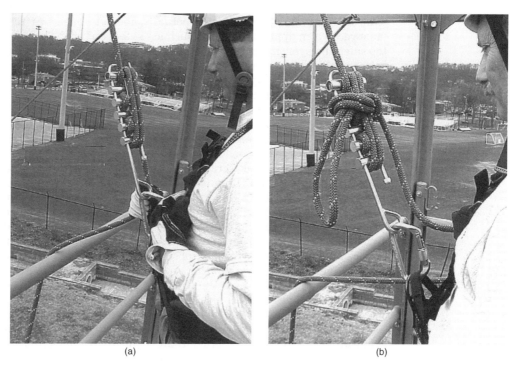

(a)   (b)

Figure 17.5 (*a*) Lock off the newer-style brake bar rack by using the control hand to wrap the rappel line twice around both the sixth bar and the "tie off" extension of the first bar; (*b*) after the newer-style brake bar rack is locked off, it can be tied off by using a bight of the control rope to tie an overhand knot around the rack.

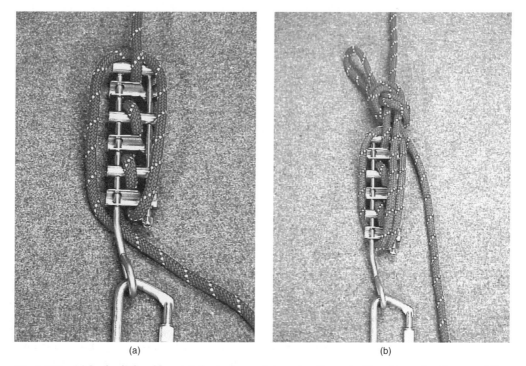

(a)   (b)

Figure 17.6 (*a*) Lock off the older-style brake bar rack by using the control hand to wedge the rappel line between the first bar and the loaded rappel line, then wrapping the line around the sixth bar; (*b*) after the older-style brake bar rack is locked off, it can be tied off by using a bight of the control rope to tie an overhand knot around the loaded rappel line just above the rack.

team members that the rappel is actually about to begin. At this point the rappeller must receive an appropriate reply such as "safety on" or the proper hand signal from the top safety belayer, indicating that she is ready to use the safety line to arrest a fall. The bottom belayer issues a reply, such as "belay on" or the proper hand signal, indicating that he is in position to perform ground belay if needed to arrest a fall or control the rappeller's descent. The rappel team leader conducts a full safety check. This includes checking for proper rigging, ensuring that all carabiners are locked, being sure that all required safety equipment is being used, ensuring that harnesses are properly donned, and ensuring that all personnel are alert and in position to proceed. The rappel team leader then gives the rappeller the go-ahead to go over the edge and load the rappel line.

The rappeller must take great care when going over the edge to avoid losing control and shock-loading the system. Never jump over the edge to load the rappel line! The idea is for the rappeller's weight to be smoothly loaded onto the line. This is easiest to perform when a high anchor point is used. When a low anchor point is used, the rappeller carefully climbs out over the edge and slowly shifts the load from his hands and feet to the rappel line. Take care not to mash fingers or hands between the rappel line or rack and the edge. For the sake of simplicity, we will assume that the edge to be negotiated in our example is a handrail.

Once the rappeller is over the edge and positioned with both feet against the structure and is leaning back, all his weight should be on the rappel line (see Fig. 17.7a). His knees are slightly bent with feet about a shoulder width apart. One of the most difficult aspects of learning to rappel is learning to trust the system. Continuing to clutch a handrail serves no purpose other than tiring the rappeller's arms before the descent begins. A good practice is for the rappeller to initially hold both hands shoulder high after loading the system. This provides confidence in the brake bar rack rather than the rappeller's hands for support.

Unlocking the rack and beginning the rappel.   Before beginning the rappel, the rappeller unties and unlocks the rack and drops one bar out of the rack, leaving five bars roped in as appropriate for a one-person load. The rappeller then cups the left hand around the two bottom bars in the rack and holds the rappel line in the right hand positioned behind the right hip, as shown in Fig. 17.7a. To begin the descent, the rappeller reduces the friction on the rope by moving the two bottom bars down, or toward the rappeller, and loosening the grip on the rope to allow it to begin sliding through the right hand.

Making a controlled descent with the rack.   Once the rappel is under way, the rappeller can vary the speed of descent in one of two ways. Pushing the bottom two brake bars up slows the descent, and spreading them further apart makes it faster. Putting more tension on the rappel line below the rack with the right hand slows the descent, and decreasing the tension on the line increases the speed of the descent.

The key word in rappelling is control. Long, fast drops punctuated by swinging into and rebounding off of the side of the structure should be avoided in rescue operations and training. Instead, to the maximum extent possible, the rappeller tries to walk backward down the side of the structure. He tries to maintain a horizontal orientation, with a 90° angle between his body and the side of the structure (see Fig. 17.7b). This minimizes the chance that his feet will slide downward and out from under him, causing a "pendulum swing" into the side of the structure that can result in loss of control of the rappel or injury to the rappeller.

At any point in the rappel the rappeller can stop the descent by simply pushing the bars on the rack together and gripping the rope tightly with the right hand. The bar can then be secured to free up both the rappeller's hands by placing the sixth bar back into the rack and locking and tying off the brake bar rack, as previously shown. After doing so the rappeller issues the command "locked off" to notify the bottom belayer that he can relax. Before untying and unlocking the rack, the rappeller issues the command "unlocking" to notify the bottom belayer to resume his duties.

Negotiating edges during the descent.   One tricky situation that the rappeller must deal with occurs while encountering an edge, such as an opening or recessed surface in the

(a)

(b)

(c)

Figure 17.7 (*a*) Once over the edge, the rappeller's weight is on the rappel line, not the arms—both hands can be used to control the rate of descent when the brake bar rack is used; (*b*) during the controlled descent, the rappeller walks backward down the side of the structure while keeping his body horizontal or perpendicular to the side of the structure; (*c*) negotiating edges during a descent may require the rappeller to invert to avoid a pendulum swing into the side of the structure.

side of the structure. If the rappeller continues to walk backward down the side of the structure with his body perpendicular to it, he will pendulum swing into the opening when his feet cross the edge. This can result in loss of control of the rappel or personal injury.

To deal with edges during a rappel, the rappeller stops walking when his feet reach the edge, but continues to lower the upper part of his body with the rack. This process is sometimes referred to as "inverting" because the rappeller's upper body actually drops below the level of his feet (see Fig. 17.7*c*). The feet are centered on the edge, which serves as a pivot point for the rappeller. As the rappeller begins to invert, he bends the knees and spreads the feet further apart. All this has the effect of moving the rappeller's body below the edge and closer to the side of the structure, which reduces the possibility of swinging into the structure. Once in an almost inverted position, the rappeller can remove his feet from the edge and safely rotate into an upright position without slamming into the side of the structure.

In many cases rotating out of the inverted position brings the rescuer's feet into contact with a lower component of the structure so that he can reassume the horizontal orientation and continue walking down the side of the structure. In other cases, as in the case of negotiating a cantilevered ledge, the rappeller may have to continue all the way to the ground without ever being able to get his feet back in contact with any part of the structure. This process is referred to as a "free rappel."

Going offline at the end of the rappel. Once the rappeller reaches the ground, he bends his knees and squats while continuing to feed rope through the brake bar rack. Then, when he stands up there will be plenty of slack in the rappel line, making unroping the rack easier. He gives a hand signal or a verbal command such as "off rappel" to let everyone know that the rappel is over. The ground belay operator then assists the rapeller in unfastening the safety belay line and disconnecting from the rappel line.

Rappelling with two-person loads using the rack. In the event that a second person needs to be loaded onto the brake bar rack, the rappeller would simply leave all six bars in the rack after untying it to begin or resume a descent. This versatility is part of the reason that the brake bar rack is preferred over the figure 8 descender for operations where descent control is required.

Rappelling with a two-person load is not a typical operation for confined space rescue. It may be required for some structural rope rescue operations, such as rescuing workers suspended by fall-arrest lines from the side of a structure. Examples of these operations include pickoff and line transfer techniques. These techniques are beyond the scope of this text, but are well covered by other sources that specifically address industrial structural rescue (Roop et al. 1998) or general rope rescue (Frank 1998).

Rappelling using the figure 8 descender for descent control

Rappelling using the figure 8 descender is very similar to rappelling using the brake bar rack; however, there are a few significant differences. As discussed in Chap. 11, the figure 8 with ears is the preferred type of figure 8 for use in rescue operations. The following discussion will assume that a figure 8 with ears is being used.

Roping in the figure 8 descender. Rope in the figure 8 descender as shown in Fig. 17.8a for a one-person load. The descender is double-wrapped for a two-person load as seen in Fig. 17.8b. The descender must be roped in for either a one- or two-person load before a rappel begins. Unlike the brake bar rack, the figure 8 descender does not allow the rescuer to make this sort of adjustment once the device is loaded.

Going over the edge. In preparing to go over the edge when a high-point anchor is used, hold the figure 8 descender locked off as shown in Fig. 17.8c, as the edge is negotiated. If a low-point anchor is used, this method may cause the fingers or hand to be mashed between the rappel line and the edge when the line is loaded. When using a low-point anchor, a better option may be to simply hold the rappel line tightly with the control hand (i.e., the strong-side hand) while using the other hand and both feet to negotiate the edge and load onto the rappel line. It is important to keep a strong grip with the control hand. If you release the grip on the rappel line while negotiating the edge with the figure 8, you will plummet. This is one of the disadvantages of using the figure 8 compared to using the brake bar rack with all the bars roped in.

Making a controlled descent with the figure 8 descender. Once the edge has been negotiated, the rappeller should be positioned for comfort and control. A common practice is to place the control hand behind the strong-side hip with the other hand loosely gripping the rappel line above the descender for extra stability (see Fig. 17.8d). This posi-

(a)   (b)

(c)   (d)

Figure 17.8 (*a*) Wrap the figure 8 descender once when roping in for a single-person load; (*b*) wrap the figure 8 descender twice when roping in for a two-person load; (*c*) the figure 8 descender can be held locked off as shown while negotiating the edge for a high-anchor-point rappel; (*d*) in rappelling with the figure 8 descender, the rate of descent is controlled by the control hand only.

tion provides maximum friction at the descender. The rappeller begins the descent by reducing the friction on the rappel line and allowing it to begin sliding through the control hand. The off-side hand maintains a loose grip above the descender for balance and slides down the rope during the descent.

It is important to realize that the rate of descent is controlled with the control hand alone. The grip of the right hand on the rappel line is loosened to make the descent go faster and tightened to slow or stop the descent.

One factor to consider when using the figure 8 descender is that the weight of the rope in the rappel line below the control hand actual helps the rappeller control the rate of descent. This is especially true when a significant length of rope is involved, such as at the beginning of a long rappel; however, this additional control is progressively lost as the rappeller descends the rope. Because of this, the rappeller may have a natural tendency to go faster and faster throughout the rappel. In some cases this has caused the rappeller to lose control. For this reason, and because of the potential for significant heat buildup, the brake bar rack is preferred over the figure 8 descender for long rappels. Some authorities recommend that the use of the figure 8 descender be limited to rappels and lowers of less than 100 ft (30 m).

Locking off and tying off the figure 8 descender. Lock off the figure 8 descender as shown in Fig. 17.9*a, b*. Tie off the descender as shown in Fig. 17.9*c* to hold the load in order to free both hands. Untie and unlock the descender by simply reversing these

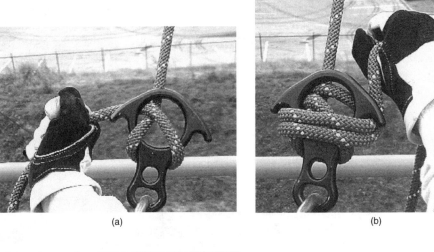

(a) (b)

(c)

Figure 17.9 (*a*) To begin locking off the figure 8 descender, the control hand wedges the control rope between the descender and the loaded rappel line at the top of the descender (the rope must "pop" into place for a good lock off); (*b*) to complete the lockoff process, wrap the control rope 2 or 3 three times around both ears and the top of the descender; (*c*) after locking off, tie off the figure 8 descender by securing a bight of the control rope to the loaded rappel line with an overhand knot.

steps. Since the figure 8 descender doesn't provide as much control as the brake bar rack, it is very easy to move further down the rappel line inadvertently while attempting to lock off and after unlocking the descender. For this reason, it is recommended that the control hand be allowed to move with the rappel line to a position near the figure 8 descender in preparation for locking off the descender and being positioned close to the descender on the rappel line when the descender is unlocked.

Team member functions during rappels

So far we have devoted most of our attention to the activities of the rappeller during rappels; however, the actions of the other team members are very important for safe and efficient rappels. These actions are described below.

Operating the bottom belay. During rappels the bottom belayer is located back from the base of the structure just beyond the landing zone of the rappeller. The belayer stands facing the rappeller with the feet a shoulder width apart and the weak-side foot closer to the rappeller. The rappel line is positioned underneath the belayer's arms and held with the slack end held in the weak-side hand and the loaded side held in the strong-side hand, as shown in Fig. 17.10a. About 18 in of slack should be maintained in the belay line. This provides enough slack to prevent any interference with the rappeller's operation of the descent control device, but also allows the bottom belayer to tension the rope immediately if needed.

In the event that the rappeller loses control, the bottom belayer must react immediately. The belayer puts both hands together, grasps both ropes, and pivots his entire body 180° while pulling down on the rappel line as seen in Fig. 17.10b. The bottom belayer can pivot to either side. This technique has the effect of wrapping the rappel line around the belayer's body to immediately tension the line, and is often referred to as a "body belay." Tensioning the bottom of the rappel line increases the friction at the brake bar rack or figure 8 descender. This arrests the fall of a rappeller who has lost control of the rappel, but only if it is performed instantly. The bottom belayer must be attentive constantly. The belayer must be prepared for a hard pull through the rappel line when the fall is arrested. Immediately after arresting the rappeller's fall, the bottom belayer can lower the rappeller all the way to the ground by varying the tension on the rappel line.

The bottom belay technique may be effective in arresting the fall of a rappeller who has lost control of the rappel; however, it is worthless if a fall occurs as a result of failure of the rappel line or the rappel line anchor. In such a case the only way to arrest the rappeller's fall is through proper use of the top safety belay.

Operating the top safety belay. Throughout the rappel the top safety belay operator feeds slack through the hitch or belay device that serves as the "emergency brake" if the rappeller falls (see Fig. 17.10c). Applicable belay techniques were described in Chap. 13. The top safety belayer provides enough slack to avoid interfering with the rappel but must be sure not to allow excessive slack, which will cause a shock load if the rappeller suddenly loads onto the safety line. A maximum of 18 in (46 cm) of slack in the safety line is recommended. The use of shock absorbers between the rapeller's harness and the end of the safety line is recommended as a way to prevent shock loads, especially during training. The top safety belay operator must be ready to take action to stop a falling rapeller at any time during rappel operations. This was described in Chap. 13.

Rappel team leader. In addition to overseeing the safety of the operation, the team leader is responsible for ensuring that the operation runs smoothly. One important task is to direct the top safety belay operator so that a proper amount of slack is maintained in the safety line throughout the rappel, since the safety line operator usually cannot actually see the amount of slack in the safety line once a rappel is under way. This is accomplished using voice commands such as "more slack on safety" or "less slack on safety." In situations where voice commands are not effective, simple hand signals can be used (see Fig. 17.16).

(a)

(b)

Figure 17.10 (*a*) The bottom belayer must remain alert and ready to act immediately during rappels. (*b*) If the rappeller loses control of the descent, the bottom belayer immediately tensions the rappel line to arrest the rappeller's fall.

Rapeller self-rescue during emergencies

As in all activities, emergencies can occur during rappels. We have already discussed the roles of the top and bottom belayers in preventing and arresting falls of the rappeller. We also need to look at a couple of options the rappeller has to avoid trouble.

Emergency leg wrap. The leg wrap is a technique borrowed from ironworkers. The rappeller can use it in a last-ditch effort to arrest an uncontrolled descent down the rappel line. If all the safety precautions mentioned above are utilized, this should never happen; however, such an event could occur. For example, if the rappeller began racking in the rappel rack before the top safety line was attached and fell or was pulled over the edge before a sufficient number of bars were racked in, no descent control or bottom belay would be possible.

To perform a leg wrap, the rappeller simply extends the strong-side foot and swings the strong-side leg in a clockwise circle around the rope as he falls. The hope is that the

Figure 17.10 (*Continued*) (*c*) The top safety belay line provides a completely independent fall-arrest system in the event that the rappeller or the rappel line fails.

(c)

rope will wrap around the rapeller's leg so that the leg can serve as a friction device. Bending the leg at the knee may provide additional friction. The leg wrap must be practiced in advance so that it can be performed effectively in an emergency. If successfully performed in an actual emergency, the leg wrap will produce severe rope burns, but it can slow or stop an uncontrolled descent and may save the rappeller's life.

Ascending self-rescue. Another emergency occurs when the descent control device becomes jammed during a rappel. This has happened when the rappeller's shirt or glove was pulled into the device, leaving it hopelessly jammed. In such a case the rappeller must resist any temptation to attempt to cut the material out of the device. Contacting a loaded rope with the edge of a knife can sever the rope completely.

Clearing a jam from the rack requires that the rappeller's weight be unloaded from the rack. This is accomplished by ascending the rope just enough to unload the rack, using a rope-grabbing device installed on the rappel line above the rack or descender. The rope-grabbing device must be equipped with a long loop into which the rappeller's foot is placed after the device is installed on the rope. The rappeller uses his leg muscles to take his weight off the rack, leaving his arms free to deal with the problem.

Ascending self-rescue can be accomplished with a Gibbs ascender on the rappel line above the rack. The ascender should either be equipped with an etrier or a webbing loop roughly equal in length to the distance between the rappeller's feet and chest. The etrier is a length of webbing with four or five sewn loops that serve as steps (see Fig. 17.11*a*). Once the ascender is installed, the rappeller places a foot into the webbing loop and straightens his leg to unload the rack, allowing it to be cleared.

A convenient alternative to this procedure can be performed using loops formed from lengths of 7- or 8-mm-diameter accessory cord tied end to end with the double fisherman's knot described in Chap. 12. A loop of this type can be installed on the rappel line above the rack using a double-wrapped Prusik hitch. The length of the Prusik loop should be roughly equal to the distance between the rappeller's feet and chest. For convenience, two loops of roughly half that length can be used, with the first loop attached to the rappel line with a Prusik hitch and the second loop attached to the first loop with a girth hitch (see Fig. 17.11*b*.); 6- to 7-ft (1.8- to 2.1-m) lengths of accessory cord are recommended for this (Roop et al. 1998, p. 133).

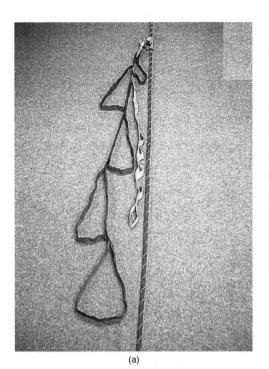

(a)   (b)

Figure 17.11 (*a*) An etrier attached to an ascender can be used to ascend a fixed line; (*b*) Ascending self-rescue can be performed using loops of accessory cord attached to the rappel line with a Prusik hitch.

Either of these self-rescue techniques could be used to completely ascend a fixed line if two ascending devices are available (one for each foot). By shifting the weight to one ascending device, the other device can be reset further up the rope. The weight is then shifted to the higher device, allowing the lower one to be reset further up the rope. This process basically allows one to "walk up the rope." Techniques for ascending fixed lines are not typically required in confined space rescue and are therefore outside the scope of this text; however, such techniques are well covered by other sources devoted to general rope rescue (Frank 1998).

Lowering Systems

Confined space rescue operations sometimes require that loads be lowered for a significant distance. For instance, we may wish to lower rescuers into a space rather than requiring them to rappel into it, in order to make the operation safer and easier for them. In some cases rescuers may remove a patient packaged in a litter from a confined space only to be faced with getting the litter down the side of a tower or off the top of a silo. The simple mechanical-advantage systems covered in Chap. 13 may be adequate for lowering loads for short distances; however, those techniques require too much rope to be useful for long lowers. In this section we will explore rigging and using rescue systems designed specifically for lowering human loads.

The lowering techniques we will examine are similar to rappelling in that a descent control device and rope will be used. Lowering systems are significantly different because the descent control device is anchored and used to control the rate at which the rope bearing the load passes through it. This provides excellent control of the descent of the load, especially when the brake bar rack is used. For the sake of simplicity, in this chapter we will assume that the load to be lowered is a rescuer. In the next chapter we will apply the same techniques to lower rescuers and patients packaged in litters.

Rigging for lowering operations

As in all rescue operations, before rigging for lowering operations can begin, the area of operations must be safe. Basic safety issues, such as proper use of safety gear and fall-protection provisions, must be addressed, as noted earlier while discussing rigging for rappel operations.

Adequate anchors or anchor systems, as described in Chap. 13, must be established for all fixed components of the system. These components include at least one descent control device (and two in some cases) and at least one safety belay location. As always, a completely separate anchor for the safety belay is strongly recommended. It may also be desirable to establish high anchor points for directional pulleys above edges that must be negotiated with a load. As we will see later, these "high directionals" can offer real advantages in certain operations, such as when we need to take a basket litter over a handrail. Always apply padding to rough surfaces or sharp edges that ropes or webbing may contact. Figure 17.12 shows a simple lowering station.

Establish a lowering line by attaching a descent control device to the selected anchor. Tie a double-loop figure 8 on a bight at the end of the rope to be used to lower the load. As we will see in Chap. 18, some operations, such as double-line litter lowers, require two separate lowering lines to be rigged. For the purposes of this section, we will assume that a single-line lowering system will be used to lower one rescuer, a single-person load.

Rig a safety belay line as described in Chap. 13. Any of the techniques described there can be used to belay the load being lowered.

Personnel roles in lowering operations

Like all rescue procedures, operating a lowering system requires teamwork and organization to be done safely and efficiently. Obviously the stakes are high whenever lowering is required. Team members involved in these operations include the following:

- The lowering team leader is responsible for overseeing the safety of the lowering operation and directing the operation so that things run smoothly.

- The lowering line operator uses the descent control device on the lowering line to lower the load.

- The safety belayer is responsible for tending the safety belay line attached to the load and using it to stop any uncontrolled descent of the load that might occur during the lower.

- The rescuer is actually the load in our example and will be an active participant in the operation. In some operations, such as litter lowers described in Chap. 18, the load may be a patient packaged in a litter. In some of these operations the load may also consist of one or two attendants in addition to the packaged patient.

- Other team members may be needed to perform supporting functions such as helping to move a litter containing a patient over a handrail at the beginning of a lower or operating tag lines from the ground to keep the load clear of the side of the structure as it is being lowered.

Preparing to lower

Before the rescuer nears the edge, attach the safety belay line to a suitable location on the rescuer's harness. A high point of attachment such as the dorsal D ring is recommended. The safety belayer maintains just enough slack in the belay line to allow the rescuer enough freedom of movement to get into position for the lower to begin. The safety belayer must be able to arrest any fall that might occur from then on.

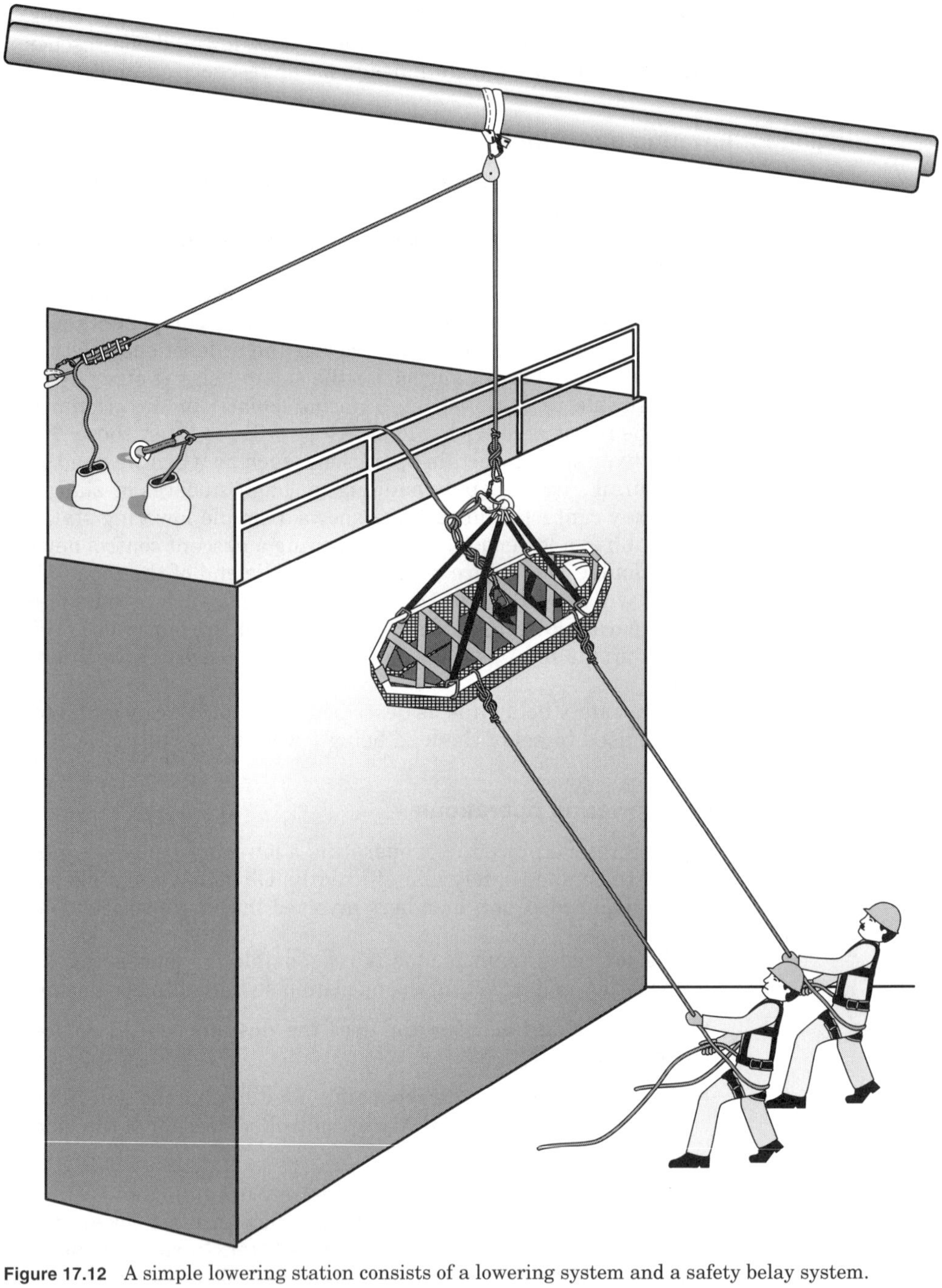

Figure 17.12   A simple lowering station consists of a lowering system and a safety belay system.

In preparation for lowering, attach the end of the lowering line to a suitable point of attachment on the rescuer's harness, such as the front waist D ring, and rope the lowering line into the descent control device. Install the carabiner used to attach the line to the rescuer so that the gate is facing toward the rescuer and the locking ring screws downward to lock. This reduces the likelihood of the locking ring working into the unlocked position and the gate being pushed open during the lower.

Before the descent control device is roped into the lowering line, it is very important that the end of the lowering line be properly positioned in relation to the edge to be negotiated. This is critical because edge problems must be dealt with in lowering operations, just as in rappels and all other operations where edges are encountered. Excessive slack in the lower line can cause a shock load when the system is loaded.

Preparing for a high-point lower. High-directional pulleys should always be rigged above edges to be negotiated, provided suitable anchor points are available (see Fig. 17.12). Remember that in our example we are preparing to lower a single rescuer. In this case, once the lowering line has been run through the high-directional pulley, we can simply attach the end of the lowering line to the rescuer, pull all slack out of the line, and rope the descent control device into the line. This "high point" setup provides the rescuer a significant advantage in going over the edge and loading the lowering system. This type of rigging offers even greater advantages when the load going over the edge is a litter containing a patient, as we will see in Chap. 18.

Preparing for a low-point lower. In many instances it is not possible to rig a high-directional pulley over the edge to be negotiated. In such a case the lowering team is required to use a low-directional point. In our example, assume that we are lowering from the top of a tower. The highest point available in the area of operations is the top of a sturdy round handrail with a 2-in outside diameter. In this case the handrail is used as a low change-of-direction point by simply allowing the lowering line to feed over the top of it during the lower.

In preparing for a low-point lower, properly index the end of the rope in relation to the edge before the descent control device is roped into the lowering line or the lowering line is attached to the load. In our example this is accomplished by positioning the knot in the end of the lowering line at or just beyond the outside edge of the handrail (see Fig. 17.13). The lowering line operator then pulls all slack out of the line, ropes in, and locks off. The goal is to have the entire knot, including the safety knot, fall just beyond the handrail when the line is fully loaded. To achieve this, the knot may need to be "fudged" a bit to the anchor side of the edge to compensate for the weight of the load and stretch in the system.

When using the brake bar rack, one method for indexing the knot to the edge is to begin with the rack roped in with three bars in place and the end of the rope well short of the handrail. It should be possible to pull the rope through the rack with a strong pull, which will have the effect of pulling all slack out of the system. In this way the end of the lowering line is pulled to the proper index point as described above (see Fig. 17.13); then all six bars are racked in and the rack locked off. Once the index point is established and the lowering device is locked off, attach the end of the lowering line to the load.

Roping in the descent control device. Roping in a descent control device for lowering is basically the same as roping in the device for rappelling. If a brake bar rack is used, rope it into the lowering line as described previously. Initially all six bars are roped in. Hold the rack in the locked-off position (Fig. 17.14a), or tie it off if needed to free the operator's

Figure 17.13 In preparation for a low-point lower, index the lowering line in relation to the edge before the descent control device is roped in or the lowering line is attached to the load.

(a)   (b)

Figure 17.14 (*a*) After locking off, the brake bar rack can be held in the locked-off position by grasping the control line and the loaded line just ahead of the rack; (*b*) after locking off, the figure 8 descender can be held in the locked off position by grasping the control line and the loaded line just ahead of the descender.

hands. If a figure 8 descender is used, rope it in and lock it off as described earlier. The figure 8 can also be held in the locked-off position (Fig. 17.14*b*), or tied off if need be. As always, the descender is double-wrapped for a two-person load.

Operating the lowering system

When all team members are in position and ready, the lowering operation can begin. Prior to this the team leader conducts a full safety check. This includes checking all rigging, being sure that all carabiners and other fasteners are locked, and checking that safety gear is properly used by all personnel.

Initial lowering team commands. The team leader issues the command "prepare to lower" to notify everyone on the team that the operation is about to begin. Each team member makes a reply so that the team leader knows that they are on alert status and ready to proceed. For example, the safety belayer replies "safety on" or "ready on safety," indicating that he is ready to belay a falling load. The lowering line operator replies "lower line ready," indicating that he is holding the lowering line in the locked-off position at the descent control device. The rescuer replies "rescuer ready," indicating that she is ready to go over the edge and load the lowering line whenever the team leader instructs her to do so. Hand signals or radio equipment may be required to issue the commands and replies in some situations.

In our example, when everyone is ready, the team leader instructs the rescuer to go over the edge. After receiving that order, the rescuer climbs over the edge and carefully loads her weight onto the lowering line. In other operations, such as litter lowers, additional team members will place the load over the edge very carefully, as we will see in Chap. 18. Once the lowering line is loaded, the team leader issues commands to begin the lower, as described below. As the operation proceeds, the team members carry out their duties.

Operating the lowering line. Operating a descent control device during lowering operations is basically the same as operating it during a rappel. The only difference is that the device is fixed to an anchor and the rope moves through it for lowering, whereas in rappelling the line is fixed and the device moves down it. When the brake bar rack is used, lower one-person loads with five bars in the rack and two-person loads with all six bars in the rack. As previously noted, when the figure 8 descender is used, it is single-wrapped for one-person loads and double-wrapped for two-person loads.

In our example assume that a brake bar rack is used. Once all preparations have been made and the rescuer has loaded onto the lowering line, the team leader issues a command such as "lower on five." This notifies the lowering line operator to drop one bar out of the system and begin lowering the load on five bars.

If the team leader or anyone else on the team issues the command "stop!", the lower line operator must immediately stop the descent of the load. When using the rack, the operator can quickly and easily stop the descent by moving the right hand ahead of the rack, as shown in Fig. 17.15. The load can be held in position simply by gripping both the loaded and unloaded lines together just ahead of the rack. All team members should repeat the command "Stop!" to verify that they heard the order.

If the command "Stop!" is followed by the command "lock off," the lower line operator adds the sixth bar to the rack and holds it in the "locked-off" position (see Fig. 17.14). If lowering is interrupted for some time, the team leader may issue the order "tie off," in which case the rack is tied off with all six bars in place, as shown in Fig. 17.5b. Once the rack is tied off, the lower line operator is no longer obligated to hold the rope to ensure that the load will not drop.

During some lowering operations the team may begin lowering a one-person load, stop part way through the lower so that a second person's load can be loaded on, and then complete the lower. In such an operation, the team leader issues a command such as "lower on six" to the lowering line operator. This notifies him to leave all six bars in the rack when he unlocks it to resume the lower. As noted in discussing rappels, this type of procedure is required most often for pickoffs and line transfers, which are not typical confined space rescue operations and therefore are outside the scope of this text.

Operating the safety belay line.   After being notified that the lowering operation is about to begin, the safety belayer must be ready to stop a falling load at any time throughout the lowering operation. Any belay technique described in Chap. 13 can be used. The same procedures and concerns described above for providing top safety belay for a rappeller also apply here.

The belayer remains constantly alert and acts immediately to arrest any fall that may occur for any reason. A slight amount of slack in the belay line is required to prevent interference with the lowering operation; however, no more than 18 in (46 cm) of slack should be allowed. The safety belayer receives directions as needed from the team leader regarding the amount of slack in the safety line. Orders such as "more slack on safety" or "less slack on safety" are used.

Rescuer's function in lowering operations.   In our example the load being lowered is a rescuer. After the lower begins, the rescuer's actions are similar to those described above for a rappeller, except that in lowering operations the rescuer is not controlling her own descent. The rescuer is free to concentrate on dealing with problems such as negotiating edges during the descent and performing assigned tasks. Rappels allow the rescuer to react almost immediately to the situation at hand. For example, a rappeller can stop almost immediately after deciding to do so. In contrast, if a rescuer on a lower line decides to stop, the command must be directed to the team leader, who relays it to

Figure 17.15 When a brake bar rack is used in a lowering system, the lowering line operator can stop the descent of the load quickly by moving the control hand ahead of the rack. The load can be held fast by holding both the control line and the loaded line in the control hand.

the lower line operator, who then reacts to it. The resulting reaction lag may cause the rescuer to stop at a lower point than she actually intended.

When acting as litter tenders during litter lowers, rescuers are lowered along with the litter and guide the litter down the side of the structure. We will describe litter-lowering operations involving both one and two tenders in Chap. 18.

Support functions during lowering operations. In addition to the team member roles already described, a number of additional supporting roles may be required for some lowering operations. For example, extra team members may be needed to assist rescuers in smoothly negotiating difficult edges at the beginning of a lower. Tag-line operators may be required to guide a load to keep it clear of the side of a structure as it is being lowered (Fig. 17.12), as described in Chap. 18.

Rescue team leader's function in lowering operations. The rescue team leader oversees the safety and efficiency of the lowering operation. This includes overseeing the safety of all team members from the time that basic rigging begins until the operation is completed. The team leader must conduct a full safety check, as described above, before anyone goes on line.

The team leader directs the lowering operation to ensure that everything goes safely and efficiently. This includes issuing the commands mentioned in describing the various team member roles. Because operations may be carried out in situations in which voice contact is not possible, simple hand signals may also be used as needed. While no set of signals or commands is universally accepted, any signals or commands can be used as long as they are clear and everyone on the team understands what they mean. Common hand signals for lowering and hauling operations are shown in Fig. 17.16.

Using Rescue Systems in Hauling and Lowering Operations

Many rescue scenarios require hauling operations to be used to move victims and/or rescuers. In some rescue operations the simple mechanical-advantage (MA) systems described in Chap. 13 may be adequate; however, those simple systems have significant limitations for other rescue operations. For example, assume that we have 150 ft of rope available to build an MA system. Based on the load and the number of haul team members available, we need a 4:1 mechanical advantage. In rigging a conventional hauling system, we will achieve an effective hauling/lowering distance of about 35 ft (10.7 m) after allowing for rope required for the haul team and for the knot in the end of the rope. Conventional MA systems are not very versatile. For example, if we are lowering with an MA system and run short of rope, there is no convenient option for extending the range of the system. Likewise, if we build an MA system, take the traveling pulley into the space to attach to the patient, and find that it is too short, we are at a similar disadvantage.

In this section we will examine several options for overcoming the limitations of conventional MA systems. These options involve rescue systems that are slightly more complicated to build and operate but significantly more versatile than conventional systems.

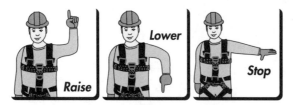

Figure 17.16 Examples of simple hand signals that may be used to direct lowering and hauling operations are shown here.

MA system piggybacked onto a mainline

One approach to overcoming the limitations of a conventional MA system is shown in Fig. 17.17. In this "piggyback" application, a simple MA system is attached to the anchor at one end and attached with a rope grab to the main line at the other end. The MA system actually functions as a 5:1 system in this application. The rope grab serves as a haul cam and can be attached at any point desired on the main line. The working end of the main line is attached to the load, while the running end passes through an anchored figure 8 descender. The figure 8 serves as a ratchet during hauling operations and a lowering device during lowering operations.

Overview of system operation. To use the system, the haul cam is pulled along the main line as far as possible toward the load and released or "set." This extends the MA system. The MA system is used to pull the load on the main line as far as possible toward the anchor. As this is done, the main line is pulled through the figure 8 descender in order to keep slack out of the main line. When the MA system reaches the end of the haul, the main line is locked off at the figure 8 and the weight is released from the 4:1 system and loaded onto the figure 8. This allows the haul cam to be released, slid further down the main line, and reset closer to the load, thus setting the stage for another haul and starting the process over again.

Personnel roles. Team members required to operate this system include the following:

■ The haul team leader oversees the overall safety of the hauling operation and directs the activities of everyone on the team.

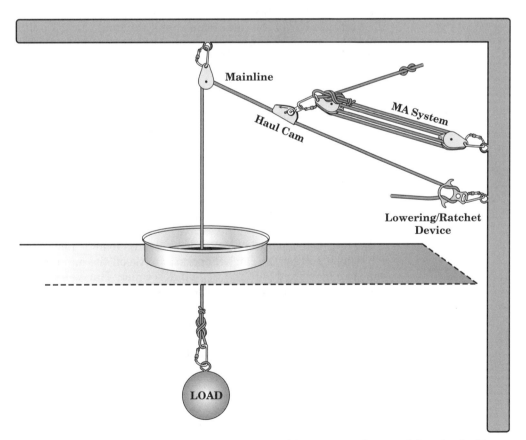

Figure 17.17 A very versatile system for hauling and lowering can be rigged by piggybacking a mechanical-advantage system onto a main-line package.

- The ratchet operator pulls any slack in the main line through the figure 8 to keep the main line taut and holds the load while the MA system is reset at the end of each haul.

- The haul team hauls on the MA system to move the load.

- The haul cam operator is a haul team member who resets the haul cam at the end of each haul.

- The attendant or hole watch observes the patient or victim being moved with the system at all times to be sure that it is safe for the operation to continue.

- The safety belayer operates the safety belay line to belay any falls that may occur.

Procedures and commands. After the system is rigged and ready for operation, the team leader performs a standard safety check and issues a command such as "prepare to haul." All team members respond so that the haul team leader knows that they are ready to proceed.

The haul team leader then issues the command "haul" and the haul team responds by pulling on the haul line, thereby raising the load. As the MA system shortens, the team leader issues the command "Stop!" before the two sets of pulleys foul out against each other. This notifies the haul team to stop hauling and hold the haul line stationary.

Next the haul team leader issues a command such as "set," which notifies the ratchet operator to be sure that all slack is out of the main line and to lock off the main line at the figure 8 and hold it in the locked-off position. The ratchet operator replies "set" to confirm that this has been done.

The haul team member issues a command such as "slack" to notify the haul team to slowly release the load onto the ratchet. The haul team leader then issues a command such as "reset" to notify the haul team to relax their grips on the haul line, after which the first member of the haul team resets the haul cam by pulling it as far as possible down the main line toward the load. After this is done, the first haul team member issues the reply "reset" to notify the team leader that the system is ready for another haul.

The haul team leader then issues the command "prepare to haul" to start the procedure over again. Note that during these operations the safety belayer is in position at all times to belay any fall that might occur. The attendant constantly views the patient or rescuer being moved to be sure that no hazardous conditions develop. The person being moved could be severely injured if a limb or the head hangs up during the haul and the haul team continues to pull. Damage to components of the hauling system could result if any part of the litter hangs up and the team continues to haul. This is especially true if mechanical devices such as the Gibbs ascender are used as a haul cam. Haul team members must note any undue resistance in the haul line. Any member of the team can stop the operation at any time by simply yelling "Stop!" should an unsafe condition develop.

If a protracted delay in hauling is likely, the team leader may issue the command "tie off," which notifies the ratchet operator to tie off the main line at the figure 8. The load can then be shifted onto the tied-off descender. This will free the ratchet operator and the haul team from having to hold the load constantly.

Use considerations for the piggyback system. The piggyback system is very versatile because it can be used to lift or to lower. The functional length is limited only by the length of rope available for the main line, not the length of the MA system used for hauling as in conventional systems.

A useful item for use with this type of system is a small, prebuilt 4:1 MA system. Such a system can be kept handy during lowering operations in case problems develop that require the load to be raised. An example of such a problem might be a load that becomes hung up during a lower and can be cleared only by being raised back up for several feet. In that case a prebuilt MA system could simply be piggybacked to the lowering line, allowing the lowering system to be converted to a lifting system very quickly. The prebuilt four-to-one could also be attached to an entrant's tag line, allowing the same type of system to be set up quickly for a nonentry rescue. The tag line must be life-safety-rated if the entrant will be lifted or lowered vertically.

The piggyback system (Fig. 17.17) is used for short lowers by simply reversing the hoisting procedure described above. For lowers longer than the maximum extended length of the MA system, the haul cam is held open or the 4:1-MA system removed from the main line and the figure 8 descender used to lower the load. For significantly long lowers, it may be desirable to replace the figure 8 with a brake bar rack before shifting the load onto the descent control device before beginning the lowering operation. This is due to the advantages of the brake bar rack over the figure 8 descender for lowering operations, as discussed in Chap. 11.

In rigging this system it is important either to use one bombproof anchor for both the main-line ratchet and the MA system, or else use anchors that are very close together. Any significant angle between the two anchors hurts the efficiency of the system. Directional pulleys are recommended to direct the main line past edges to prevent excessive wear to the rope. This is especially true in hauling operations.

Using inchworm systems

Prerigged mechanical-advantage systems with an extended length of only 10 to 20 ft (3 to 6 m) are sometimes very valuable in maneuvering a packaged patient out of a confined space. In some cases an even shorter, more compact prebuilt system is very useful. One example of this is the inchworm system shown in Fig. 17.18, which is used to move a load along a fixed line through a series of repeated short hauls.

The inchworm system is rigged using a short MA system, most frequently a four-to-one (4:1) system that is a few feet long when fully extended. Note that a simple figure 8 knot in the end of the haul line on the MA system keeps the line from being pulled through the pulley when the system is extended.

Rigging and using the inchworm for vertical applications.   To rig the system for vertical applications, attach a rope grab to the fixed line above the load to act as a haul cam. Attach the MA system to the rope grab as shown in Fig. 17.18. Attach another rope grab to the fixed line at the level of the load to serve as a ratchet cam. Attach the load to the ratchet cam, and attach the traveling pulley of the MA system to the load.

The vertical inchworm system can be used to good advantage in situations that would prevent a haul team outside the space from being able to hoist the load. One example is a situation in which a patient must be precisely maneuvered past a series of offset baffles or trays in a vertical vessel in order to be removed through the top. Move the load up the fixed line by pushing the haul cam as far as possible up the fixed line, extending the MA system, and pulling on the haul line of the MA system to move the load up the line until the MA system is fully shortened. At that point, slowly release the tension on the haul line until the load shifts to the bottom or ratchet cam. Reset the top or haul cam, and the process is ready to start over again. As in all vertical situations, belay the load continuously.

Rigging and using the inchworm for horizontal applications.   Configure the inchworm system as described above for use in horizontal situations, except that the ratchet cam is not needed and the load does not need to be attached directly to the fixed line. The inchworm in this configuration could be used to maneuver a patient through a series of tight turns through a horizontal piping or tubing system along a fixed line anchored outside the space. In such a case a life-safety-rated rope would not be required since it is a strictly horizontal scenario. A utility-rated rope might be more appropriate to use because of the likely wear and tear on the loaded rope stretched around tight, rough bends in this scenario.

Converting a lowering system to a Z rig under load.

It is possible that we may begin a rescue operation using a lowering system to lower a load, only to find that we must raise a load in order to continue or complete the operation.

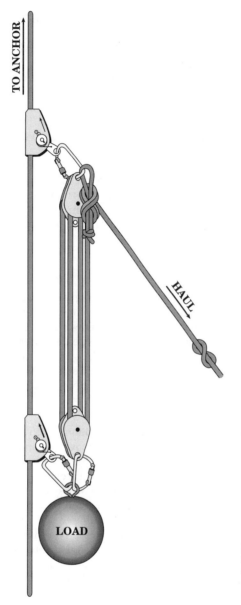

Figure 17.18 The inchworm system uses a short mechanical-advantage system to move a load along a fixed line through a series of short hauls.

In some cases we may lack a separate MA system and the other components needed to convert the lowering system to the piggyback system. In such cases another option is to covert the lowering line to a Z-rig hauling system. This can be done with the lowering line under load, as shown in Fig. 17.19 and described in the following steps.

Note that the use of an anchor plate or rigging plate makes the entire procedure much easier. A suitable safety belay is used to belay the load throughout the entire process so that the load will be supported by at least two systems at all times. In our example a brake bar rack is used, but the same steps can be used when a figure 8 descender is used as the lowering device.

Step 1: prepare for the changeover. Beginning with a typical lowering line setup (Fig. 17.19a), lock off and tie off the brake bar rack. Attach a load-releasing device, such as a mariner's knot or load release hitch, to the anchor plate. Creating and using load-releasing devices is described below. Attach a rope-grabbing device to the end of the load-releasing device. Either soft or mechanical rope grabs can be used as long as the precautions previously mentioned related to the use of mechanical rope grabs are duly

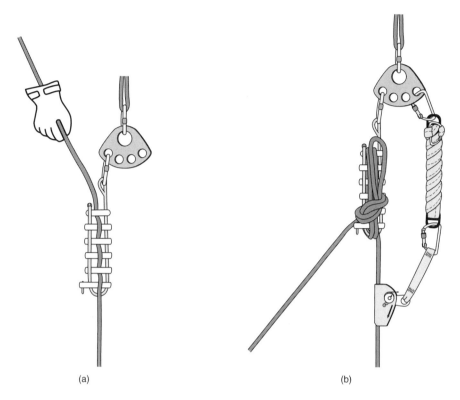

(a)   (b)

Figure 17.19 (*a*) Three basic steps are involved in converting a typical lowering line setup (shown) to a hauling system under load; (*b*) the first step in converting a lowering system to a Z-rig hauling system is to lock off and tie off the brake bar rack and attach a rope-grabbing device to the lowering line, then attaching the rope-grabbing device to the anchor with a load-releasing device.

noted. Attach the rope-grabbing device to the lowering line below the brake bar rack (Fig. 17.19*b*).

Step 2: unloading the brake bar rack. Slide the rope-grabbing device as far as possible down the haul line to remove all slack from the load-releasing device (Fig. 17.19*b*). This rope-grabbing device will serve as the ratchet cam in the hauling system. Carefully untie and unlock the brake bar rack and slowly release it to shift the load onto the rope grab and load-releasing device. Remove the rope from the brake bar rack. The rack can be removed from the anchor plate if it will not be needed again (Fig. 17.19*c*).

Step 3: building the Z rig. Attach a single-sheave pulley to the end of the load-releasing device or to the anchor plate so that it rests just above the rope-grabbing device, and thread the rope through it. Install a second rope-grabbing device below the ratchet cam to serve as the haul cam. Attach a single-sheave pulley to the haul cam, thread the rope through it, and the Z rig is complete (Fig. 17.19*d*). Once the Z rig is built, it can be used as described in Chap. 13 to haul the load.

Converting a Z rig to a lowering system under load.

After hoisting duties are completed using the Z-rig system, it may be desirable to convert it to a lowering system. This is accomplished by reversing the previous steps, as shown in Fig. 17.20 and described below. Like the previous conversion, this one can be done with the system loaded. The load is belayed with a suitable safety belay during the entire procedure.

Step 1: removing the Z rig. Slowly release the weight of the load onto the ratchet cam (Fig. 17.20*a*). Remove the haul line from the pulley on the haul cam, and remove the

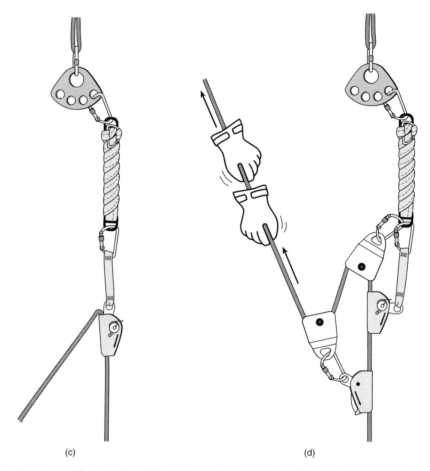

(c)   (d)

Figure 17.19 (*Continued*)   (*c*) The second step in converting the lowering system to a Z rig is to slowly and carefully transfer the load from the brake bar rack to the rope-grabbing device and remove the rope from the rack; (*d*) the third step in converting the lowering system to a Z-rig is to build the Z-rig hauling system.

pulley and haul cam from the rope. Remove the rope from the anchor pulley, and remove the anchor pulley from the load-releasing device.

Step 2: roping in the brake bar rack.   Attach the brake bar rack to the rigging plate, pull all slack out of the rope, and rope in the rack. Rope in all six bars, then tie off the rack (Fig. 17.20*b*).

Step 3: transferring the load to the brake bar rack.   Carefully unfasten the load-releasing device and slowly extend it to transfer the load to the brake bar rack (Fig. 17.20*c*). Remove the rope grab and the load-releasing device. Retie the load-releasing device, as described below, at the earliest opportunity. The load is now ready to be lowered with the brake bar rack (Fig. 17.20*d*).

Load-releasing devices

In some confined space rescue operations it may be desirable to shift or release the weight of a load from one system to another. One fast and convenient way to accomplish this is by using load-releasing devices such as the mariner's knot or the load-release hitch. Load-releasing devices can basically be thought of as a way of providing an extendable anchor system. They function as a simple combination friction control and mechanical-advantage device, allowing smooth load transitions.

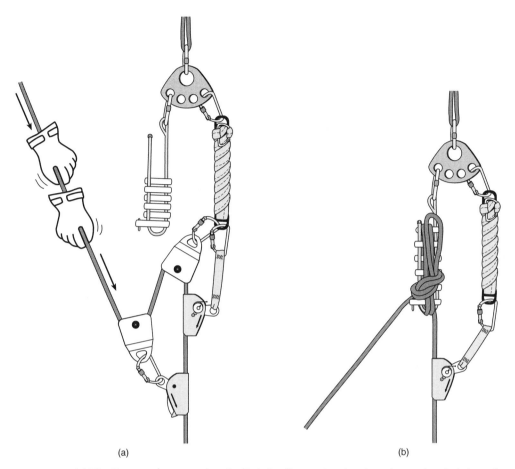

(a)   (b)

Figure 17.20 (*a*) The first step in converting the Z-rig hauling system to a lowering system is to transfer the load to the rope-grabbing device that serves as a ratchet for the hauling system, and disassemble the hauling system; (*b*) the second step in converting the Z rig to a lowering system is to attach a brake bar rack to the anchor, rope in the rack, pull out all slack between the rack and the load, and lock off and tie off the rack.

Rigging load-releasing devices. Several techniques can be used to rig load-releasing devices. All use two or three carabiners. The CMC mariner's knot is rigged using webbing slings or straps. The load release hitch is rigged using accessory cord.

The CMC mariner's knot shown in Fig. 17.21 was tied using a large ProSeries Rescue Runner™ by CMC. The large Rescue Runner is essentially a 60-in (152-cm) sling of flat webbing with an eye sewn in each end. Other similar webbing items, such as a tied loop sling, can also be used.

The mariner's knots shown in Fig. 17.21*e* were each rigged with two general-use steel carabiners and one light-use aluminum carabiner. The general-use steel carabiners are used at the anchor end and the load end of the device, and the aluminum carabiner serves as the safety carabiner used to secure the end of the sling when the mariner's hitch is completed. This prevents confusion about which carabiner connects to what when the device is rigged and used. Some rescuers use colored carabiners as the safety carabiner, and others use a nonlocking aluminum carabiner. Some rescuers complete the mariner's knot by attaching the safety carabiner, directly to the anchor carabiner. However, some authorities warn against this because of concerns that the safety carabiner may become loaded, making it impossible to remove (Frank 1998, p. 60).

As compared to rope, webbing has very little elasticity and offers very little in the way of shock absorption if shock-loaded. Webbing used to rig load-releasing devices has been reported to fuse under high-impact loads (Lipke 1997, p. 47). CMC Rescue designed the ProSeries Load Release Strap™ (Fig. 17.21*e*) to decrease that problem while offering ease of rigging and use. As always, when working with load-releasing devices take care

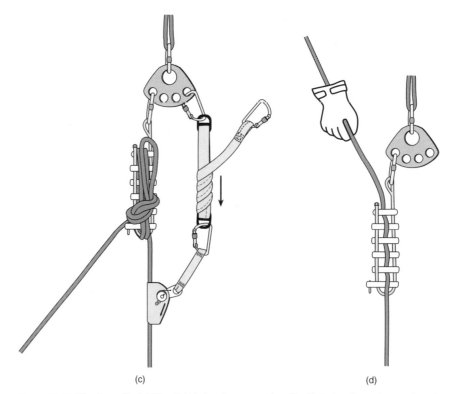

(c)   (d)

Figure 17.20 (*Continued*) (*c*) The third step in converting the Z rig to a lowering system is to unfasten and slowly extend the load-releasing device in order to transfer the load from the rope-grabbing device to the brake bar rack; (*d*) after the load has been shifted back to the brake bar rack, untie and unlock the rack to conduct lowering operations.

to avoid shock loads by using standard precautions such as minimizing slack in systems and using shock absorbers.

The load-release hitch is very similar to the mariner's knot but is tied using accessory cord rather than webbing. The accessory cord has more shock-absorbing capability than does webbing. The load-release hitch can be tied as shown in Fig. 17.22 using 32 to 33 ft (9.7 to 10 m) of 8- or 9-mm accessory cord. Several techniques are commonly used to tie the load-release hitch. The technique shown in Fig. 17.22 was included in the third edition of CMC Rescue's *Rope Rescue Manual Field Guide*. CMC Rescue credits it to "Rigging for Rescue" (CMC, 1998, p. 17).

Using load-releasing devices.   In the previous section, we saw an excellent example of the use of load-releasing devices in changing a hauling system to a lowering system under load. The use of the mariner's knot in that application is shown and described in Fig. 17.20*a* through *c*.

As noted in Chap. 13, load-releasing devices are very useful when tandem Prusik belay (TPB) systems are used to belay a load being lowered with a device such as the brake bar rack or the figure 8 descender. During these lowering operations, if the Prusik hitches "set" or lock onto the belay line, the weight of the load must be removed from the Prusik hitches before they can be unlocked. This can be a problem, since devices such as the brake bar rack and figure 8 descender can lower a load but cannot haul a load. Load-releasing devices provide a quick and convenient way to shift the weight of the load off the Prusik hitches and back onto the lowering device so that normal lowering can resume.

As an example, when using the mariner's knot to unload the TPB, prepare to extend the load-releasing device by removing all slack from the lowering line and securing the lowering device. Next, reverse the process you followed in tying the mariner's knot (Fig. 17.23).

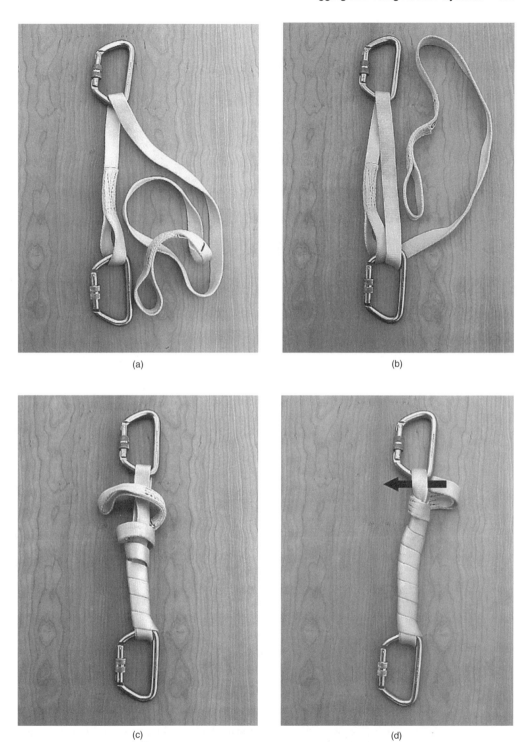

(a)   (b)

(c)   (d)

Figure 17.21 (*a*) Begin the CMC mariner's knot by clipping one of the eyes in a large CMC ProSeries Rescue Runner™ into one of the steel carabiners. This carabiner will be attached toward the load when the release device is used. Pass the unattached end of the Rescue Runner through the other steel carabiner. This carabiner will be attached to the anchor when the device is used. (*b*) Pass the unattached end of the rescue runner through the load carabiner. (*c*) Begin wrapping the unattached end of the rescue runner around the lengths of webbing between the two carabiners. (*d*) Wrap the Rescue Runner from the load carabiner to near the anchor carabiner. Push the eye at the end of the Rescue Runner between the two lengths of webbing at the anchor knot, as indicated by the arrow.

(e)

Figure 17.21 (*Continued*) (*e*) To complete the mariner's knot, use the aluminum carabiner as the safety carabiner to secure the end of the Rescue Runner as shown. A similar device rigged with the CMC Load Release Strap™ is shown on the right.

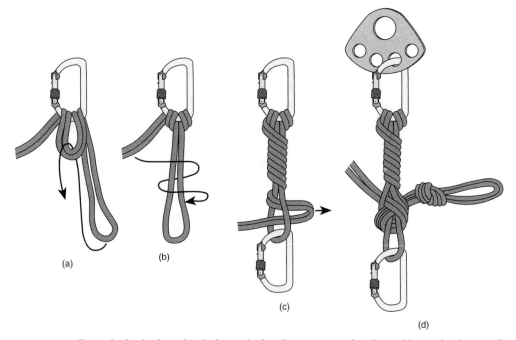

(a)   (b)   (c)   (d)

Figure 17.22 To tie the load release hitch, form a bight of accessory cord and tie a Munter hitch around a carabiner with the bight. Let about 8 in of the bight extend beyond the carabiner (*a*). Wrap the bight several times with the other end of the accessory cord (*b*). Push a loop of the cord through the bight as shown (*c*). To finish, fasten the loop with a half-hitch and secure the half-hitch with an overhand knot (*d*).

Remove the safety carabiner and begin slowly unwrapping the webbing along the side of the mariner's knot. This will gradually reduce the friction holding the load. Slowly allow the device to extend, thus shifting the weight of the load from the belay line back onto the lowering line. Retie the load-releasing device and rerig the safety belay before resuming normal operations

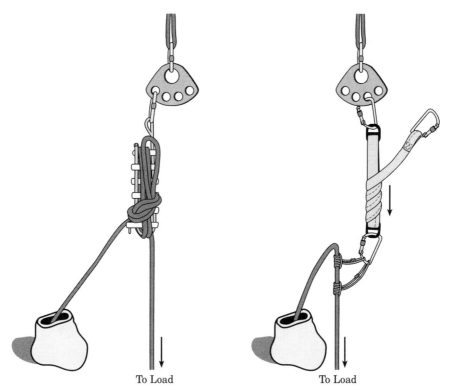

To Load   To Load

Figure 17.23 If a tandem Prusik belay (TPB) system "locks" while a load is being lowered with a brake bar rack or figure 8 descender, load-releasing devices provide a convenient way to shift the load from the TPB back to the lowering device. This allows the tandem Prusik hitches to be unlocked so that normal lowering operations can resume.

Summary

Moving human loads can be one of the greatest challenges of confined space rescue. This challenge includes getting rescuers into the space and getting packaged patients and rescuers back out of the space. In this chapter we covered rigging and using systems intended specifically for moving human loads during rescue operations. This chapter built on the information previously provided on building and using basic systems in Chap. 13. In Chap. 18 we will build on the information provided in this chapter and Chap. 13 as we cover procedures for packaging and transferring patients during confined space rescues.

Packaging and Transferring Patients for Confined Space Rescue

Confined space rescue operations frequently require that rescuers enter the space and remove an incapacitated patient. This requires that the patient be packaged in order to make removal possible. In some cases full spinal immobilization will be required. In some instances a packaged patient must be lowered or raised in order to remove him from a space. In other cases, specialized rope rescue techniques are required to transfer a patient to the ground after the patient is removed from an elevated space.

Various techniques are available for accomplishing rescue objectives. This chapter includes some of the available techniques for patient packaging and transfer during confined space rescues. No one technique is best for all rescues in all situations. Choosing which of the available techniques is best requires professional judgment based on the specifics of the rescue situation and the personnel, equipment, and expertise available. In most cases, the best choice is the simplest and easiest technique that will work to safely solve the problem at hand.

Patient Packaging and Litter Rigging

Anyone who has tried to move an unconscious adult human being knows that such a "dead weight" is extremely difficult to move. This is especially true in cramped confined spaces.

Packaging the victim for removal from the space makes the job much easier and safer for both rescuers and the victim. Packaging provides "handles" on the victim's body that makes moving the victim much easier. Packaging provides points of attachment for equipment such as mechanical-advantage systems that allow rescuers to move heavy loads with minimal effort. Proper packaging also prevents the patient from being injured while being moved.

Considerations for Spinal Immobilization

Spinal injury can occur when a patient sustains any sort of impact to the body, even a short fall to the ground. Paralyzing or fatal damage to the spinal cord may result if a patient with a spinal injury is moved without the proper precautions. For this reason, special precautions must be taken in all cases where the patient may have suffered spinal damage. These precautions require that spinal immobilization be established and maintained throughout all patient handling.

Spinal immobilization requires specialized equipment and techniques, as commonly used by emergency medical services (EMS) personnel. For the most part these techniques are beyond the scope of this text. We will include reference to spinal immobilization techniques in covering the information in this chapter. The assumption is that the readers are either already familiar with these techniques or will get the proper EMS training required to use them before handling patients in a rescue situation.

Using harnesses in victim removal

Putting a harness on the victim may not be patient packaging in a true sense, but it may be the fastest way, and in some cases the only way, to establish a point of attachment for a rope system to the victim. This approach should not be used in situations in which there are any reasons to suspect spinal cord damage, since it provides no spinal immobilization.

Putting a typical full-body harness on an unconscious victim can be very difficult. Some harnesses are specifically designed to be easy to put on an unresponsive patient and are marketed specifically as "victim harnesses." In many cases rescuers do not have a victim harness or any other type of harness for use in patient removal. In these cases an improvised harness may be the only means available for patient removal.

Using improvised harnesses in rescue operations

A number of simple harnesses can be tied or rigged with webbing. These harnesses can be applied to patients in tight spaces in which no other means of packaging, attaching a system to the patient, or "getting a handle" on the patient is possible. One-inch tubular webbing, or other webbing of at least equivalent strength, should be used to rig the improvised harnesses described below.

Rescuers can use the same techniques to rig improvised harnesses onto themselves in order to use rope systems for escape in situations where harnesses are not available. For this reason, many firefighters carry webbing loops in their bunker gear for use as a hose tool with the knowledge that they can also be used to rig an escape harness in a "do or die" situation.

Swiss seat or hasty harness. The *Swiss seat,* also known as the "hasty harness," is tied using a 20-ft (6.1-m) or longer length of webbing, as shown in Fig. 18.1. Although the Swiss seat has been used as an effective seat harness for years, great care must be taken to ensure that a properly tied square knot with safety knots is used to finish the harness. If a granny knot or thieve's knot is tied instead and no safety knots are used, the Swiss seat can fail during a lower or rappel. As the introductory scenario in Chap. 12 illustrates, this can result in severe personal injury.

Another consideration in using the Swiss seat is that it provides little in the way of vertical stability when used alone because the point of attachment it offers is below the center of body mass for most people. For this reason, the person in the harness must remain alert to avoid flipping over and falling out of the seat. In order to maintain the vertical stability of a patient who is not fully alert, the use of an improvised chest harness is required if the Swiss seat is used.

Improvised seat harness. The improvised seat harness differs from the Swiss seat in that a pretied loop of webbing is used to rig it. This allows the use of the water knot, which is generally considered a better knot than the square knot for tying webbing. It also allows the knot to be tied under calm, nonemergency conditions when errors are less likely.

The improvised seat harness can be tied using a 12- to 16-ft (3.6- to 4.9-m) length of webbing tied into a sling with the water knot. The length of webbing required varies depending on the size of the person to be packaged. Figure 18.2 shows the procedure for tying an improvised seat harness using a sling created from a 12-ft (3.6-m) length of webbing. Like the Swiss seat, the improvised seat harness does not provide adequate vertical stability unless an improvised chest harness is used also.

Improvised chest harness. The improvised chest harness is rigged using an 8- to 10-ft (2.4- to 3.0-m) length of webbing tied into a loop using the water knot. Figure 18.3 shows the procedure for tying an improvised chest harness using a sling created from an 8-ft (2.4-m) length of webbing. It can be attached with a carabiner to the rope running to the

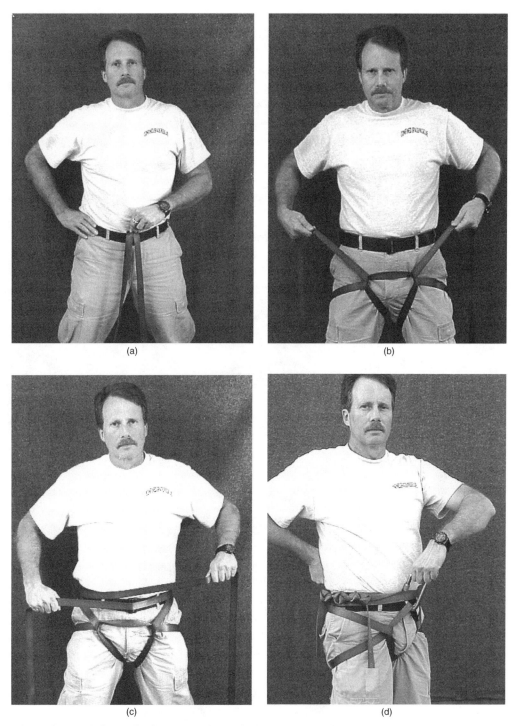

(a)   (b)

(c)   (d)

Figure 18.1   (*a*) To begin the Swiss seat, form a bight by grasping the webbing at a point 1 ft (30 cm) on either side of the midpoint. Tuck the bight behind the belt buckle and allow the two lengths of webbing to hang between the legs. (*b*) Reach behind the knees and grasp a length of webbing in each hand. Pull the lengths of webbing behind the thighs then push them through the bight below the belt buckle. Pull tight. (*c*) Wrap the two lengths of webbing tightly in opposite directions around the waist. Continue until the two lengths meet on one hip with enough webbing left over to tie a square knot with double safeties. At least two full waist wraps are required. (*d*) Finish the Swiss seat by connecting the webbing ends with a square knot and adding at least two overhand safety knots on each side of the square knot. Hook a carabiner through the bight and all the waist wraps to create the point of attachment to the system.

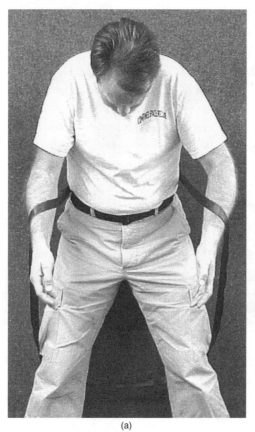

(a)

(b)

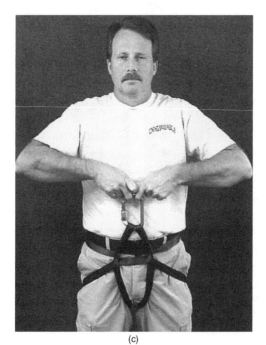

(c)

Figure 18.2 (*a*) To begin the improvised seat harness, place the webbing sling behind the back and reach through the sling on either side so that the sling is supported by the forearms; (*b*) reach through the sling and grasp the lower end of the sling with both hands; (*c*) finish the improvised seat harness by pulling both hands up to form two bights of webbing in front of the waist and place a carabiner through both bights of webbing to form the point of attachment for the system.

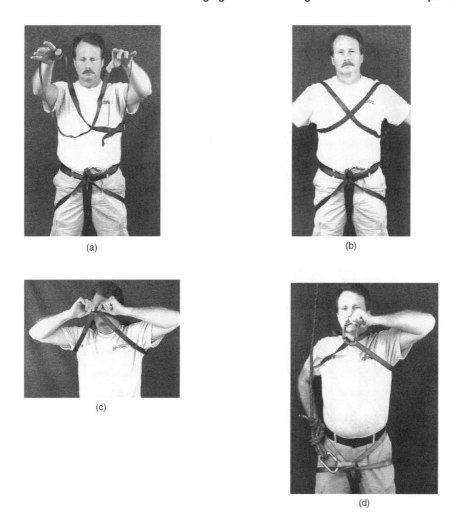

Figure 18.3 (*a*) Begin the improvised chest harness by holding the webbing sling at arm's length and flipping one side over to form a horizontal figure 8 in the sling; (*b*) place one arm through each side of the figure 8 and slide it down the arms so that the point where the webbing crosses to form the figure 8 is in front of the chest; (*c*) grasp the point where the webbing crosses in front of the chest and move it over the head and back between the shoulder blades; (*d*) to complete the improvised chest harness, hook a carabiner through both lengths of the webbing extending in front of the chest to establish the point of attachment for the system.

seat harness to provide more vertical stability for the wearer. In some cases the two loops of webbing that serve as the handles for an improvised seat harness are long enough to allow the improvised chest harness to be clipped into the same carabiner used to attach the rope to the improvised seat harness.

Improvised full-body harness. The improvised full-body harness is very similar to the improvised seat harness. The only difference is that the horizontal length of webbing is positioned under the arms instead of around the waist, so the improvised body harness provides better vertical stability. It can be rigged as an escape harness using a pretied 16- to 20-ft (4.9- to 6.1-m) length of webbing made into a sling by connecting the ends with a water knot. Figure 18.4 shows the procedure for rigging the improvised full-body harness using a 16-ft (4.9-m) length of webbing.

The same harness can be rigged in a different way on a standing victim using a 20- to 24-ft (6.1- to 7.3-m) length of webbing tied into a sling with a water knot. The steps in rigging this harness are shown in Fig. 18.5. A sling created from a 24-ft (7.3-m) length of webbing is shown in the figures. A full-body harness can be rigged on a patient lying supine, as shown in Fig. 18.6.

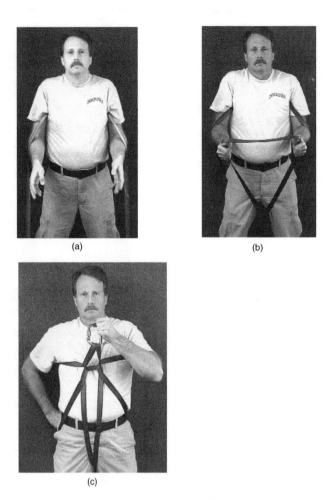

(a)

(b)

(c)

Figure 18.4 (*a*) To rig the improvised full-body harness on yourself, place the webbing sling behind your back, clamp the webbing under each arm, and place your hands through the sling; (*b*) reach between your feet and grasp the bottom of the sling with both hands; (*c*) finish the improvised full-body harness by pulling both hands up to form two bights of webbing in front of your chest and place a carabiner through both bights of webbing to form the point of attachment for the system.

When the improvised full-body harness is rigged to a victim who is not alert, the webbing sling should be long enough to place the carabiner attached to the harness well above the patient's head. This will help prevent the patient from being struck by the carabiner.

The rescue knot.   The rescue knot, also known as the "life knot," can be used to improvise a full-body harness when nothing but a rope is available to effect escape or rescue from a dangerous situation. Figure 18.7 shows how to tie the rescue knot.

Using the long backboard

The long backboard is considered the mainstay of patient transfer for situations involving trauma-related injuries. The use of proper patient handling and packaging techniques with a long backboard can prevent existing injuries from worsening. This is especially important in situations involving spinal injuries. For this reason, the long backboard is sometimes called a "spine board."

The long backboard is also useful for situations in which no trauma has occurred but a patient is incapacitated and must be carried out of the space by rescuers. In such a case, securing the patient to a long backboard may be one way of establishing "handles" on the patient and a good means of conveyance.

(a)

(b)

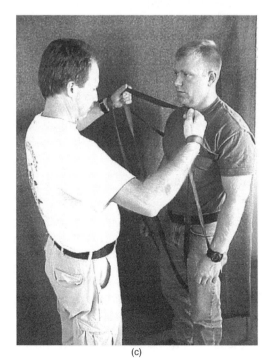

(c)

Figure 18.5   (*a*) Begin rigging the improvised full-body harness on a standing victim by standing in front of the victim with the upper end of the sling tucked under your chin. Pass the lower end of the sling between the victim's legs. (*b*) Reach around both sides of the victim and grasp the lower end of the sling with both hands. Pull the lower end of the sling up the victim's back and underneath the victim's arms, so that you are holding a bight of webbing in each hand in front of the victim's chest. (*c*) Pass each hand from the outside in through the bight of webbing on either side of the victim's chest and grasp the bight of webbing under your chin with both hands.

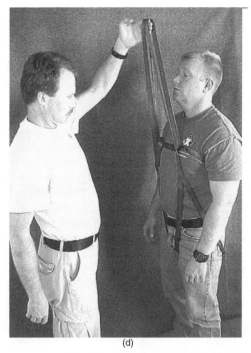
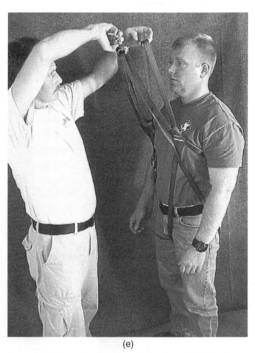

(d)   (e)

Figure 18.5 (*Continued*) (*d*) Pull both hands up to form two bights of webbing above the victim's chest. Place a carabiner through both bights of webbing to form the point of attachment for the system. (*e*) To complete rigging the improvised full-body harness on the standing victim, pull the length of webbing across the victim's chest up and clip it into the carabiner.

Considerations for backboard design. Although all long backboards are quite similar in function, they are available in a variety of specific configurations. For confined space rescue, boards with a slim profile are recommended. Boards that taper at the foot end may fit better into basket litters. If a long board is to be used with a basket litter that has a leg divider, the board must have a cutout between the patient's legs to fit over the leg divider in the litter. One such device is the Miller Full Body Splint/Litter™ by Life Support Products.

Maintaining cervical spinal stabilization. Whenever spinal injury is possible, rescuers must establish cervical spinal immobilization as soon as possible and maintain it throughout the transfer procedure. This begins with one rescuer supporting the patient's head and neck region to maintain in-line manual spinal stabilization while another applies a cervical spine immobilization collar (C-spine collar). The rescuers then "log roll" the patient and position the patient on the long backboard while the rescuer at the head maintains continuous C-spine stabilization and directs the operation. The rescuers fasten the patient's thorax and legs to the backboard with the head and neck region stabilized using a cervical spine immobilization device (Fig. 18.8). Only then can manual in-line C-spine stabilization be released.

Addressing these patient handling techniques in detail is beyond the scope of this text. The reader should either be proficient in these techniques through previous emergency medical training or get the appropriate training before packaging and handling patients in rescue operations. For our purposes, we will assume that the patient is already in a supine position on a long backboard and that all required precautions for C-spine immobilization have been taken if spinal injury is possible.

Using diamond lashed webbing to secure the patient to the board. One means of securing the patient is to lash the patient to the board with a 30-ft (9.1-m) length of one-inch tubular webbing by using the diamond lash technique shown in Fig. 18.9. If no trauma to the patient is involved and the backboard is being used merely to make the patient easier to move, wrap the webbing around the feet, as seen in Fig. 18.9*b*, to bind the

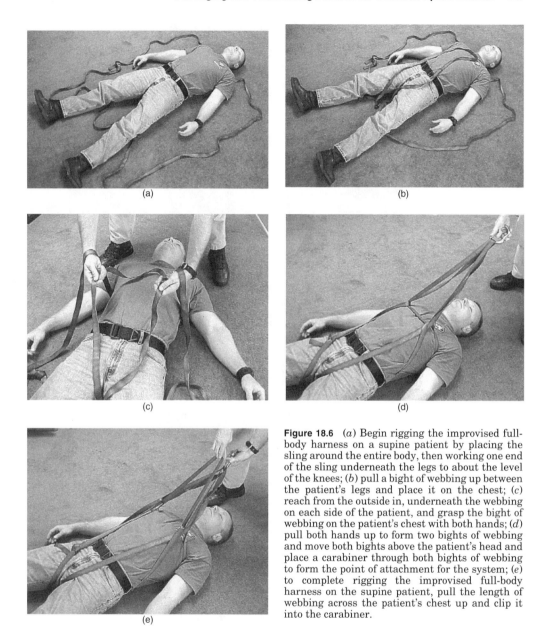

Figure 18.6 (*a*) Begin rigging the improvised full-body harness on a supine patient by placing the sling around the entire body, then working one end of the sling underneath the legs to about the level of the knees; (*b*) pull a bight of webbing up between the patient's legs and place it on the chest; (*c*) reach from the outside in, underneath the webbing on each side of the patient, and grasp the bight of webbing on the patient's chest with both hands; (*d*) pull both hands up to form two bights of webbing and move both bights above the patient's head and place a carabiner through both bights of webbing to form the point of attachment for the system; (*e*) to complete rigging the improvised full-body harness on the supine patient, pull the length of webbing across the patient's chest up and clip it into the carabiner.

patient more securely to the board. Avoid this if spinal injury is suspected, since it may put pressure on the lower extremities.

Using straps to secure the patient to the board.   As an alternative to diamond lashing with webbing, patients may be secured to the long backboard using straps designed for the purpose, as shown in Fig. 18.8. Backboard straps are available with both buckle and Velcro closures.

Considerations for packaging with the backboard.   Whenever the backboard and patient become vertically oriented during the transfer process, the patient may tend to slide down the board, compressing the feet and lower extremities. This may aggravate existing spinal injuries or injuries to the lower extremities and can be very painful to the patient. This is one reason why vertically oriented transfer packages are not recommended unless horizontal packages are not an option. This can be prevented in a couple of ways when vertically oriented packages are used. If the patient is wearing a harness,

(a)

(b)

(c)

Figure 18.7 (*a*) Begin the rescue knot by tying a double-loop figure 8 on a bight with loops large enough to fit over the thighs. Leave at least 3 ft (0.9 m) of the working end of the rope extending beyond the knot. Pass one loop of the double loop figure 8 up each leg to the top of the thigh. (*b*) Tie a half hitch around the chest under the arms with the standing part of the rope extending over the right shoulder. (*c*) Form a bight in the standing part of the rope above the right shoulder and pull the bight down and push it behind the length of rope running across the left breast.

(d)

(e)

(f)

Figure 18.7 (*Continued*) (*d*) Notice the loop that the previous movement formed around the vertical length of rope above the figure 8 knot. Pass the bight upward through the loop. (*e*) Place the working end of the rope through the bight and pull all the slack out. (*f*) Pull on the standing part of the rope and "set" the knot. A safety knot can be tied around both vertical lengths of rope below the rescue knot with the working end of the rope.

attach the harness to the backboard or some other part of the package or lowering system with a utility strap, a webbing sling, or some other suitable connector to support the vertical patient. It may be convenient to use an adjustable connection, such as a utility belt or Prusik hitched accessory cord. If the patient is not wearing a harness, rig an improvised seat harness with a 16-ft (4.9-m) length of one-inch webbing (see Fig. 18.10) and connect it to the backboard or the package.

Bind the patient to the board securely enough to allow the board to be rotated perpendicular to the long axis into a vertical position to allow the patient to vomit without undue movement of the patient's spine. Do not bind the patient so tightly that he is sub-

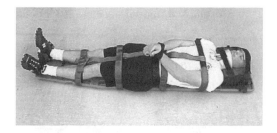

Figure 18.8 Complete cervical spine immobilization is required to prevent spinal injuries from being worsened during patient transfer.

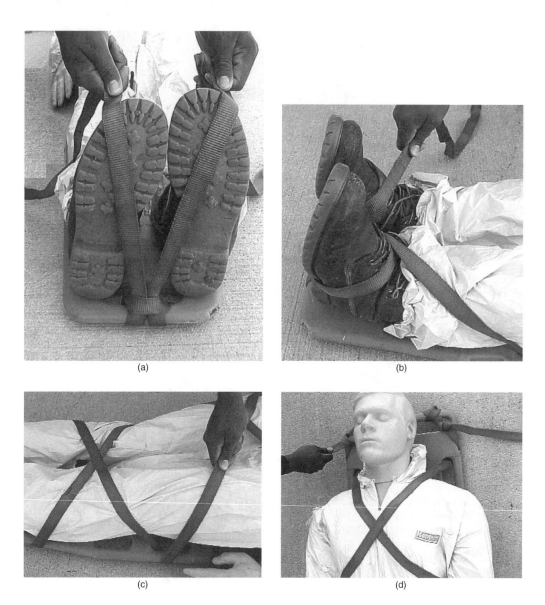

(a)

(b)

(c)

(d)

Figure 18.9 (*a*) To begin diamond-lashing a supine patient to a long backboard, locate the midpoint of the length of webbing and form a girth hitch around lifting handles at the foot end of the backboard. (*b*) If no spinal injuries are suspected, the patient's feet can be wrapped with the webbing for greater security. (*c*) Pass the webbing ends through handle openings above the feet on opposite sides of the backboard. Cross the webbing ends and pass them through handle openings higher up on opposite sides of the board. Continue this process to form the diamond-lash pattern up the board. (*d*) Continue diamond lashing to a location above the patient's shoulders, then pull all slack out of the webbing along the entire length of the backboard. Tie a clove hitch through a handle at the head end of the board with each of the webbing ends. Work all slack out through the clove hitches and add safety knots to complete the diamond lashing procedure.

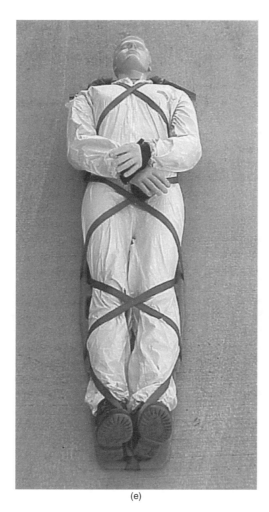

(e)

Figure 18.9 (*Continued*) (*e*) This photograph provides an overview of the patient diamond lashed to the backboard.

jected to undue discomfort. Use padding as needed. Avoid tightening webbing or straps directly over the patient's knees or any other parts of the body that may be painful if compressed. Do not package in a way that puts excessive pressure on the thorax, thus restricting the patient's breathing ability.

Some rescuers place the webbing or straps underneath the arms when binding the patient's torso. This is based on the idea that bindings placed over the arms are subject to loosening if the patient moves his or her arms. Placing the webbing or straps directly over the torso leaves the patient's arms accessible for activities such as checking blood pressure and starting intravenous medications. If the patient's arms are not bound by the webbing or straps binding the patient to the board, they must be separately restrained. This can be accomplished by attaching the wrists to the package using straps, webbing, or tape (see Figs. 18.8 and 18.9*e*).

Using short immobilization devices

Several types of short spinal immobilization devices are available for use in confined space rescue. These devices most commonly are used in removing patients from vehicles following motor vehicle accidents. They have proved very useful for packaging and moving patients in situations that are too cramped to allow a long backboard to be brought into play. This same characteristic makes them very useful in certain confined space rescues.

The simplest example of a short immobilizer is the short backboard, which is merely a torso-length version of the long board. Other short immobilizers currently available include the Kedrick Extrication Device (KED™) by Ferno, the Oregon Spine Splint II™

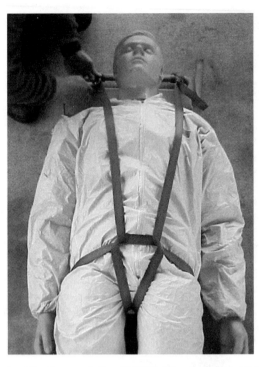

Figure 18.10 Harnesses worn by victims or improvised harnesses rigged to them can be used to prevent the patient from sliding downward in a vertically oriented package.

by Skedco, and the LSP Halfback Device™ by Life Support Products. All three devices are similar in concept and function, but have some differences in design.

Kendrick Extrication Device.   The KED is basically a splint that can be fastened around the upper body to immobilize the back and the head and neck region. It also has straps that support the upper legs. It has several handles attached to allow rescuers to grasp it for moving the patient once packaged. The KED is used very commonly in motor vehicle extrication operations.

Oregon Spine Splint II.   The Oregon Spine Splint II (OSS) was developed by Skedco specifically for use with the SKED stretcher. In design and function it is similar to the Ferno KED and the LSP Halfback devices. Skedco has approved the use of the Oregon Spine Splint II for spinal immobilization in conjunction with the SKED stretcher (described below) and when a full-body harness is placed around both the OSS and the patient (Roop et al. 1998, p. 163).

LSP Halfback.   The LSP Halfback is similar to both the OSS and the KED in design and function. It has six lifting handles for patient transfer when used as a short immobilizer. In addition, it can be used as a lifting harness in situations in which spinal injuries are not suspected. Toward that end, a lifting bridle provided by the manufacturer can be attached to D rings built into the Halfback device. As an example of the general way in which packaging is done with a short immobilizer, Fig. 18.11 shows the procedure for packaging a patient with the Halfback. Conduct training before using the Halfback in rescue operations, using specific instructions provided by the manufacturer.

Use considerations for short immobilizers.   At best, short spinal immobilizers run a poor second to long backboards for spinal immobilization. Use the long board whenever possible for patient packaging when spinal injuries are suspected; however, situations may be encountered during confined space rescues in which there is not enough room for the long board to be used. In such cases the short immobilizer, along with extremely careful handling by rescuers, may allow the patient to be safely moved to a location where full packaging is possible. Whenever a patient with suspected spinal injuries is moved

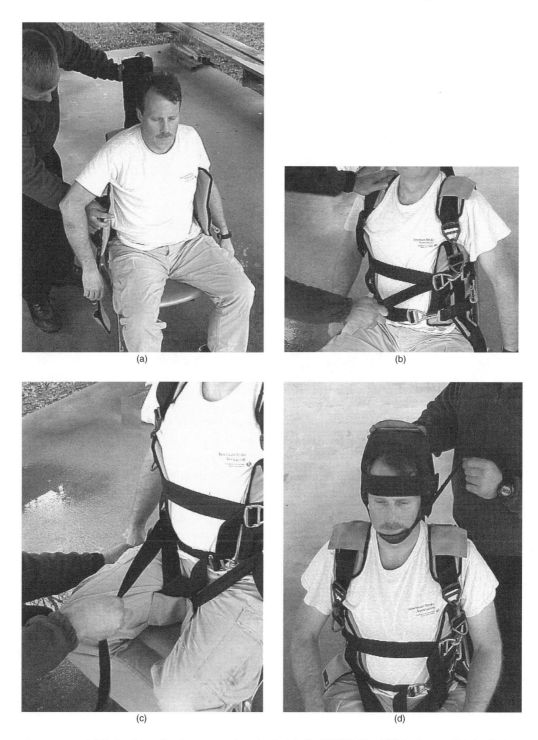

Figure 18.11 (*a*) To begin packaging a seated patient with the LSP Halfback™ device, unfold the device and place it behind the patient's back. (*b*) Next, fasten the straps across the torso and the shoulder straps. Tighten the shoulder straps to pull the Halfback up against the patient's underarms, then tighten the torso straps. (*c*) Pass the leg straps underneath the knees from the outside in, work them underneath the thighs to the groin area, then fasten and tighten them. (*d*) Complete packaging with the Halfback by using the Velcro-mounted hood to immobilize the patient's head to the spine plate. (*Note:* Cervical spine immobilization was omitted for clarity of the photographs.)

Figure 18.11 (*Continued*) (*e*) A lifting bridal is provided with the Halfback.

(e)

with a short immobilizer, fully immobilize the patient on a long backboard at the earliest feasible opportunity.

It is critical for rescuers to be aware of proper procedures and use limitations of short spinal immobilizers. All manufacturers provide use information for their equipment. Use short immobilizers in a manner approved by the manufacturer. For example, the lifting handles on short immobilizers are intended only for minor movements required to position a patient on a backboard. If rescuers attach a hauling system to the lifting handles and haul the patient by the handles, spinal immobilization may be compromised and the patient injured further as a result.

Using the SKED litter

The SKED™ stretcher by Skedco is constructed of a flexible high-strength plastic material that is rolled up around a patient and secured with straps to form a "cocoon" around the patient (Fig. 18.12). This provides a very small package profile with numerous handles. In some cases it may be the only type of litter that will allow a patient to be passed through a small portal in order to be removed from a confined space. Because of this, the SKED has been used with great success in confined space rescue operations. The SKED also can be pulled and lifted or lowered in horizontal and vertical orientations using equipment provided with the SKED basic rescue system.

Spinal immobilization concerns with the SKED. The SKED stretcher alone provides no spinal immobilization, so it is important to establish spinal stabilization before the patient is placed in the SKED. This can be accomplished by securing the patient to a long backboard with all the C-spine precautions previously mentioned before the patient is packaged in the SKED.

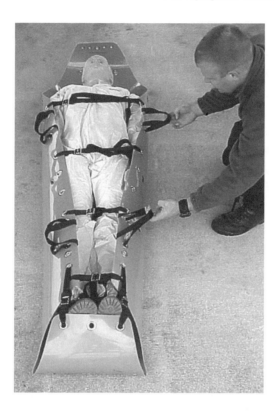

Figure 18.12 To package a patient in the SKED, the patient is placed on the SKED and straps are fastened across the patient to pull the SKED into a tubular shape. According to the standard instructions for rigging the SKED, the foot straps are tightened to pull the flap of material at the bottom of the SKED up to the patient's feet.

The use of the long backboard also provides good longitudinal rigidity to the package. For this reason it is a good idea to use the long board with the SKED, even in cases where no spinal damage is suspected.

Patient packaging with the SKED. The general procedure for packaging a patient in the SKED involves placing the patient on the SKED and fastening the straps across the patient's body to pull the SKED into a tubular shape (Fig. 18.12). According to the standard instructions provided for rigging the SKED, the foot straps are used to pull the flap of material at the foot end of the SKED up to the patient's feet, as shown in Fig. 18.12. Before using the SKED in a rescue operation, consult the specific instructions available from Skedco. Written directions are supplied with each SKED, and an instructional video is available from Skedco.

Rigging the SKED stretcher for transfer in the horizontal orientation. The SKED can be readily rigged for hauling or lowering in the horizontal orientation as shown in Fig. 18.13. The two bridle straps provided with the SKED are of different lengths; the shorter of the two is marked "head" because it is used at the head end of the stretcher. This gives the SKED a slightly head-up orientation during horizontal lowers and lifts. Specific instructions on rigging the SKED for horizontal operations are available from Skedco. Additional information on rigging horizontal litters is included later in this chapter in the section on using basket litters.

A safety belay line can be attached to the horizontally rigged SKED, as shown in Fig. 18.14. A discussion of rigging safety belays is included below in the section on litter lowering operations. Operation of safety belay systems was discussed in Chap. 13.

Standard SKED rigging for transfer in the vertical orientation. The SKED also can be rigged readily for hauling or lowering in the vertical orientation. Figure 18.15 shows a SKED rigged for these types of operations according to the user instructions provided by Skedco.

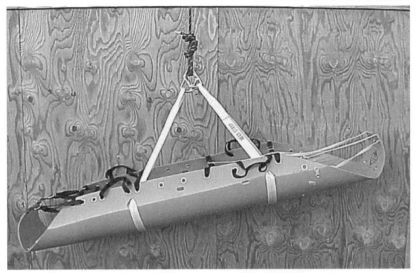

Figure 18.13 This SKED litter is rigged for hauling or lowering in the horizontal orientation.

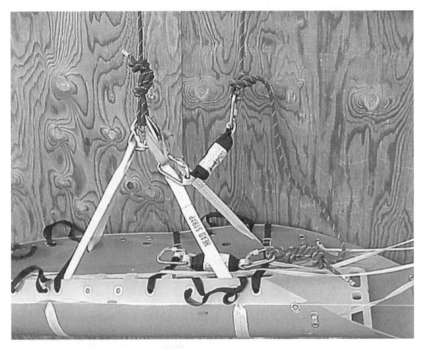

Figure 18.14 This horizontal SKED is equipped with a safety belay line rigged to attach to both the SKED bridle and the trainee/patient's harness.

To rig the SKED for vertical operations according to Skedco's instructions, use the 30-ft (9.1-m) length of ³/₈-in (10-mm)-diameter static kermantle rope provided with the SKED to rig a bridle. Locate the midpoint of the rope and tie a figure 8 on a bight into it. Use a double-loop figure 8, since this is the point where the bridle will attach to the lowering or lifting line.

To begin rigging the bridle to the SKED, run the lengths of rope on either side of the knot through grommets in the head end of the SKED. Weave both lengths of rope through the grommets and carrying handles down either side of the SKED, and then run them through the grommets at the foot of the SKED.

Finish the bridle by tying a square knot in the rope between the grommets at the foot end, then passing the two remaining lengths of rope up through the lifting handles above the foot. Tie another square knot between the lower lifting handles and add safety knots on both sides of the square knot (Fig. 18.15). Note that this same bridle configuration could also be used to pull the SKED horizontally, as when removing a patient through pipes or tubing.

A safety belay line can be attached to the vertically rigged SKED, as shown in Fig. 18.16. A discussion of rigging safety belays is included below in the section on litter lowering operations. Operation of safety belays was discussed in Chap. 13.

Another way to enhance the safety of the vertical SKED involves the use of a webbing sling. Form the sling by connecting the ends of a 16-ft (4.9-m) length of one-inch webbing using the water knot. Loop the sling around the board between the middle and the head end of the board by passing it through two handles on opposite sides of the board, as shown in Fig. 18.17. Attach the ends of the sling to both the haul line and the safety belay line. The sling provides an extra margin of safety in the event that the litter bridle fails.

Alternative SKED rigging for transfer in the vertical orientation. When the SKED is used in the vertical orientation with the standard rigging, the flap at the foot of the stretcher keeps the patient from sliding out the bottom of the SKED. There may be a tendency for the patient's body weight to shift onto the feet and legs in the bottom of the

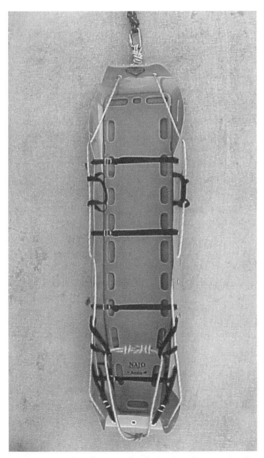

Figure 18.15 In the standard rigging for use in the vertical orientation, the foot flap of the SKED prevents the patient from sliding out through the bottom when the litter is vertical.

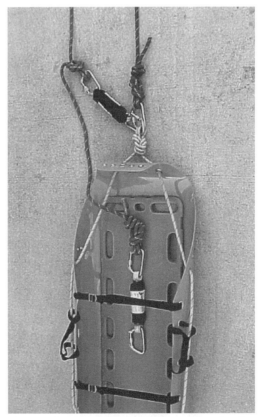

Figure 18.16 This vertical SKED is equipped with a safety belay line rigged to attach to both the rope bridle and the trainee/patient's harness.

Figure 18.17 As an extra safety measure, a sling can be attached to the backboard and the safety belay line when the SKED is rigged for use in the vertical orientation.

vertical stretcher. This may worsen spinal injuries and can be extremely painful to patients with significant injuries to the lower extremities. An alternative method of vertical rigging that allows the foot flap of the SKED to be left loose was developed by ROCO Rescue and approved by Skedco (Roop et al. 1998, p.190). This procedure assumes that the patient is firmly secured to a backboard before being packaged in the SKED.

As shown in Fig. 18.18, the alternative procedure follows the standard rigging instructions from Skedco up to the point at which the bridle ropes are being woven through the grommets and handles down each side of the SKED. To use the alternative rigging, run the ropes through the head-end grommet, the head-end lifting handles, and the middle grommet just as you would for the standard rigging. Pass the ropes from the outside to the inside through the middle grommet.

Loop each rope at least twice around the nearest backboard carrying handle toward the foot end of the backboard. Pass the ropes through the foot-end carrying handles and the unused grommets at the foot end of the SKED. Pass the ropes from the outside to the inside through the lower grommets. Tie the ends of the rope temporarily in a loose square knot between the lower grommets.

To complete the alternative rigging, pull all the slack out of the bridle ropes and fully tighten the lashing straps on the SKED. Retie the square knot between the lower set of grommets to eliminate all the slack pulled out of the bridle and finish with appropriate safety knots on each side of the square knot.

Other considerations for vertical SKED rigging. Anytime a package is vertical, the patient may tend to slide toward the foot end and may suffer potential additional injury or pain as a result. In rigging the SKED for use in the vertical orientation, use the same steps described earlier in reference to the backboard to prevent this (see Fig. 18.10).

Using basket litters

Basket litters have been used widely in elevated rescue operations for decades. Basket litters are not nearly as versatile as the SKED stretcher for transferring patients through small passages because they are bulkier and more rigid; however, these same

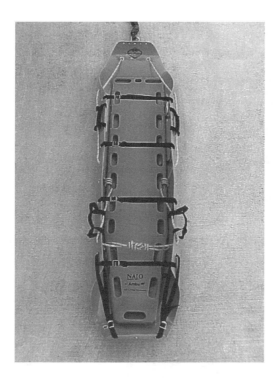

Figure 18.18 An alternative for rigging the SKED for use in the vertical orientation allows the foot flap to remain loose.

characteristics make basket litters excellent devices for operations such as lowering a patient to the ground after removal from an elevated space. In some cases it may be possible to use a "best of both worlds" approach by packaging the patient in the SKED for removal from an elevated confined space, then securing the SKED in a basket litter to be lowered to the ground.

Considerations for basket litter design and construction.   Basket litters are available in various designs and construction materials. It is critical that basket litters used in elevated rescue operations be of superior strength and structural integrity. Litters of lesser quality might be acceptable when rescuers are carrying a patient on the ground, but rope rescue operations put much more stress on the litter and the stakes are much higher.

One approach to ensure basket litter adequacy is to use a litter such as a steel and wire Stokes litter designed to meet relevant military specifications. Some currently available plastic and aluminum basket litters also are adequate. If a basket litter with a leg divider is used, any long backboard used with the litter must have a cutout between the legs to fit over the divider when the long backboard is placed in the litter.

Securing the patient to the basket litter.   The techniques used to secure the patient must be determined based on the mechanism of injury or illness. The key distinction will be whether or not spinal injury to the patient is possible.

If significant trauma can be ruled out, simply place the patient into the basket litter and secure him using the diamond lash technique with 30 ft (9.1 m) of one-inch tubular webbing, as shown in Fig. 18.19. Figure 18.41c shows the complete diamond lashed package. Extremely large patients may require a longer length of webbing. Since no trauma is suspected, wrap the patient's feet as shown in Fig. 18.19b to better secure the patient to the litter. Avoid placing webbing directly over the knees or putting undue pressure on the thorax that may restrict the patient's breathing.

If trauma-related injuries are possible, secure the patient to a long backboard using spinal immobilization precautions before placing him into the litter. Diamond-lash the packaged patient into the basket litter. Do not wrap the feet if spinal injury or trauma to the lower extremities are suspected.

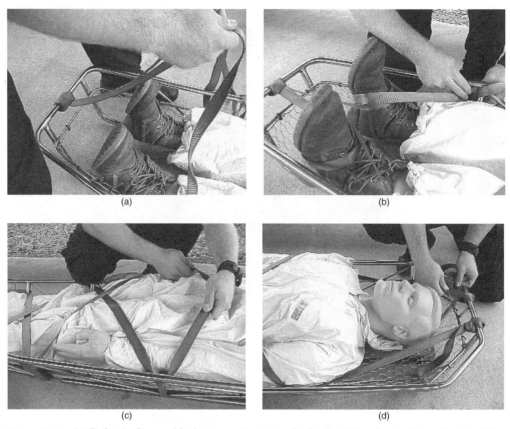

Figure 18.19 (*a*) To begin diamond-lashing a patient into the basket litter, locate the midpoint of the length of webbing and form a girth hitch with it around the top rail at the foot end of the litter. (*b*) If no spinal injuries are suspected, wrap the patient's feet with the webbing for greater security. Pass the webbing ends around vertical support members on opposite sides of the litter. (*c*) Cross the webbing ends and pass them around vertical support members higher up on opposite sides of the litter. Avoid wrapping the webbing around the top rail of the litter as it may be subject to abrasion as a result. Continue this process to form the diamond-lash pattern up the litter. (*d*) Continue diamond-lashing to a location above the patient's shoulders, then pull all slack out of the webbing along the entire length of the litter. Tie a clove hitch around the top litter rail at the head end of the litter with each of the webbing ends. To complete the diamond-lashing procedure, work all slack out through the clove hitches and tie safety knots for both.

Rigging the basket litter for patient transfer in the horizontal orientation.   If it is feasible, a horizontal litter orientation is preferable during patient transfer. The basket litter can be quickly and easily rigged for either single-line or double-line hauls or lowers in the horizontal orientation (see Figs. 18.31 and 18.32).

A variety of prefabricated litter bridles are commercially available for use in the single-line horizontal application. These bridles are convenient and easy to use. One example is the CMC ProSeries Vertical Evac Stretcher Harness, which features a fully adjustable litter bridle and a preattached tender line. In this chapter we will focus on rigging equivalent bridles in the field using common rescue equipment components.

Rigging horizontal single-line litter bridles.   One example of rigging for this type of operation is shown in Fig. 18.20. The setup shown in the photograph can be rigged using four utility straps, eight large locking steel carabiners, and a steel rigging ring, as follows.

1. Clip four of the carabiners through the litter rails on opposite sides with two toward the head end and two toward the foot end, as shown. Position the carabiners so that the gates face inward toward the patient and the locking nuts screw downward into the locked position. This will reduce the chance that the locking nuts will work into the unlocked position and the gates will be pushed open during use.

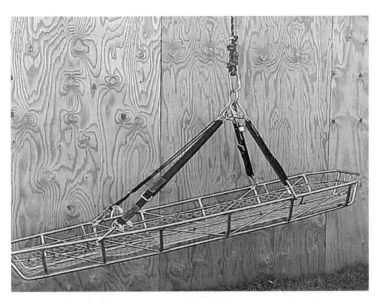

Figure 18.20   A basket litter can be rigged in various ways for single-line lowers or hauls in the horizontal orientation. In this example, the rigging ring provides a point of attachment for the bridle hardware.

2. Fully shorten all four belts to a length of 3 ft.

3. Double a utility belt through one of the carabiners at the head end and bring both D rings together. Attach both D rings to the rigging ring with a carabiner. Repeat this process for the other head-end carabiner using another utility strap.

4. Attach the D ring at one end of one of the remaining belts to one of the foot-end carabiners. Attach the D ring at the other end of the strap to the rigging ring. Repeat this process for the other foot-end carabiner using the fourth utility belt. Note that the effective length of the foot-end belts is twice that of the head-end straps, so the head end of the litter will be elevated.

5. Check all rigging. Be sure that all carabiners are locked. Orient carabiners connecting the belts to the rigging ring so that the gates screw downward into the locked position.

A belay line can be attached to the horizontally rigged basket litter, as shown in Fig. 18.21. A discussion of rigging safety belays is included in the section on litter lowering operations below. Operation of the belay system was described in Chap. 13.

The rigging shown in Fig. 18.20 is just one way of rigging the basket litter for a horizontally oriented single-line lower. Many variations that will work well are possible. For example, Fig. 18.22 shows a similar system without the rigging ring, in which the carabiners at the top of the bridle connect directly to the main lowering/hauling line. Two carabiners are used to attach the shorter belts toward the head of the litter to the main line. This is done because a single carabiner would be subjected to a three-way pull or triload in that application. The two belts toward the foot end of the litter are long enough that a single carabiner can to be used to attach them to the main line without triloading the carabiner. To attach a belay line to the bridle shown in Fig. 18.22, pass a doubled sling of webbing through all hardware at the top of the bridle to create the point of attachment (Fig. 18.23). As another alternative, replace the carabiners connecting the belts to the lowering/hauling line with two trilinks, with one trilink holding the D rings for the foot-end straps and the other trilink holding the D rings for the head-end straps. This eliminates any potential for three-way loads on carabiners in the bridal. It also significantly shortens the effective length of the bridle, and is actually the bridle connection to the lowering line for the short-rigged bridle described below.

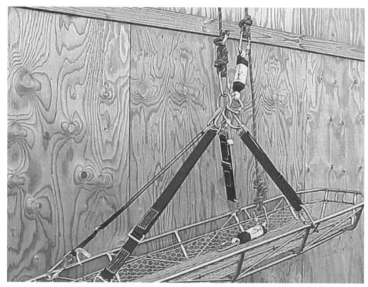

Figure 18.21 This horizontal basket litter is equipped with a safety belay line rigged to attach to both the rigging ring in the litter bridle and the trainee/patient's harness.

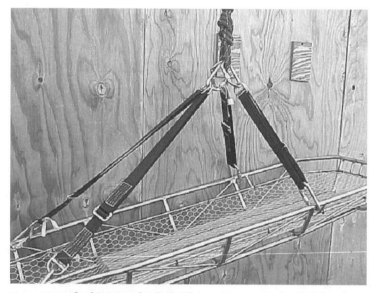

Figure 18.22 In this example, the bridle carabiners are attached directly to the end of the lowering/hauling line.

To rig for stairwell lowers, incorporate a suitable swivel as shown in Fig. 18.24. The swivel prevents twisting of the lowering line as the litter is lowered down the stairwell, as described later in this chapter.

Other alternatives to the bridle rigging described above include using loops of webbing when utility straps are not available. Another option is to use a rope bridle that has been prerigged with Prusik hitched loops of accessory cord to provide universal adjustment of all four legs of the bridle (see Fig. 18.25).

Short-rigged single-line horizontal litter bridles. In some cases it is advantageous to shorten the bridle as much as possible. This makes edge transitions much easier by shortening the distance that the litter will have to be lowered before loading onto the lowering system. The longer this distance, the more difficult it is to negotiate the edge. The chance of losing control of the litter and shock-loading the system is also increased.

Figure 18.23 In this example, a doubled sling of webbing is passed through all the bridle hardware to create a point of attachment for the belay line at the litter bridle.

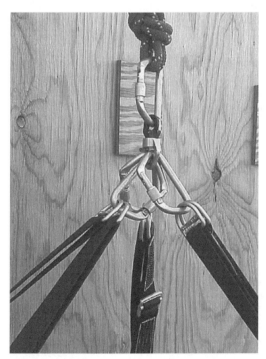

Figure 18.24 The Petzl swivel is used in this single-line horizontal litter bridle to prevent twisting of the lowering line during stairwell lowers.

Figure 18.25 This rope bridle has legs that are universally adjustable using Prusik hitched loops of accessory cord. It also has a tender line that provides convenient positioning of a litter tender.

Some commercially available litter bridles are fully adjustable and can be shortened as much as needed. When fully adjustable bridles are not available, other means must be used to shorten the bridle. The bridles shown in Figs. 18.20 through 18.23 could be "short-rigged" by omitting the carabiners from the litter rails and doubling the utility straps directly around the rails, as seen in Fig. 18.26. The D rings of the straps are connected directly to the knot using trilinks (see Fig. 18.26). When this rig is used, the straps on the foot end should be longer than those on the head end to give the litter a head-up orientation. This is not the ideal bridle rigging because the utility straps are subject to damage at the point where they wrap around the litter rail. Care must be taken to prevent them from hanging up or rubbing along sharp or abrasive surfaces during patient transfer operations. The advantage of a shorter bridal in edge negotiations may outweigh this disadvantage in some cases.

Rigging horizontal double-line litter bridles.   One example of rigging for this type of operation is shown in Fig. 18.27. The setup shown can be rigged using four utility straps, four large locking steel carabiners, and two $^{11}/_{42}$-in steel trilinks, as follows.

1. Clip all four carabiners through the litter exactly as described for the horizontal single-line bridle.

2. Fully shorten all four utility belts to a length of 3 ft.

3. Double one belt through one of the carabiners at the head end and bring both D rings together. Repeat this process for the other head-end carabiner using the other utility strap. Use one of the trilinks to connect the D rings from both head-end straps to the knot at the end of the head-end lowering line.

4. Repeat step 3 for the foot end of the basket and connect the foot-end bridle to the foot-end lowering line. The foot-end straps are of the same length as the head-end straps, and the attitude of the basket is controlled with the two lowering lines.

5. Check all rigging. Be sure all trilinks are closed and all carabiners are locked. Orient the carabiners connecting the straps to the basket so that the gates face inward and the locking nuts screw downward into the locked position.

Safety belays are seldom used on double-line systems because the two lowering systems back each other up. For extra safety, connect a fall-arrest lanyard and shock absorber from the trainee/patient's harness to the lowering system as shown in Fig. 18.28. An optional belay line rigged to both lowering systems is also shown in Fig. 18.28.

Figure 18.26 The basket litter can be "short-rigged" to make edge negotiations easier by omitting nonessential components when rigging the bridle.

Figure 18.27 The basket litter is rigged to provide points of attachment for two separate systems for double-line lowers or hauls.

A discussion of using the fall-arrest lanyard and optional safety belay is included in the section below on litter lowering operations.

Rigging the basket litter for patient transfer in the vertical orientation. Various techniques can be used to rig a basket litter to be lifted or lowered in the vertical orientation. One effective technique is to use 30 ft (9.1 m) of $\frac{31}{48}$-in (10-mm) or larger static kermantle rope to create a bridle. One example of how this can be done is shown in Fig. 18.29. The complete bridle is shown in Fig. 18.41c.

As noted in discussing the use of the backboard and the SKED litter, whenever a package is vertically oriented, the patient has a strong tendency to slide toward the foot end. This also applies to the basket litter. If the patient has leg or spinal injuries, the resulting compression of the legs can be very painful to the patient and may worsen the injuries. Apply the remedies previously discussed to prevent this when using the basket litter (see Fig. 18.10).

A safety belay line can be attached to the vertically rigged litter as shown in Fig. 18.30. For extra safety, a webbing sling can be looped through the backboard, if one is used, and attached to the main line and the safety belay line, as previously described in reference to the SKED and shown in Fig. 18.17. A discussion of rigging safety belays is included in the section below on litter lowering operations. Operation of safety belays was described in Chap. 13.

Litter Lowering Operations

In some cases, after rescuers have properly packaged a patient, they must face the major challenge of lowering the patient from an elevated confined space to complete the rescue. In this section we will examine several ways of meeting that challenge.

Basic considerations and assumptions regarding litter lowers

The techniques described below assume that we are beginning with a patient who is properly packaged in a stretcher or litter that is properly rigged for the type of lower to be performed. We will then attach the litter or stretcher to an appropriate lowering system and conduct the lower as described below. We will rig and operate the lowering system as

Figure 18.28 Safety precautions for double-line operations include attaching a fall-arrest lanyard with a shock absorber from one of the lines to the trainee/patient's harness and attaching a safety belay line to both lines.

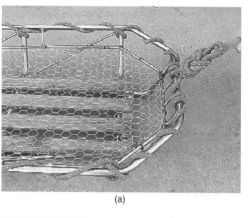
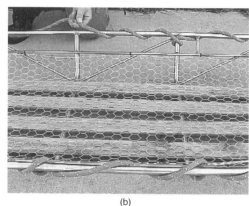

(a)   (b)

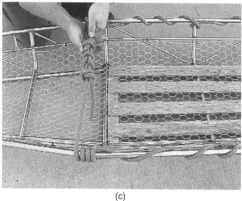

(c)

Figure 18.29 (*a*) To rig a basket litter for patient transfer in the vertical orientation, begin by tying a double-loop figure 8 on a bight in the middle of the bridle rope. Keep the loops short. Wrap each end of the rope at least twice around the top rail of the litter at the head end as shown. Pull all slack in both ropes through, so that the figure 8 knot is directly above the head end of the litter. (*b*) Work one of the ropes down each side of the litter, making several wraps around the top rail on each side as you go. Also make a full wrap around each vertical support member along the way. Continue wrapping the rope around the litter rails and support members about two-thirds of the way toward the foot of the litter. (*c*) To complete the bridle, make four full wraps with the rope around the top rail on each side of the litter at a point just below a vertical support member. Pull any remaining slack out of the bridle and tie a square knot with the rope ends at the bottom of the bridle as shown. Tie a safety knot on each side of the square knot.

described in Chap. 17, including personnel roles and operational commands. In the examples below, we will assume that the lowers are being conducted in a training session, with a trainee posing as the patient, and that the trainee/patient is wearing a full-body harness. We will also assume that the edge to be negotiated is a strong handrail.

Consider the following factors in choosing lowering systems. In most cases, the simplest and easiest option for completing the rescue safely will be the best choice.

Horizontal versus vertical litter lowers.   As a general rule, it is preferable to lower the patient with the litter in the horizontal orientation rather than the vertical orientation. If a horizontal lower is carried out properly, the patient should never see anything but sky during the entire lowering operation. In comparison, a vertical lower gives the patient a bird's eye view of just how high he is during the operation. The patient is more likely to panic or become distressed during a vertical lower.

When a litter is vertically oriented, the patient has a strong tendency to slide down to the foot end. If the patient has leg or spinal injuries, the resulting compression of the legs can be very painful to the patient and may worsen the injuries. Practices intended to avoid this when vertically oriented litters are used were discussed in describing rig-

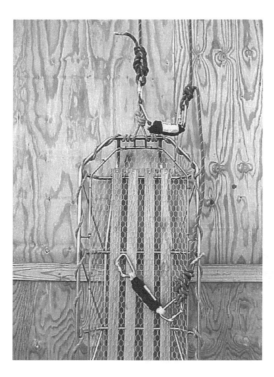

Figure 18.30 This vertical basket litter is equipped with a safety belay line rigged to attach to both the rope bridle and the trainee/patient's harness.

ging options for both the SKED and the basket litter. The best way to avoid it is to lower the litter in the horizontal orientation whenever it is feasible to do so.

Using tag lines versus litter tenders for lowers. Lowering operations are all about controlling the litter. The rate of descent must be carefully controlled. In addition, the litter must be guided during the lower to prevent it from hanging up on obstructions or rubbing along the side of the structure from which the lower is being made. The litter can be guided in one of two ways: (1) by the use of *tag lines*, ropes that are attached to the litter and used by team members on the ground to guide the basket during the descent (Fig. 18.31) and (2) by utilizing *litter tenders*, who ride along with the basket and guide it as it is lowered (Fig. 18.35).

As a general rule, tag lines are preferred over litter tenders for guiding the litter, if this is feasible. Tag lines allow team members guiding the litter to do so from a safe location. The use of attendants requires team members to load onto the system after the basket is over the edge, which can be awkward for them and unsettling for patients. In some cases the use of tag lines is not feasible because of the height involved, the location, the configuration of the structure, or other factors. Litter attendants have the advantage of being able to comfort and care for the patient as they guide the litter.

Single-line versus double-line litter lowers. Some rescuers prefer a single-line litter lower (Fig. 18.31) because it requires less equipment, less rigging time, and less manpower than does the double-line system (Fig. 18.32). The single-line system is also simpler to operate, since the double-line system requires the actions of two separate lowering systems to be coordinated. However, two attendants may be required for difficult lowers so that one can guide the basket while the other cares for the patient. Such a case necessitates two lowering lines, since a three-person load would exceed the working load limit of a single rope. In some cases, rescuers select a double-line system, even though tag lines or a single attendant will be used to control the load. These rescuers believe that the double-line system is significantly safer than the single-line system because two separate lowering systems are used and each system backs the other up.

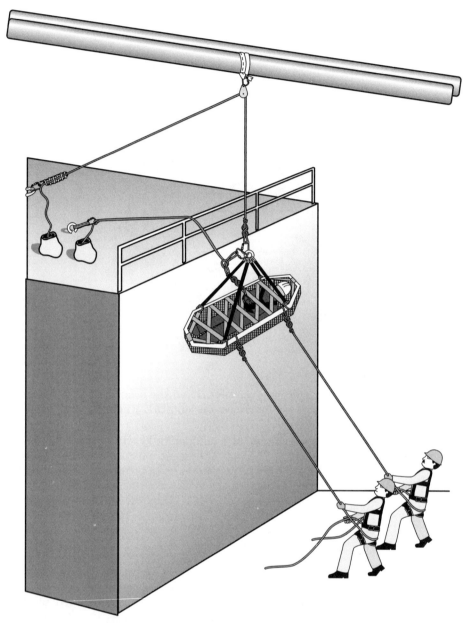

Figure 18.31 Single-line lowering operations utilize a single lowering system.

High-point versus low-point lowers. There are significant advantages to rescuers in negotiating edges with a litter if descent control devices are anchored at a level above the one from which the litter is lowered. Using high-directional pulleys conveys the same advantage when the descent control device is anchored at the same level from which the litter is being lowered (Fig. 18.31). For this reason, high-point lowering operations (Fig. 18.31) are always preferred over low-point lowers (Fig. 18.32) if they are feasible.

Use of safety belays during litter lowers. Always use belay lines in training. They are also strongly recommended for use in actual rescue operations. For litter lowering operations, a belay line should be attached to both the litter bridle and the patient's harness, if the patient is wearing one, before the litter is moved over the edge. Trainees posing as victims during training evolutions should always wear harnesses to provide a point

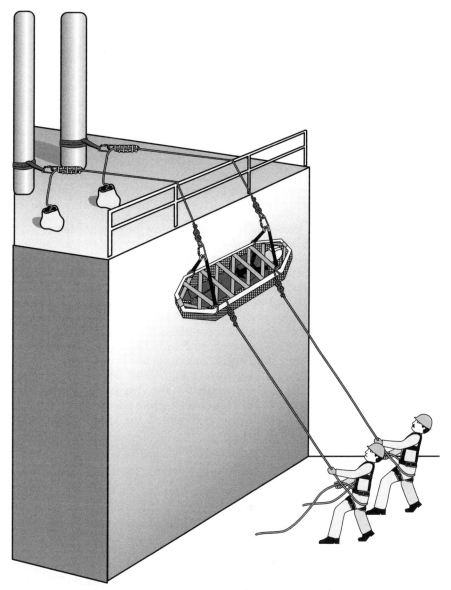

Figure 18.32 Double-line lowering operations utilize two separate lowering systems.

of attachment for the belay system. If litter tenders are used, they should also be attached to a belay line.

Safety belays for single-line lowers. Examples of belay lines rigged for single-line lowering systems are shown in Figs. 18.14, 18.16, 18.21, and 18.23, as well as several other figures in this chapter. To rig the belay lines shown, tie a double-loop figure 8 into the end of a suitable rope, then tie a double-loop butterfly above the knot at the end of the line. Equip the end of the safety line with a shock absorber and attach it to the trainee/patient's harness. Equip the butterfly knot with a shock absorber and connect it to the litter bridle.

When safety belay lines are rigged in this way, the distance between the butterfly knot and the end of the safety line should have additional slack equal to the length of the shock absorber when fully deployed or extended. This is usually around 3.5 ft (1.1 m). If any part of the lowering system fails, the litter could shock-load the belay line. If shock loading occurs, the length between the figure 8 knot and the butterfly loop must be sufficient to allow the shock absorber at the litter bridle to fully extend without the patient's harness loading onto the belay line. With the belay system rigged in this way, the harness should load onto the belay line only in the event the litter fails or the patient falls out of the litter.

If an attendant uses the same belay line as the patient, the rope used for the belay line must be rated for general use. For rescues in which the patient is not wearing a harness, the attendant can attach the end of the safety belay line to his harness. In some rescues the victim will wear a harness, and in all training evolutions the trainee posing as patient should wear a harness, so that a point of attachment for the end of the belay line will be available. In these cases, the attendant can attach one end of a fall-arrest lanyard to the double-loop butterfly knot in the safety line and attach the other end of the lanyard, fitted with a shock absorber, to his harness (see Fig. 18.35). If no fall-arrest lanyard is available, substitute a sling created by connecting the ends of a length of one-inch webbing with a water knot. Another option is to provide a separate belay system for the attendant if extra equipment and an extra belayer are available.

Safety belays for double-line lowers. Safety belay lines usually are not used during double-line lowers because two separate lowering systems are in place; however, if one lowering system fails, the other system can receive a three-person load and be overloaded. This is highly unlikely because each system backs up the other system; however, a safety belay line can be rigged easily by tying a double-loop figure 8 with very large loops. Rig one loop to the head-end lowering system and the other loop to the foot-end lowering system. The loops should pass through the hardware connecting the bridles to the knots in the ends of the lowering lines, as seen in Fig. 18.28.

As an optional safety measure during training, attach a fall-arrest lanyard, anchor strap, or webbing sling and shock absorber from the trainee/patient's harness to one of the lowering lines as shown in Fig. 18.28. The same optional measure can be used in an actual rescue if the patient is wearing a harness. It is recommended that this device be attached to the foot lowering line (Roop et al. 1998, p. 182). Rescuers should have at least two points of attachment to the system during double-line lowers as discussed below.

Conducting horizontal single-line high-point lowers

In horizontal single-line high-point lowers the litter is lowered to the ground in a horizontal orientation using a single lowering line. Either the descent control device is anchored at a level above the edge that must be negotiated to begin the lower, or a high-directional pulley is rigged above the edge (see Fig. 18.31).

The horizontal single-line high-point lower will serve as the model lowering operation for this section of the text. The other types of lowering operations described have numerous similarities to this operation. To simplify coverage of the other types of litter lowering operations, reference will be made to this operation to avoid redundancy, so it is important to fully understand the procedures for this litter lowering technique before going on to the other types of litter operations.

Rigging the litter. In preparation for the lower, package the patient in a suitable litter and rig the litter for a horizontal single-line lower. Attach the lowering line to the litter bridle as described earlier and as seen in Figs. 18.20 through 18.23. If an attendant is used, the lowering line must be rated for general use. If tag lines are used, rig them to the litter before the operation begins, as shown in Fig. 18.31 or 18.33.

Lifting the litter. Before issuing orders to begin the operation, the rescue team leader conducts a safety check of all rigging and the status of the patient and verifies that everyone involved in the operation is ready to proceed. The team leader gives a command such as "prepare to lift" to notify the lifting team to prepare to lift the litter. The safety belayer issues a reply such as "safety on" or "ready on safety" to confirm that he is on alert status and ready to belay a falling litter. At this point the person at the head of the stretcher assumes the role of lifting team leader and directs the lifting team. The lifting team leader issues a command such as "lift." In response the lifting team lifts the litter up and holds it with the top surface of the litter approximately even with the top of the handrail, or as close to that height as they are able to lift it. To avoid injury due to lifting strain, use adequate numbers of personnel and good lifting techniques, such as lifting with the legs instead of the back, whenever lifting or handling loaded litters.

Roping in the lowering line.   Once the litter has been lifted, the lowering line operator ropes in the descent control device. If a brake bar rack is used, he ropes in all six bars and holds the rack locked off, or else ties it off, as described in Chap. 17. If a figure 8 descender is used, he ropes it in, using a single wrap (with tag lines) or a double wrap (with a basket tender) and locks or ties it off (see Table 18.1). The lowering line operator informs the team leader of the status of the line, for example, by saying "locked off." This lets the team leader know that it is safe to move the litter over the edge.

Moving the litter over the edge.   The team leader makes a quick final safety check and directs the lifting crew to place the litter on the outside of the rail. When a high-directional pulley is used, this process is greatly facilitated if an extra team member grasps the haul line with both hands midway between the descent control device and the high-directional pulley and pulls down with his body weight. This maneuver utilizes a compound force to lift the litter higher than the handrail so that the lifting team can easily push the litter beyond the edge. The extra team member slowly releases the lowering line so that the litter comes to rest on the end of the lowering line on the outside of the handrail. During this maneuver, lifting team members must take care to avoid having their fingers mashed between the edge and the litter. The way the litter lower is conducted from this point on depends on whether tag lines or an attendant are used to guide the litter to the ground.

Horizontal single-line high-point lowers using tag lines.   To conduct the lower using tag lines to guide the load, attach two tag lines to the litter when the litter is initially rigged. Attach tag lines to a horizontally rigged basket litter as shown in Fig. 18.31 or 18.33. Attach tag lines to carrying handles on a horizontally rigged SKED litter in the same way. To attach a tag line to a litter, tie a figure 8 on a bight and attach it with a carabiner (Fig. 18.33), or attach the rope directly to the litter using a figure 8 followthrough (Fig. 18.31). A single rope can be used for both tag lines by attaching both ends to the litter or by tying two butterfly loops near the midpoint and attaching them to the litter (Fig. 18.33). After the litter is over the edge, feed the tag lines down to the two tag-line tenders on the ground.

The tag-line tenders each take a tag line, back away from the structure, and assume the position shown in Figs. 18.31 and 18.34. Each tag-line tender grasps the line coming from the litter in one hand, passes the rope just below the buttocks, and

TABLE 18.1   Using Descent Control Devices for Litter Lowers

Type of litter lower to be performed	Number of bars roped in on brake bar rack	Number of wraps on figure 8 descender
Single-line lower with tag lines	5	Single wrap
Single-line lower with litter tender	6	Double wrap
Double-line lower with tag lines	4	Single wrap
Double-line lower with one litter tender	5	Single wrap
Double-line lower with two litter tenders	6	Double wrap
Stairwell lower	5	Single wrap

Figure 18.33 Tag lines attached to litters allow them to be guided by tag-line tenders on the ground during lowers.

Figure 18.34 Tag-line tenders guide the litter to keep it clear of the side of the structure during lowers.

holds the rope taut with the other hand. This position allows the tender's entire body to be used like an anchor post to hold the litter off the side of the building during the lower.

Once the tag-line tenders are in position, the team leader issues an order such as "lower" to begin the lowering phase of the operation. In response, the lowering line operator unlocks the brake bar rack, removes one bar, and begins lowering the litter on five bars (see Table 18.1). If a figure 8 descender is used, it should be single-wrapped. During the descent the tag-line tenders keep the litter clear of the side of the structure so that it does not rub or hang up. As the lower proceeds, the team leader directs the operation using voice commands or hand signals. Figure 17.16 shows examples of simple hand signals. Coordination of factors such as rate of descent and the amount of slack in the safety belay line is controlled by signals from the team leader. If any team member recognizes an unsafe condition at any time during the lower, that team member issues the command "Stop!" immediately to warn everyone to stop the lower until the hazard is abated. In reply, all team members should repeat the command "Stop!" to verify that they heard the order.

As the litter approaches the ground, the tag-line tenders guide it to a smooth landing. Once the litter is on the ground, a command such as "on the ground" or "off line" is issued to notify everyone that the lower is over. The rigging is removed from the litter, and preparations begin for further treatment and transport of the patient.

Horizontal single-line high-point lowers with a litter tender. Conducting a litter lower with an attendant requires the attendant to climb out, load onto the lowering line, and take position after the litter has been moved over the edge. Before even approaching the handrail, the attendant should be attached to a safety belay system as discussed previously.

Rigging the litter tender's connection to the lowering system. In preparation for going over the edge, the attendant rigs a connector that will be used to attach the front waist D ring of his harness to the knot in the end of the lowering line. It is recommended that the length of the connector be quickly and easily adjustable. One way to do this is by using a doubled utility belt with a carabiner in each end, as shown in Fig. 18.35. Adjust the connector so that it reaches from the bottom of the knot in the lowering line to the nearest point on the top rail of the litter. The waist D ring of the litter tender's harness should be supported a few inches below the top rail of the basket when the tender is suspended from the lowering system by the connector (Fig. 18.35). If a utility belt is not available, rig a sling by connecting the ends of a length of one-inch webbing using a water knot.

Once the connector is rigged and adjusted, the attendant attaches one end of the connector to the knot in the end of the lowering line and attaches the other end to the front waist D ring of his harness. When the attendant is attached to both the lowering system and safety belay line, the team leader does a safety check of the rigging and gives him the go-ahead. The attendant then climbs over the handrail and loads onto the lowering system.

One alternative to the utility belt or webbing sling connector is to use a tender line. The *tender line* is a static kernmantle rope of at least 14 ft (4.3 m) in length that is attached to the end of the lowering line as shown in Fig. 18.25. Use either a Gibbs ascender or a Prusik hitched loop of accessory cord as a rope-grabbing device to form the attendant's point of attachment on the tender line. All rope and hardware used in the tender line must be at least light-duty-rated. Tie a stopper knot at the end of the tender line to keep the tender from sliding off the end of the line if the rope-grabbing device fails to grab. Like the utility belts, the length of this connector must be properly adjusted before the attendant loads onto the system.

The tender line may be rigged to allow the attendant to adjust his position on the tender line during a lower. To do this, attach a second rope-grabbing device to the tender line above the attendant's point of attachment. Fit the upper rope-grabbing device with an etrier or a loop of webbing or rope. This allows the attendant to bend one leg at the knee, put the foot into a properly positioned loop suspended from the upper rope grabbing device, and straighten the leg, unloading his weight from the rope-grabbing device attached to his harness. This is very similar to the ascending self-rescue procedure for rappelling described in Chap. 17. Once the rope-grabbing device attached to the attendant's harness is unloaded, he can move it to any position desired, then load his weight back onto it.

Positioning the litter tender for a horizontal single-line lower.   Going over the edge and loading onto the lowering system can be accomplished in different ways. When the edge is a handrail, the attendant may be able simply to climb over the handrail just beyond the head of the stretcher and maneuver into the attendant position by moving his feet lat-

Figure 18.35 For horizontal single-line lowers, attach the litter tender to the lowering line using a connector adjusted so that the front waist D ring of the tender's harness is supported a few inches below the top rail of the litter.

erally along the edge of the decking and moving the hands along the outer side of the litter. This is probably the easiest method for both the attendant and the patient, but will not work in some situations. Another option is for the attendant to climb out, over, and around the head end of the litter itself. When a Stokes basket is used, the attendant grasps the lowering line with his hands, steps around the top rail of the litter to the proper location, then slowly lowers himself into position, taking care to step on the litter rail instead of the patient's face.

After the attendant is in position (Fig. 18.35), he can relax with his weight supported by the connector rigged between his harness waist D ring and the end of the lowering line. He grips the top rail of the litter with both hands and places his feet against the side of the structure. When the lower begins, he uses his legs to keep the litter clear of the side of the building. Some attendants like to attach the waist D ring of the harness to the top rail of the litter with an additional carabiner to assist in controlling the litter and provide an additional point of attachment to the system. A second attendant cannot be used in a single-line lower, because the rope and other system components used are not rated for a three-person load.

Lowering the litter and tender.   Once the attendant is in position, the team leader issues an order to begin the lowering phase of the operation. If a brake bar rack is used, the lowering line operator unlocks the rack but leaves all six bars racked in and begins lowering the litter and attendant. If a figure 8 descender is used, it should be double-wrapped (see Table 18.1). The team leader directs the lowering operation using voice commands or hand signals as needed. Using information provided by the attendant, the team leader coordinates factors such as the rate of descent and the amount of slack in the safety belay line. In some cases, radio communication between the attendant and the team leader or a team member assigned to relay messages may be required. If any team member recognizes an unsafe condition at any time during the lower, he issues the command "Stop!" immediately to warn everyone to stop the lower until the hazard can be abated. Each team member should reply "Stop!" to verify that the order was received by all.

As the litter approaches the ground, the attendant guides it to a smooth landing. Once the litter is on the ground, a command such as "on the ground" or "off line" is issued to notify everyone that the lower is over. The rigging is removed from the litter, and preparations begin for treatment and transport of the patient.

Conducting horizontal double-line high-point lowers

Horizontal litter lowers also may be conducted using two lowering lines, as seen in Figs. 18.36 through 18.39. Except for the use of the double lowering lines, these operations are quite similar to the equivalent single-line lower already described. The incorporation of a second lowering line offers the option of utilizing two litter tenders. Two attendants may be required to control the litter during difficult lowers or so that one attendant can care for the patient while the other controls the litter. The procedures used in double-line operations vary depending on whether tag lines, a single attendant, or two attendants are used.

Double-line high-point lowers using tag lines.   Horizontal high-point double-line lowers using tag lines, as seen in Fig. 18.36, are carried out following the same steps described above for the single-line high-point lower using tag lines, except that two lowering lines must be rigged. Lower the single-person load with at least four bars in both brake bar racks for operations in which brake bar racks are used (see Table 18.1). If figure 8 descenders are used, both descenders are single-wrapped. Other than that, the only significant difference is the way the lowering lines are attached to the litter, as described previously and seen in Fig. 18.27.

Double-line lowers with two-litter tenders.   Other than the fact that two lowering lines must be rigged, the procedure for preparing to do a horizontal double-line high-point lower with two attendants is basically the same as described above for an equivalent single-line system. Once the litter has been lifted and moved to the outside of the rail and the attendants are ready to load onto the system, the procedures used for this operation differ significantly.

Rigging the litter tenders' connections to the lowering system. Each attendant rigs a connector to use in attaching to the system, as described above for single-line litter lowers. The connectors are adjusted so that they will reach from the bottom of the knot in the lower line to the nearest point on the top rail of the litter. This allows each tender to be supported so that the waist D ring of his harness is positioned a few inches below the top rail of the basket when the tender is suspended from the lowering system by the connector, as shown in Fig. 18.37.

Positioning two-litter tenders for the horizontal double-line lower. In preparation for loading onto the system, the foot attendant attaches the connector between the front waist D ring of his harness and the knot in the end of the lowering line nearest the foot of the litter. The team leader conducts a safety check of all rigging. The foot attendant moves around the foot end of the litter and into position. This is done as described earlier for single-line lowers, except that the attendant goes around the foot end of the litter. Once in position, the foot attendant uses a carabiner to connect the front waist D ring of his harness to the top rail of the litter to provide a second point of attachment to the system. The foot attendant begins calming the patient while the head attendant loads onto the other lowering line.

Once the foot attendant is in place, the head attendant attaches the connector between the front waist D ring of his harness and the knot in the end of the lowering line nearest the head of the litter. The team leader conducts a rigging safety check before giving the head attendant the go-ahead. The head attendant moves around the head of the litter and into position, as described previously. The head attendant uses a carabiner to connect the waist D ring of his harness to the top litter rail to provide a second point of attachment to the system (see Fig. 18.37).

When attendants are loading onto the lowering lines, they have only one point of attachment to the system: the connector strap or loop. It is recommended that a second system be used for additional fall protection while attendants are loading on. This is accomplished by attaching a separate safety belay line or a fall arrest lanyard to each rescuer during the procedure.

Safety belays for the double-line double-tender lower. Safety belay lines are not commonly used during double-line lowers because two separate lowering systems are in place; however, the failure of one system could place a three-person load on the other system and overload it. For this reason safety belay lines are sometimes used during double-line lowers. The rigging and use of a safety belay line with the double-line system was

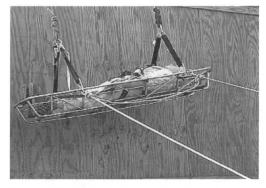

Figure 18.36 In this horizontal double-line lowering operation, tag lines are used to guide the litter during the lower.

Figure 18.37 Two litter tenders can be used for horizontal double-line lowers.

discussed earlier and is shown in Fig. 18.28. Another safety precaution is to attach a fall-arrest device between the patient's harness and one of the lowering lines, as previously discussed and shown in Fig. 18.28.

Lowering the litter and tenders.   After both attendants are in place and everyone is ready, the double-line lower proceeds. When the order to lower is issued for operations in which brake bar racks are used, both lowering line operators unlock their racks and begin to lower the basket with all six bars in the rack. If figure 8 descenders are used, they should be double-wrapped (see Table 18.1). Lowering proceeds as described above for the equivalent single-line system.

During the lower the foot attendant concentrates on guiding the litter and communicates with the team leader. The head attendant concentrates on comforting and caring for the patient and assists the foot attendant as needed in handling the litter.

The team leader directs the lowering line operators using voice commands and/or hand signals to maintain a reasonable rate of descent and a slightly head-up attitude of the litter. Ideally the team leader is positioned between the two lowering lines and uses the right hand to direct the lowering line operator on his right and the left hand to direct the other lowering line operator (see Fig. 18.38).

Double-line lowers with a single-litter tender.   The single-attendant double-line lower is conducted much like the double-attendant lower just described. Only one attendant is used, and the attendant is attached to both lowering systems as shown in Fig. 18.39.

Rigging the litter tender's connection to the lowering system.   In preparation for attaching to the lowering systems, the attendant rigs two connectors as described earlier. Adjust each connector so that when the carabiner at one end is hooked into the knot in the end of one of the lowering lines, the carabiner in the other end will reach to a point just below the midpoint of the top rail of the litter. Ideally the front waist D ring of the attendant's harness is positioned a few inches below the litter rail when the attendant is attached to both lowering lines (see Fig. 18.39).

Positioning the litter tender for the horizontal double-line lower.   The attendant has a couple of options for getting into the proper position on the litter. One option is to attach one of the connectors from the harness waist D ring to the knot in the foot lowering line, climb over and around the foot of the stretcher, connect the other connector from the waist D ring to the knot in the head-lowering line, then slowly lower into the attendant position, as seen in Fig. 18.39. Another option is to attach the first connector from the waist D ring to the knot in the head-lowering line, go around the head end of the litter, assume the attendant position, then have the foot-lowering line operator lower the foot of the litter enough to allow the other connector to be attached to the knot in the foot-lowering line. It may not be possible to put a third carabiner through the harness waist D ring if the attendant desires a direct attachment to the

Figure 18.38 During double-line lowering operations, the rescue team leader coordinates the actions of both lowering line operators to maintain an appropriate rate of descent and a slightly head-up attitude of the litter during the lower.

Figure 18.39 When using a single litter tender for horizontal double-line lowering operations, use connectors to attach the tender to both lowering lines.

top rail of the litter. In that case a webbing sling or some other means to connect to the D ring is required, if such a connection is desired. A secondary fall-protection system should be used while the attendant is loading on, as discussed above for two attendant operations.

Safety belays for the double-line single-tender lower.   A safety belay is seldom used during single-attendant double-line lowers, since each lowering system backs up the other and there is no significant likelihood for overloading one system should the other system fail. A safety belay line with a large double-loop figure 8 may be attached to the system, as described above and shown in Fig. 18.28. A fall-arrest lanyard or similar improvised item with a shock absorber can be connected from the patient's harness to one of the lowering lines, as seen in Fig. 18.28.

Lowering the litter and tender.   Once the attendant is in position, the lower begins. After receiving the order to lower from the team leader, both the lowering line operators unlock their brake bar racks, remove one bar, and lower the load to the ground on five bars (see Table 18.1). If figure 8 descenders are used, both are single-wrapped. The attendant functions as described earlier to control the litter and comfort the patient. The attendant communicates with the team leader who directs the lowering line operators using voice commands and/or hand signals to maintain a reasonable rate of descent and a slightly head-up attitude of the litter.

Conducting horizontal low-point litter lowers

Horizontal litter lowers using a low change-of-direction point (Fig. 18.32) are conducted much like equivalent high-point lowers. The main difference in low-point operations is that moving the basket over the edge is more challenging than in equivalent high-point lowers. Single-line or double-line low-point lowering systems can be used. In our examples we will assume that the edge to be negotiated is a strong, 2-in outside-diameter round handrail that serves as the change of direction point.

Preparing to lower the litter. In preparation for the lower, properly rig the litter for either a single-line or double-line lower. Take all safety precautions described previously before the team proceeds. Index the lower line knot(s) to the edge (the top of the handrail in our example) as described for low-point lowers in Chap. 17. If brake bar racks are used, rope them in on all six bars and lock and/or tie them off. If figure 8 descenders are used, rope them in and single-wrap them for a one-person load, and double-wrap for greater loads (see Table 18.1). Attach the lowering line(s) to the litter as described earlier for either single- or double-line lowers. Use all fall-protection precautions and safety belay lines for all attendants and patients.

Lifting the litter. After the team leader completes a safety check and orders the operation to proceed, the lifting crew lifts the litter to the handrail. The lifting crew should consist of at least four members, with members positioned at the foot and head of the litter, plus at least two along the inner side of the litter. The lifting team members along the side of the litter lift with one hand on the inner rail and one hand on the outer rail of the litter (see Fig. 18.40*a*). When the lifting team leader gives the order to lift, the team lifts the stretcher, being careful to lift with their legs and not their backs. The team members along the sides of the litter must be very careful, as they are in an awkward position for lifting. As the lift is made, the rail of the litter closest to the edge is rotated upward. As the initial lift is completed, the litter is placed on the handrail with an inward tilt of about 45° (Fig. 18.40*b*), so that the patient is looking in toward the rescuers, not out toward the edge. The litter is then rotated into a fully vertical position, with the lower side of the litter resting on the handrail and the lower rail of the litter positioned just inside the handrail (see Fig. 18.40*c*).

Moving the litter over the edge. Team members can make loading the litter onto the lowering system much smoother and easier by using lengths of webbing to control the load. In the current example, before the litter goes over the edge, team members at each end of the litter double 12-ft (3.7-m) lengths of one-inch webbing around the upper litter rail and wrap a turn of webbing around the handrail (see Figs. 18.40*b, c*). A lifting team member along the side of the litter doubles a length of one-inch webbing of similar length around the midpoint of the lower litter rail and makes a wrap around the handrail with the webbing.

Next the lifting team moves the litter over the edge. The lifting team members along the side of the litter push the lower litter rail over the handrail with open hands. Lifting team members must be very careful not to get their fingers mashed between the litter and the handrail.

As soon as the litter is on the outside of the handrail, the team member at the middle of the litter uses the webbing that he has wrapped around the handrail to support most of the weight of the litter. Team members at the head and foot of the litter hold the litter in a nearly vertical position so that the patient continues to look in toward the platform. The team member at the middle of the litter uses the webbing wrapped around the handrail to slowly lower the litter until the weight of the litter loads onto the near side of the litter bridle. The team members at the head and foot of the litter then use the pieces of webbing that they have wrapped around the handrail to lower the upper rails of the litter slowly until the far side of the litter is also loaded onto the bridle. As this is done, the litter gradually rotates from the nearly vertical position into a horizontal position (see Fig. 18.40*d*).

Lowering the litter. After the litter has been placed over the edge and loaded onto the lowering system(s), the lower proceeds just as described above for equivalent high-point lowers. The deployment and use of tag lines or attendants is the same. Remember that two attendants are used only on a double-line lowering system.

Figure 18.40   (*a*) In preparing to negotiate the edge for a horizontal low-point litter lower, the lifting team rotates the litter as it is lifted so that the patient faces in toward the team and away from the edge; (*b*) the initial lift places the litter on the handrail with an inward tilt of 45° toward the lifting team; (*c*) the lifting team rotates the litter into a nearly vertical orientation with the lower side of the litter resting on the handrail before moving the litter over the edge; (*d*) lengths of webbing doubled around the litter frame and wrapped around the handrail provide good control for loading the weight of the litter onto the lowering system(s).

Conducting vertical litter lowers

Some situations require that a patient be lowered to the ground in a vertically oriented litter. A common example is a situation in which the litter must pass through a narrow opening prior to the lower (see Fig. 18.42*a*). Another example is a situation in which the litter must pass through such an opening during the lower. Vertical lowers are conducted using a single lowering line. Either tag lines or an attendant guide the litter during the lower. These operations are very similar to the single-line horizontal operations described above.

To prepare for a vertical lower, properly package the patient in a litter rigged with a bridle configured for a vertical lower as described previously. Rig a single-line lowering system as described above for a single-line horizontal lower and attach it to the bridle. Use a suitable safety belay line and all other safety precautions described earlier.

Conducting high-point vertical litter lowers.   If a high point of attachment is used for the descent control device or a high-directional pulley is used, then the higher, the better. Most of the length of the litter must clear the edge before the lower can begin. One approach to doing a vertical high-point lower is to have a lifting team lift the litter as high as possible in the vertical position while the team member operating the descent control device pulls all slack through the device, ropes it in, and locks it off. The lifting crew then moves the litter over the edge and the litter is in position to be lowered. If tag lines are used, attach them to the litter before it is placed over the edge. If an attendant is used, he loads onto the system before the lower proceeds.

Conducting low-point vertical lowers.   In many instances when litters are lowered in the vertical orientation it is not possible to use the high-point technique. In these cases a low-point vertical lower is required. This technique is similar to the low-point single-line horizontal lower described previously.

As an example of how this technique can be applied, assume that a litter must be lowered vertically but must be taken over a handrail horizontally because there is not enough headspace for the litter to clear the rail using the high-point technique. The team prepares for the lower by indexing the knot at the handrail, just as they do for all low-point lowers, roping in and locking off the descent control device and attaching the knot in the end of the lowering line to the litter bridle.

Negotiating edges for low-point vertical Lowers.   At least four lifting team members are used to move the litter over the handrail. Place the litter on top of the handrail with the long axis of the litter at about a 45° angle to the rail. Place the foot end of the litter on the outside of the rail and the head end on the inside of the rail (see Fig. 18.41a). To aid in controlling the litter, loop a 24-ft (7.3-m) length of one-inch webbing around the litter rail at the foot end and wrap it around the handrail (Fig. 18.41a). The lifting team tilts the head of the litter up and the feet down (Fig. 18.41b) as the litter is slowly lowered into a vertical position (see Fig. 18.41c). The webbing wound around the railing bears most of the weight as the litter is slowly lowered and loaded onto the lowering line.

Passing the litter through small openings for low-point vertical lowers.   As another example of a low-point vertical lower, assume that the litter must be passed endwise through a small window-like opening. The team prepares for the lower by indexing the knot at the lower edge of the opening, roping in and locking off the descent control device, and attaching the lowering line to the litter bridle. Loop a 24-ft (7.3-m) length of one-inch webbing through the foot of the litter and use it to control the weight of the litter for a smooth transition onto the lowering system. Lift the litter and pass it feet first through the opening. Place the litter with its approximate balance point on the bottom edge of the opening, then lower the feet and elevate the head as you pass the litter the rest of the way through the opening (Fig. 18.42a) and slowly lower it onto the lowering system (Fig. 18.42b). Take care to avoid mashing the patient's face into the structure above the opening.

Using tag lines in a vertical lower.   When tag lines are used in vertical lowering operations, attach them to the basket litter as shown in Fig. 18.41c, or the SKED as shown in Fig. 18.42b, before the litter is moved over the edge. Once the litter is in position to be lowered, the tag-line tenders assume their positions as shown in Fig. 18.34. Operate the tag lines as described above in reference to horizontal high-point single-line lowers. When the order to lower is given, the lowering line operator unlocks the brake bar rack, removes one bar, and lowers the litter to the ground on five bars. If a figure 8 descender is used, it should be single-wrapped (see Table 18.1). As the litter nears the ground, the tag-line operators move in close to it and guide it into a horizontal position as it comes to rest.

Using an attendant in a vertical lower.   If an attendant is used for a vertical lower, he should rig a suitable connecting device, as described above for equivalent horizontal single-line lowers. Adjust the connector so that the attendant's feet are slightly below the bottom of the litter when he is suspended by it (see Fig. 18.43). The attendant's feet

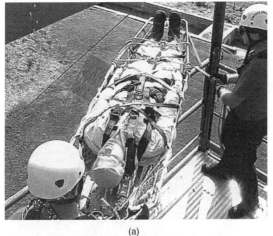

(a)

(b)

(c)

Figure 18.41   (*a*) In preparing to negotiate the edge for a vertical low-point litter lower, the lifting team places the litter on the handrail with the foot end angled out and the head end angled in. (*b*) A length of webbing doubled around the litter frame at the foot end and wrapped around the handrail can help control the weight of the litter as the edge is negotiated. (*c*) Slowly load the weight of the litter onto the lowering system to avoid a shock load as the litter assumes the vertical orientation. Attach tag lines to the litter frame as shown when used in vertical lowers.
(*Note:* The lifting team should consist of at least four team members. Two team members were omitted for clarity of the photograph.)

(a)   (b)

Figure 18.42 (*a*) Webbing can be used to ease the transition onto the lowering system when a litter must be passed endwise through a small opening, as illustrated by this low-point lowering operation with a vertical SKED litter; (*b*) attach tag lines to the lower lifting handles of the SKED litter for use in vertical lowering operations.

must reach the ground before the foot of the basket. Any of the options previously described for attaching the attendant to a belay system for a horizontal single-line lower can also be used in a vertical lower.

After the rescue team leader has done a full safety check and given her approval, the attendant climbs over the rail and slowly loads onto the lowering line. He positions himself to guide the litter during the descent as seen in Fig. 18.43*a*. The rescue team leader gives the order to lower. The lowering line operator unlocks the rack but leaves all six bars racked in and lowers the load as directed by the team leader. If a figure 8 descender is used, it should be double-wrapped (see Table 18.1).

During the lower the attendant holds the litter clear of the side of the structure, guides it past any edges encountered, communicates with the team leader, and comforts the patient. The attendant's feet reach the ground just before the foot of the litter, allowing him to guide the litter into a horizontal position on the ground as the lower is completed (see Fig. 18.43*b*).

Stairwell lowering systems

Completing rescue operations often requires that a patient be brought down a stairway after being removed from a confined space. In many cases stairwell lowering systems are utilized to make the procedure much easier and safer than carrying the litter down the stairway.

Figure 18.43 (*a*) A litter tender guides the litter during this vertical lowering operation; (*b*) the tender must be attached to the lowering system so that his feet contact the ground before the foot of the litter, so that he can guide the litter to a horizontal position as the lowering operation is completed.

To conduct a stairwell lower, package the patient in a suitable litter rigged for a horizontal single-line lower. Anchor a descent control device above the landing where the patient will be loaded onto the system, or run the lowering line through a high-directional pulley as shown in Fig. 18.44. The procedure is very similar to that described for performing a horizontal single-line high-point lower. Four team members guide the litter as it is lowered down the stairwell. These team members do not bear any significant part of the weight of the litter; that is what the descent control device is for. Lower the litter on five bars when a brake bar rack is used (see Table 18.1). If a figure 8 descender is used, it should be single-wrapped.

All team members keep a firm grip on the litter as it is lowered, guiding it down and around the turns in the stairwell. They minimize the pendulum motion that could have a "sawing" effect on the lowering line. The progressive turning motion as the litter descends the stairway will have a twisting effect on the system. Prevent this by placing a suitable swivel between the end of the lowering line and the litter bridle (Fig. 18.24).

In order for the stairwell lower to work well, there must be an adequate "keyhole" space down the center of the stairwell to provide a space for the rope to move through. Check the keyhole space for rough edges and use padding to cover them. During the lower it may be advisable for an extra crew member to move down the stairwell, remaining on a landing above the litter, to guide the lowering line and keep it from hanging up on anything.

Safety belays are rarely used in stairwell lowers. The team guiding the litter down the stairwell constitutes the safety system, since they should be able to control the litter if the lowering system fails. If fewer than four team members are available to guide the

Figure 18.44   Stairwell lowering systems use a single lowering line to lower the weight of the litter down a stairwell as several team members guide the litter.

litter, it may be desirable to use a safety belay. Rig the belay as described above for horizontal, single-line high-point litter lowers.

Summary

Confined space emergencies usually occur because someone is incapacitated within a permit space. In this chapter we addressed one of the most important aspects of confined space rescue: packaging the victim for safe removal from the space. When confined spaces are elevated, the packaged patient must be lowered to a safe location. We covered several procedures for safely lowering packaged patients in order to complete rescue operations.

Both types of operation discussed in this chapter warrant a high level of training, practice, and skill. Improper patient packaging can result in further injury to the patient during transfer procedures. The use of unsafe procedures during lowering operations threatens the safety of both patient and rescuer.

Permit-Required Confined Space Decision Flowchart (1910.146, Appendix A)

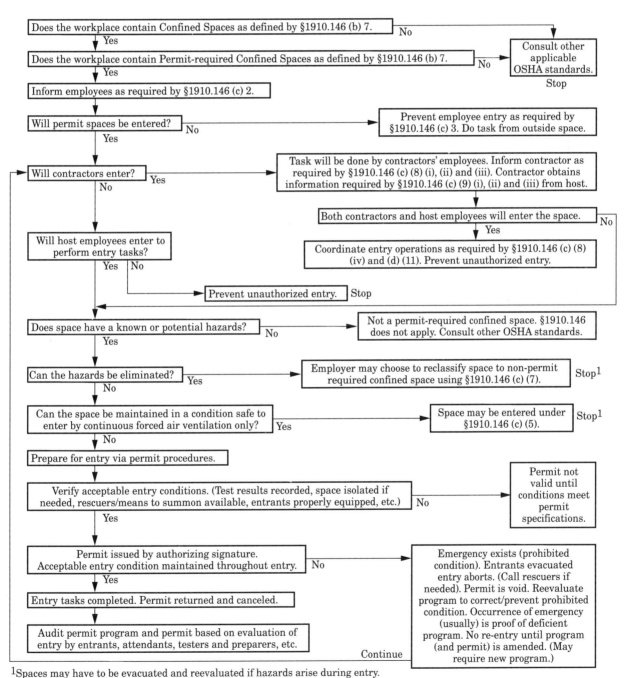

Does the workplace contain Confined Spaces as defined by §1910.146 (b) 7. — No → Consult other applicable OSHA standards. Stop

Yes ↓

Does the workplace contain Permit-required Confined Spaces as defined by §1910.146 (b) 7. — No →

Yes ↓

Inform employees as required by §1910.146 (c) 2.

Will permit spaces be entered? — No → Prevent employee entry as required by §1910.146 (c) 3. Do task from outside space.

Yes ↓

Will contractors enter? — Yes → Task will be done by contractors' employees. Inform contractor as required by §1910.146 (c) (8) (i), (ii) and (iii). Contractor obtains information required by §1910.146 (c) (9) (i), (ii) and (iii) from host.

No ↓

Both contractors and host employees will enter the space. — No

Yes ↓

Coordinate entry operations as required by §1910.146 (c) (8) (iv) and (d) (11). Prevent unauthorized entry.

Will host employees enter to perform entry tasks? — Yes / No

Prevent unauthorized entry. Stop

Does space have a known or potential hazards? — No → Not a permit-required confined space. §1910.146 does not apply. Consult other OSHA standards.

Yes ↓

Can the hazards be eliminated? — Yes → Employer may choose to reclassify space to non-permit required confined space using §1910.146 (c) (7). Stop[1]

No ↓

Can the space be maintained in a condition safe to enter by continuous forced air ventilation only? — Yes → Space may be entered under §1910.146 (c) (5). Stop[1]

No ↓

Prepare for entry via permit procedures.

Verify acceptable entry conditions. (Test results recorded, space isolated if needed, rescuers/means to summon available, entrants properly equipped, etc.) — No → Permit not valid until conditions meet permit specifications.

Yes ↓

Permit issued by authorizing signature. Acceptable entry condition maintained throughout entry. — No → Emergency exists (prohibited condition). Entrants evacuated entry aborts. (Call rescuers if needed). Permit is void. Reevaluate program to correct/prevent prohibited condition. Occurrence of emergency (usually) is proof of deficient program. No re-entry until program (and permit) is amended. (May require new program.)

Yes ↓

Entry tasks completed. Permit returned and canceled.

Audit permit program and permit based on evaluation of entry by entrants, attendants, testers and preparers, etc.

Continue

[1]Spaces may have to be evacuated and reevaluated if hazards arise during entry.

Examples of Permit-Required Confined Space Programs (1910.146, Appendix C)

Example 1

Workplace. Sewer entry.

Potential hazards. The employees could be exposed to the following:
Engulfment.

Presence of toxic gases. Equal to or more than 10 ppm hydrogen sulfide measured as an 8-h time-weighted average. If the presence of other toxic contaminants is suspected, specific monitoring programs will be developed.

Presence of explosive or flammable gases. Equal to or greater than 10% of the lower flammable limit (LFL).

Oxygen deficiency. A concentration of oxygen in the atmosphere equal to or less than 19.5% by volume.

A. Entry without Permit/Attendant

Certification. Confined spaces may be entered without the need for a written permit or attendant provided that the space can be maintained in a safe condition for entry by mechanical ventilation alone, as provided in 1910.146(c)(5). All spaces shall be considered permit-required confined spaces until the pre-entry procedures demonstrate otherwise. Any employee required or permitted to pre-check or enter an enclosed/confined space shall have successfully completed, as a minimum, the training as required by the following sections of these procedures. A written copy of operating and rescue procedures as required by these procedures shall be at the work site for the duration of the job. The Confined Space Pre-Entry Check List must be completed by the LEAD WORKER before entry into a confined space. This list verifies completion of items listed below. This check list shall be kept at the jobsite for duration of the job. If circumstances dictate an interruption in the work, the permit space must be re-evaluated and a new check list must be completed.

Control of atmospheric and engulfment hazards

Pumps and lines. All pumps and lines which may reasonably cause contaminants to flow into the space shall be disconnected, blinded and locked out, or effectively isolated by other means to prevent development of dangerous air contamination or engulfment. Not all laterals to sewers or storm drains require blocking. However, where experience or knowledge of industrial use indicates there is a reasonable potential for contamination of air or engulfment into an occupied sewer, then all affected laterals shall be blocked. If blocking and/or isolation requires entry into the space, the provisions for entry into a permit-required confined space must be implemented.

Surveillance. The surrounding area shall be surveyed to avoid hazards such as drifting vapors from the tanks, piping, or sewers.

Testing. The atmosphere within the space will be tested to determine whether dangerous air contamination and/or oxygen deficiency exists. Detector tubes, alarm-only gas monitors, and explosion meters are examples of monitoring equipment that may be used to test permit space atmospheres. Testing shall be performed by the LEAD WORKER who has successfully completed the Gas Detector training for the monitor he or she will use. The minimum parameters to be monitored are oxygen deficiency, LFL, and hydrogen sulfide concentration. A written record of the pre-entry test results shall be made and kept at the work site for the duration of the job. The supervisor will certify in writing, based upon the results of the pre-entry testing, that all hazards have been eliminated. Affected employees shall be able to review the testing results. The most hazardous conditions shall govern when work is being performed in two adjoining, connecting spaces.

Entry procedures. If there are no non-atmospheric hazards present and if the pre-entry tests show there is no dangerous air contamination and/or oxygen deficiency within the space and there is no reason to believe that any is likely to develop, entry into and work within may proceed. Continuous testing of the atmosphere in the immediate vicinity of the workers within the space shall be accomplished. The workers will immediately leave the permit space when any of the gas monitor alarm set points are reached as defined. Workers will not return to the area until a SUPERVISOR who has completed the gas detector training has used a direct reading gas detector to evaluate the situation and has determined that it is safe to enter.

Rescue. Arrangements for rescue services are not required where there is no attendant. See the rescue portion of section B, below, for instructions regarding rescue planning where an entry permit is required.

Entry permit required

Permits: confined space entry permit. All spaces shall be considered permit-required confined spaces until the pre-entry procedures demonstrate otherwise. Any employee required or permitted to pre-check or enter a permit-required confined space shall have successfully completed, as a minimum, the training as required by the following sections of these procedures. A written copy of operating and rescue procedures as required by these procedures shall be at the work site for the duration of the job. The Confined Space Entry Permit must be completed before approval can be given to enter a permit-required confined space. This permit verifies completion of items listed below. This permit shall be kept at the job site for the duration of the job. If circumstances cause an interruption in the work or a change in the alarm conditions for which entry was approved, a new Confined Space Entry Permit must be completed.

Control of atmospheric and engulfment hazards

Surveillance. The surrounding area shall be surveyed to avoid hazards such as drifting vapors from tanks, piping or sewers.

Testing. The confined space atmosphere shall be tested to determine whether dangerous air contamination and/or oxygen deficiency exists. A direct reading gas monitor shall be used. Testing shall be performed by the SUPERVISOR who has successfully completed the gas detector training for the monitor he or she will use. The minimum parameters to be monitored are oxygen deficiency, LFL, and hydrogen sulfide concentration. A written record of the pre-entry test results shall be made and kept at the work site for the duration of the job. Affected employees shall be able to review the testing results. The most hazardous conditions shall govern when work is being performed in two adjoining, connected spaces.

Space ventilation. Mechanical ventilation systems, where applicable, shall be set at 100% outside air. Where possible, open additional manholes to increase air circulation.

Use portable blowers to augment natural circulation if needed. After a suitable ventilating period, repeat the testing. Entry may not begin until testing has demonstrated that the hazardous atmosphere has been eliminated.

Entry procedures. The following procedure shall be observed under any of the following conditions: (1) testing demonstrates the existence of dangerous or deficient conditions and additional ventilation cannot reduce concentrations to safe levels; (2) the atmosphere tests as safe but unsafe conditions can reasonably be expected to develop; (3) it is not feasible to provide for ready exit from spaces equipped with automatic fire suppression systems, and it is not practical or safe to deactivate such systems; or (4) an emergency exists and it is not feasible to wait for preentry procedures to take effect.

All personnel must be trained. A self contained breathing apparatus shall be worn by any person entering the space. At least one worker shall stand by the outside of the space ready to give assistance in case of emergency. The standby worker shall have a self-contained breathing apparatus available for immediate use. There shall be at least one additional worker within sight or call of the standby worker. Continuous powered communications shall be maintained between the worker within the confined space and standby personnel.

If at any time there is any questionable action or non-movement by the worker inside, a verbal check will be made. If there is no response, the worker will be moved immediately. *Exception:* If the worker is disabled due to falling or impact, he or she shall not be removed from the confined space unless there is immediate danger to his or her life. Local fire department rescue personnel shall be notified immediately. The standby worker may only enter the confined space in case of an emergency (wearing the self-contained breathing apparatus) and only after being relieved by another worker. Safety belt or harness with attached lifeline shall be used by all workers entering the space with the free end of the line secured outside the entry opening. The standby worker shall attempt to remove a disabled worker via his or her lifeline before entering the space.

When practical, these spaces shall be entered through side openings—those within $3\frac{1}{2}$ feet (1.07 m) of the bottom. When entry must be through a top opening, the safety belt shall be of the harness type that suspends a person upright and a hoisting device or similar apparatus shall be available for lifting workers out of the space.

In any situation where their use may endanger the worker, use of a hoisting device or safety belt and attached lifeline may be discontinued.

When dangerous air contamination is attributable to flammable and/or explosive substances, lighting, and electrical equipment shall be Class 1, Division 1 rated per National Electrical Code and no ignition sources shall be introduced into the area.

Continuous gas monitoring shall be performed during all confined space operations. If alarm conditions change adversely, entry personnel shall exit the confined space and a new confined space permit issued.

Rescue.   Call the fire department services for rescue. Where immediate hazards to injured personnel are present, workers at the site shall implement emergency procedures to fit the situation.

Example 2

Workplace.   Meat and poultry rendering plants.

Cookers and dryers are either batch or continuous in their operation. Multiple batch cookers are operated in parallel. When one unit of a multiple set is shut down for repairs, means are available to isolate that unit from the others which remain in operation.

Cookers and dryers are horizontal, cylindrical vessels equipped with a center, rotating shaft and agitator paddles or discs. If the inner shell is jacketed, it is usually heated with steam at pressures up to 150 psig (1034.25 kPa). The rotating shaft assembly of the continuous cooker or dryer is also steam-heated.

Potential hazards.   The recognized hazards associated with cookers and dryers are the risk that employees could be:

1. Struck or caught by rotating agitator;

2. Engulfed in raw material or hot, recycled fat;

3. Burned by steam from leaks into the cooker/dryer steam jacket or the condenser duct system if steam valves are not properly closed and locked out;

4. Burned by contact with hot metal surfaces, such as the agitator shaft assembly, or inner shell of the cooker/dryer;

5. Heat stress caused by warm atmosphere inside cooker/dryer;

6. Slipping and falling on grease in the cooker/dryer;

7. Electrically shocked by faulty equipment taken into the cooker/dryer;

8. Burned or overcome by fire or products of combustion; or

9. Overcome by fumes generated by welding or cutting done on grease-covered surfaces.

Permits. The supervisor in this case is always present at the cooker/dryer or other permit entry confined space when entry is made. The supervisor must follow the pre-entry isolation procedures described in the entry permit in preparing for entry, and ensure that the protective clothing, ventilating equipment, and any other equipment required by the permit are at the entry site.

Control of hazards. *Mechanical.* Lock out main power switch to agitator motor at main power panel. Affix tag to the lock to inform others that a permit entry confined space entry is in progress.

Engulfment. Close all valves in the raw-material blow line. Secure each valve in its closed position using chain and lock. Attach a tag to the valve and chain warning that a permit entry confined space entry is in progress. The same procedure shall be used for securing the fat recycle valve.

Burns and heat stress. Close steam supply valves to jacket and secure with chains and tags. Insert solid blank at flange in cooker vent line to condenser manifold duct system. Vent cooker/dryer by opening access door at discharge end and top center door to allow natural ventilation throughout the entry. If faster cooling is needed, use a portable ventilation fan to increase ventilation. Cooling water may be circulated through the jacket to reduce both outer and inner surface temperatures of cooker/dryers faster. Check air and inner surface temperatures in cooker/dryer to assure they are within acceptable limits before entering, or use proper protective clothing.

Fire and fume hazards. Careful site preparation, such as cleaning the area within 4 inches (10.16 cm) of all welding or torch cutting operations, and proper ventilation are the preferred controls. All welding and cutting operations shall be done in accordance with the requirements of 29 CFR Part 1910, Subpart Q, OSHA's welding standard. Proper ventilation may be achieved by local exhaust ventilation, or the use of portable ventilation fans, or a combination of the two practices.

Electrical shock. Electrical equipment used in cooker/dryers shall be in serviceable condition.

Slips and falls. Remove residual grease before entering cooker/dryer.

Attendant. The supervisor shall be the attendant for employees entering cooker/dryers.

Permit. The permit shall specify how isolation shall be done and any other preparations needed before making entry. This is especially important in parallel arrangements of cooker/dryers so that the entire operation need not be shut down to allow safe entry into one unit.

Rescue. When necessary, the attendant shall call the fire department as previously arranged.

Example 3

Workplace. Workplaces where tank cars, trucks, and trailers, dry-bulk tanks and trailers, railroad tank cars, and similar portable tanks are fabricated or serviced.

A. During fabrication

These tanks and dry-bulk carriers are entered repeatedly throughout the fabrication process. These products are not configured identically, but the manufacturing processes by which they are made are very similar.

Sources of hazards. In addition to the mechanical hazards arising from the risks that an entrant would be injured due to contact with components of the tank or the tools being used, there is also the risk that a worker could be injured by breathing fumes from welding materials or mists or vapors from materials used to coat the tank interior. In addition, many of these vapors and mists are flammable, so the failure to properly ventilate a tank could lead to a fire or explosion.

Control of hazards

Welding. Local exhaust ventilation shall be used to remove welding fumes once the tank or carrier is completed to the point that workers may enter and exit only through a manhole. (Follow the requirements of 29 CFR 1910, Subpart Q, OSHA's welding standard, at all times.) Welding gas tanks may never be brought into a tank or carrier that is a permit entry confined space.

Application of interior coatings/linings. Atmospheric hazards shall be controlled by forced air ventilation sufficient to keep the atmospheric concentration of flammable materials below 10% of the lower flammable limit (LFL) [or lower explosive limit (LEL), whichever term is used locally]. The appropriate respirators are provided and shall be used in addition to providing forced-air ventilation if the forced-air ventilation does not maintain acceptable respiratory conditions.

Permits. Because of the repetitive nature of the entries in these operations, an "Area Entry Permit" will be issued for a 1-month period to cover those production areas where tanks are fabricated to the point that entry and exit are made using manholes.

Authorization. Only the area supervisor may authorize an employee to enter a tank within the permit area. The area supervisor must determine that conditions in the tank trailer, dry-bulk trailer or truck, etc. meet permit requirements before authorizing entry.

Attendant. The area supervisor shall designate an employee to maintain communication by employer-specified means with employees working in tanks to ensure their safety. The attendant may not enter any permit entry confined space to rescue an entrant or for any other reason, unless authorized by the rescue procedure and, even then, only after calling the rescue team and being relieved by an attendant or another worker.

Communications and observation. Communications between attendant and entrant(s) shall be maintained throughout entry. Methods of communication that may be specified by the permit include voice, voice-powered radio, tapping or rapping codes on tank walls, signaling tugs on a rope, and the attendant's observation that work activities such as chipping, grinding, welding, spraying, etc., which require deliberate operator control continue normally. These activities often generate so much noise that the necessary hearing protection makes communication by voice difficult.

Rescue procedures. Acceptable rescue procedures include entry by a team of employee-rescuers, use of public emergency services, and procedures for breaching the tank. The area permit specifies which procedures are available, but the area supervisor makes the final decision based on circumstances. (Certain injuries may make it necessary to breach the tank to remove a person rather than risk additional injury by removal through an existing manhole. However, the supervisor must ensure that no breaching procedure used for rescue would violate terms of the entry permit. For instance, if the tank must be breached by cutting with a torch, the tank surfaces to be cut must be free of volatile or combustible coatings within 4 inches (10.16 cm) of the cutting line and the atmosphere within the tank must be below the LFL.

Retrieval line and harnesses. The retrieval lines and harnesses generally required under this standard are usually impractical for use in tanks because the internal configuration of the tanks and their interior baffles and other structures would prevent rescuers from hauling out injured entrants. However, unless the rescue procedure calls for breaching the tank for rescue, the rescue team shall be trained in the use of retrieval lines and harnesses for removing injured employees through manholes.

B. Repair or service of "used" tanks and bulk trailers.

Sources of hazards. In addition to facing the potential hazards encountered in fabrication or manufacturing, tanks or trailers which have been in service may contain residues of dangerous materials, whether left over from the transportation of hazardous cargoes or generated by chemical or bacterial action on residues of non-hazardous cargoes.

Control of atmospheric hazards. A "used" tank shall be brought into areas where tank entry is authorized only after the tank has been emptied, cleansed (without employee entry) of any residues, and purged of any potential atmospheric hazards.

Welding. In addition to tank cleaning for control of atmospheric hazards, coating and surface materials shall be removed 4 inches (10.16 cm) or more from any surface area where welding or other torch work will be done and care taken that the atmosphere within the tank remains well below the LFL. (Follow the requirements of 29 CFR 1910, Subpart Q, OSHA's welding standard, at all times.)

Permits. An entry permit valid for up to 1 year shall be issued prior to authorization of entry into used tank trailers, dry bulk trailers or trucks. In addition to the pre-entry cleaning requirement, this permit shall require the employee safeguards specified for new tank fabrication or construction permit areas.

Authorization. Only the area supervisor may authorize an employee to enter a tank trailer, dry bulk trailer, or truck within the permit area. The area supervisor must determine that the entry permit requirements have been met before authorizing entry.

Rescue Team or Rescue Service
Evaluation Criteria (1910.146, Appendix F)

(1) This appendix provides guidance to employers in choosing an appropriate rescue service. It contains criteria that may be used to evaluate the capabilities both of prospective and current rescue teams. Before a rescue team can be trained or chosen, however, a satisfactory permit program, including an analysis of all permit-required confined spaces to identify all potential hazards in those spaces, must be completed. OSHA believes that compliance with all the provisions of §1910.146 will enable employers to conduct permit space operations without recourse to rescue services in nearly all cases. However, experience indicates that circumstances will arise where entrants will need to be rescued from permit spaces. It is therefore important for employers to select rescue services or teams, either on site or off site, that are equipped and capable of minimizing harm to both entrants and rescuers if the need arises.

(2) For all rescue teams or services, the employer's evaluation should consist of two components: an initial evaluation, in which employers decide whether a potential rescue service or team is adequately trained and equipped to perform permit space rescues of the kind needed at the facility and whether such rescuers can respond in a timely manner, and a performance evaluation, in which employers measure the performance of the team or service during an actual or practice rescue. For example, based on the initial evaluation, an employer may determine that maintaining an on-site rescue team will be more expensive than obtaining the services of an off-site team, without being significantly more effective, and decide to hire a rescue service. During a performance evaluation, the employer could decide, after observing the rescue service perform a practice rescue, that the service's training or preparedness was not adequate to effect a timely or effective rescue at his or her facility and decide to select another rescue service, or to form an internal rescue team.

A. Initial Evaluation

I. The employer should meet with the prospective rescue service to facilitate the evaluations required by §1910.146(k)(1)(i) and §1910.146(k)(1)(ii). At a minimum, if an off-site rescue service is being considered, the employer must contact the service to plan and coordinate the evaluations required by the standard. Merely posting the service's number or planning to rely on the 911 emergency phone number to obtain these services at the time of a permit space emergency would not comply with paragraph (k)(1) of the standard.

II. The capabilities required of a rescue service vary with the type of permit spaces from which rescue may be necessary and the hazards likely to be encountered in those spaces. Answering the questions below will assist employers in determining whether the rescue service is capable of performing rescues in the permit spaces present at the employer's workplace.

1. What are the needs of the employer with regard to response time (time for the rescue service to receive notification, arrive at the scene, and set up and be ready for entry)? For example, if entry is to be made into an IDLH atmosphere, or into a space that can quickly develop an IDLH atmosphere (if ventilation fails or for other reasons), the rescue team or service would need to be standing by at the permit space. On the other hand, if the danger to entrants is restricted to mechanical hazards that would cause injuries (e.g., broken bones, abrasions) a response time of 10 or 15 minutes might be adequate.

2. How quickly can the rescue team or service get from its location to the permit spaces from which rescue may be necessary? Relevant factors to consider would include the location of the rescue team or service relative to the employer's workplace, the quality of roads and highways to be traveled, potential bottlenecks or traffic congestion that might be encountered in transit, the reliability of the rescuer's vehicles, and the training and skill of its drivers.

3. What is the availability of the rescue service? Is it unavailable at certain times of the day or in certain situations? What is the likelihood that key personnel of the rescue service might be unavailable at times? If the rescue service becomes unavailable while an entry is under way, does it have the capability of notifying the employer so that the employer can instruct the attendant to abort the entry immediately?

4. Does the rescue service meet all the requirements of paragraph (k)(2) of the standard? If not, has it developed a plan that will enable it to meet those requirements in the future? If so, how soon can the plan be implemented?

5. For off-site services, is the service willing to perform rescues at the employer's workplace? (An employer may not rely on a rescuer who declines, for whatever reason, to provide rescue services.)

6. Is an adequate method for communications between the attendant, employer, and prospective rescuer available so that a rescue request can be transmitted to the rescuer without delay? How soon after notification can a prospective rescuer dispatch a rescue team to the entry site?

7. For rescues into spaces that may pose significant atmospheric hazards and from which rescue entry, patient packaging, and retrieval cannot be safely accomplished in a relatively short time (15-20 minutes), employers should consider using airline respirators (with escape bottles) for the rescuers and to supply rescue air to the patient. If the employer decides to use SCBA, does the prospective rescue service have an ample supply of replacement cylinders and procedures for rescuers to enter and exit (or be retrieved) well within the SCBA's air supply limits?

8. If the space has a vertical entry over 5 feet in depth, can the prospective rescue service properly perform entry rescues? Does the service have the technical knowledge and equipment to perform rope work or elevated rescue, if needed?

9. Does the rescue service have the necessary skills in medical evaluation, patient packaging, and emergency response?

10. Does the rescue service have the necessary equipment to perform rescues, or must the equipment be provided by the employer or another source?

B. Performance Evaluation

Rescue services are required by paragraph (k)(2)(iv) of the standard to practice rescues at least once every 12 months, provided that the team or service has not successfully performed a permit space rescue within that time. As part of each practice session, the service should perform a critique of the practice rescue, or have another qualified party perform the critique, so that deficiencies in procedures, equipment, training, or number of personnel can be identified and corrected. The results of the critique, and the corrections made to respond to the deficiencies identified, should be given to the employer to enable it to determine whether the rescue service can quickly be upgraded to meet the employer's rescue needs or whether another service must be selected. The following questions will assist employers and rescue teams and services evaluate their performance.

1. Have all members of the service been trained as permit space entrants, at a minimum, including training in the potential hazards of all permit spaces, or of representative permit spaces, from which rescue may be needed? Can team members recognize the signs, symptoms, and consequences of exposure to any hazardous atmospheres that may be present in those permit spaces?
2. Is every team member provided with, and properly trained in, the use and need for PPE, such as SCBA or fall arrest equipment, which may be required to perform permit space rescues in the facility? Is every team member properly trained to perform his or her functions and make rescues, and to use any rescue equipment, such as ropes and backboards, that may be needed in a rescue attempt?
3. Are team members trained in the first aid and medical skills needed to treat victims overcome or injured by the types of hazards that may be encountered in the permit spaces at the facility?
4. Do all team members perform their functions safely and efficiently? Do rescue service personnel focus on their own safety before considering the safety of the victim?
5. If necessary, can the rescue service properly test the atmosphere to determine if it is IDLH?
6. Can the rescue personnel identify information pertinent to the rescue from entry permits, hot work permits, and MSDSs?
7. Has the rescue service been informed of any hazards to personnel that may arise from outside the space, such as those that may be caused by future work near the space?
8. If necessary, can the rescue service properly package and retrieve victims from a permit space that has a limited size opening [less than 24 inches (60.9 cm) in diameter], limited internal space, or internal obstacles or hazards?
9. If necessary, can the rescue service safely perform an elevated (high angle) rescue?
10. Does the rescue service have a plan for each of the kinds of permit space rescue operations at the facility? Is the plan adequate for all types of rescue operations that may be needed at the facility? Teams may practice in representative spaces, or in spaces that are "worst-case" or most restrictive with respect to internal configuration, elevation, and portal size. The following characteristics of a practice space should be considered when deciding whether a space is truly representative of an actual permit space:

    (1) Internal configuration
       (a) *Open*—there are no obstacles, barriers, or obstructions within the space. One example is a water tank.
       (b) *Obstructed*—the permit space contains some type of obstruction that a rescuer would need to maneuver around. An example would be a baffle or mixing blade. Large equipment, such as a ladder or scaffold, brought into a space for work purposes would be considered an obstruction if the positioning or size of the equipment would make rescue more difficult.
    (2) Elevation
       (a) *Elevated*—a permit space where the entrance portal or opening is above grade by 4 feet or more. This type of space usually requires knowledge of high-angle rescue procedures because of the difficulty in packaging and transporting a patient to the ground from the portal.
       (b) *Nonelevated*—a permit space with the entrance portal located less than 4 feet above grade. This type of space will allow the rescue team to transport an injured employee normally.
    (3) Portal size
       (a) *Restricted*–a portal of 24 inches or less in the least dimension. Portals of this size are too small to allow a rescuer to simply enter the space while using SCBA. The portal size is also too small to allow normal spinal immobilization of an injured employee.
       (b) *Unrestricted*—a portal of greater than 24 inches in the least dimension. These portals allow relatively free movement into and out of the permit space.

    (4) Space access
        (a) *Horizontal*—the portal is located on the side of the permit space. Use of retrieval lines could be difficult.
        (b) *Vertical*—the portal is located on the top of the permit space, so that rescuers must climb down, or the bottom of the permit space, so that rescuers must climb up to enter the space. Vertical portals may require knowledge of rope techniques, or special patient packaging to safely retrieve a downed entrant.

References

AAR, *Emergency Action Guides,* Association of American Railroads, Washington, D.C., 1993.

ACGIH, *Industrial Ventilation, A Manual of Recommended Practice,* 22d ed., American Conference of Governmental Industrial Hygienists, Cincinnati, Ohio, 1995.

ANSI, *American National Standard for Industrial Head Protection,* Z89.1-1997, American National Standards Institute, New York, 1997.

ANSI, *American National Standard for Personal Protection—Protective Footwear,* Z41-1999, National Safety Council, Itasca, Ill., 1999.

ANSI, *American National Standard for Respiratory Protection,* Z88.2-1992, American National Standards Institute, New York, 1992.

ANSI, *Practice for Occupational and Educational Eye and Face Protection,* Z87.1-1989. American National Standards Institute, New York, 1989.

ANSI, *Requirements for Safety Belts, Harnesses, Lanyards, Lifelines, and Droplines for Construction and Industrial Use,* Z10-1991. American National Standards Institute, New York, 1991.

ASTM, *Guide for Inspection of Nylon, Polyester and/or Nylon/Polyester Blend Kernmantle Rope,* ASTM F1740, American Society for Testing and Materials, West Conshohocken, Pa., 1996.

Bigon, M., and G. Regazzoni, *The Morrow Guide to Knots,* Willliam Morrow and Co., Inc., New York, 1982.

CMC Rescue, Inc., *Confined Space Entry and Rescue, a Training Manual,* CMC Rescue, Inc., Santa Barbara, Calif., 1996.

CMC Rescue, Inc., *Rope Rescue Manual Field Guide,* 3d ed., CMC Rescue, Inc., Santa Barbara, Calif., 1998.

Ellis, J. N., *Introduction to Fall Protection,* 2d ed., American Society of Safety Engineers, Des Plaines, Ill., 1993.

Forsberg, K., and L. H. Keith, *Chemical Protective Clothing Performance Index,* 2d ed., Wiley, New York, 1999.

Forsberg, K., and S. Z. Mansdorf, *Quick Selection Guide to Chemical Protective Clothing,* 3d ed., Wiley, New York, 1997.

Frank, J. A., *CMC Rope Rescue Manual,* 3d ed., CMC Rescue, Inc., Santa Barbara, Calif., 1998.

Lipke, R., *Technical Rescue Riggers Guide,* rev. ed., Conterra Technical Systems, Inc., Bellingham, Wash., 1997.

NFPA, NFPA 1404, *Standard for a Fire Department Self-Contained Breathing Apparatus Program,* 1996 ed., National Fire Prevention Association, Quincy, Mass., 1996.

NFPA, NFPA 1500, *Standard on Fire Department Occupational Safety and Health Program,* 1997 ed., National Fire Prevention Association, Quincy, Mass., 1997.

NFPA, NFPA 1983, *Standard on Fire Service Life Safety Rope and System Components,* 2001 ed., National Fire Protection Association, Quincy, Mass., 2001.

NFPA, NFPA 1991, *Standard on Vapor-Protective Ensembles for Hazardous Materials Emergencies,* 2000 ed., National Fire Protection Association, Quincy, Mass., 2000.

NFPA, NFPA 1992, *Standard on Liquid Splash-Protective Ensembles and Clothing for Hazardous Materials Emergencies,* 2000 ed., National Fire Protection Association, Quincy, Mass., 2000.

NFPA, NFPA 1993, *Standard on Support Function Protective Clothing for Hazardous Chemical Operations,* National Fire Protection Association, Quincy, Mass., 1994.

NIOSH, *NIOSH Guide to Industrial Respiratory Protection,* DHHS (NIOSH) Publication 87-116, U.S. Government Printing Office, Washington, D.C., 1987.

OSHA, *Assessing the Need for Personal Protective Equipment: A Guide for Small Business Employers,* OSHA 3151, Occupational Safety and Health Administration, Washington, D.C., 2000.

Pelsue, T. A. Co., *A Pocket Guide for Confined Space Ventilation,* T. A. Pelsue Company, Engelwood, Colo., 1993.

Plog, B. A., J. Niland, and P. J. Quinlan, *Fundamentals of Industrial Hygiene,* National Safety Council, Itasca, Ill., 1996, Chaps. 19 and 20.

Roop, M., T. Vines, and R. Wright, *Confined Space and Structural Rope Rescue,* Mosby-Year Book, Inc., St. Louis, 1998.

Sargent, C., *Confined Space Rescue,* Fire Engineering Books and Videos, Saddle Brook, N.J., 2000.

Wieder, A. W., C. Smith, and C. Brackage, *Essentials of Fire Fighting,* 3d ed., Fire Protection Publications, Stillwater, Okla., 1992.

Index

ABOUT THE AUTHORS

This book was written by the staff of the Workplace Safety Training Program operated as a part of the Center for Labor Education and Research at the University of Alabama at Birmingham. The program is committed to advancing the health and safety of workers and emergency responders.

DWIGHT ALAN VEASEY is director of the Workplace Safety Training Program at the University of Alabama at Birmingham. He is a firefighter/EMT and a certified confined space and rope rescue technician. His specialty training areas are chemical safety, hazardous materials emergency response, and confined space entry and rescue.

LISA C. MCCORMICK is Curriculum Coordinator for the Workplace Safety Training Program at the University of Alabama at Birmingham. She is a confined space and rope rescue technician. Her specialty training areas include hazardous materials chemistry, hazardous materials emergency response, and confined space entry and rescue.

BARBARA HILYER is the former director of the Workplace Safety Training Program at the University of Alabama at Birmingham. She is a leading authority on training techniques for effective safety and health training. Her specialty training areas include toxicology, hazard and risk assessment, and training methods.

KENNETH W. OLDFIELD is Industrial Hygiene Services Manager for the Workplace Safety Training Program at the University of Alabama at Birmingham. He is a certified industrial hygienist. His specialty training areas include computer applications, hazardous materials air monitoring, personal protective equipment, and respirator facepiece fit-testing.

LLOYD SAM HANSEN, JR. is an Instructor for the Workplace Safety Training Program at the University of Alabama at Birmingham. He holds the rank of battalion chief in the fire service and is a certified confined space and rope rescue technician. His specialty training areas include hazardous materials emergency response, confined space entry and rescue, and incident management systems.